Fungal walls and hyphal growth

Fungal walls and hyphal growth

SYMPOSIUM OF
THE BRITISH MYCOLOGICAL SOCIETY
HELD AT QUEEN ELIZABETH COLLEGE
LONDON, APRIL 1978

EDITED BY
J.H.BURNETT & A.P.J.TRINCI

CAMBRIDGE UNIVERSITY PRESS
CAMBRIDGE
LONDON · NEW YORK · MELBOURNE

CAMBRIDGE UNIVERSITY PRESS
Cambridge, New York, Melbourne, Madrid, Cape Town,
Singapore, São Paulo, Delhi, Tokyo, Mexico City

Cambridge University Press
The Edinburgh Building, Cambridge CB2 8RU, UK

Published in the United States of America by Cambridge University Press, New York

www.cambridge.org
Information on this title: www.cambridge.org/9780521279192

First published 1979
First paperback edition 2011

A catalogue record for this publication is available from the British Library

Library of Congress Cataloguing in Publication data

Main entry under title:

Fungal walls and hyphal growth.

Includes index.
1. Fungi–Growth–Congresses. 2. Fungi–Hyphae–Congresses. 3. Plant cell walls–Congresses. 1. Burnett, John Harrison, 1922– 11. Trinci, A. P. J. 111. British Mycological Society.
QK601.F86   589'.2'0431   78–72082

ISBN 978-0-521-22499-4 Hardback
ISBN 978-0-521-27919-2 Paperback

Contents

Contributors

L. Alberghina, *Cattedra de Biochemica Comparata, Facoltà di Scienze, Università di Milano, Italy.*

B. W. Bainbridge, *Microbiology Department, Queen Elizabeth College, Campden Hill, London W8 7AH, UK.*

S. Bartnicki-Garcia, *Department of Plant Pathology, University of California, Riverside, California 92521, USA.*

B. Bowers, *National Heart, Lung and Blood Institute, National Institutes of Health, Bethesda, Maryland 20014, USA.*

C. E. Bracker, *Department of Botany and Plant Pathology, Purdue University, West Lafayette, Indiana 47907, USA.*

J. H. Burnett, *Department of Agricultural Science, University of Oxford, Parks Road, Oxford OX1 3PF, UK.*

E. Cabib, *National Institute of Arthritis and Metabolic Diseases, National Institutes of Health, Bethesda, Maryland 20014, USA.*

M. G. Costantini, *Centro del CNR di Biologia Cellulare e Molecolare delle Piante, Università di Milano, Milano, Italy.*

G. A. Cook, *Department of Biochemistry, Queen Elizabeth College, Campden Hill, London W8 7AH, UK.*

A. Duran, *National Institute of Arthritis and Metabolic Diseases, National Institutes of Health, Bethesda, Maryland 20014, USA.*

M. Fèvre, *Université Claude Bernard Lyon I, Laboratoire de Physiologie Végétale, Laboratoire de Mycologie associé au CNRS no. 44, 43 Boulevard du 11 Novembre 1918, 69621 Villeurbanne, France.*

G. W. Gooday, *Department of Microbiology, Marischal College, University of Aberdeen, Aberdeen AB9 1AS, UK.*

D. H. Jennings, *Department of Botany, The University, PO Box 147, Liverpool L69 3BX, UK.*

R. Marchant, *School of Biological and Environmental Studies, New University of Ulster, Coleraine, County Londonderry, Northern Ireland, UK.*

P. Markham, *Microbiology Department, Queen Elizabeth College, Campden Hill, London W8 7AH, UK.*

E. Martegani, *Centro del CNR di Biologia Cellulare e Molecolare delle Piante, Università di Milano, Milano, Italy.*

J. F. Peberdy, *Department of Botany, University of Nottingham, Nottingham, UK.*

A. M. S. Pope, *Research Division, G. D. Searle and Company Ltd, Lane End Road, High Wycombe, Buckinghamshire HP12 4HL, UK.*

J. I. Prosser, *Department of Microbiology, Marischal College, University of Aberdeen, Aberdeen AB9 1AS, UK.*

R. C. Righelato, *Tate and Lyle Group Research and Development, PO Box 68, Reading RG6 2BX, UK.*

R. F. Rosenberger, *Division of Genetics, National Institute for Medical Research, Mill Hill, London NW7 1AA, UK.*

J. Ruiz-Herrera, *Departamento di Genética y Biología Molecular, Centro de Investigatión y Estudios Avanzados del Instituto Polytécnico Nacional, México DF, México.*

J. H. Sietsma, *Department of Developmental Plant Biology, Biological Centre, University of Groningen, Haren, The Netherlands.*

J. L. Stirling, *Department of Biochemistry, Queen Elizabeth College, Campden Hill, London W8 7AH, UK.*

E. Sturani, *Centro del CNR di Biologia Cellulare e Molecolare delle Piante, Università di Milano, Milano, Italy.*

A. P. J. Trinci, *Department of Microbiology, Queen Elizabeth College, Campden Hill, London W8 7AH, UK.*

B. P. Valentine, *Beecham Pharmaceuticals, Clarendon Road, Worthing, Sussex BN14 8QH, UK.*

S. C. Watkinson, *Department of Botany, University of Oxford, South Parks Road, Oxford OX1 3RA, UK.*

J. G. H. Wessels, *Department of Developmental Plant Biology, Biological Centre, University of Groningen, Haren, The Netherlands.*

R. Zippel, *Cattedra di Biochimica Comparata, Facoltà di Scienze, Università di Milano, Milano, Italy.*

Preface

Ideas concerning the growth of fungal hypae have developed, appropriately enough, in a manner reminiscent of the classical growth curve. During an initial long lag period, the existence of 'true' and 'fungal' cellulose was recognized (de Bary, 1866), the structural complexity of fungal walls established (Carnoy, 1870), the universality of tip-growth established (Reinhardt, 1892), and the integration of tip-growth with that of the rest of the mycelium demonstrated (Henderson Smith, 1924). Apart from minor refinements, ideas showed little development for some 35 years until, in 1959, the papers of Robertson and Zalokar ushered in the present grand period of growth. In the former paper, growth at the apex was formulated in terms of a balance between 'extension growth' and 'wall setting', while in the latter paper a direct dependence of apical growth on materials translocated from a minimal, determinate region behind the tip was established. These ideas, together with the application of electron microscopy to fungi culminating, ten years later, with the recognition of the significance of apical vesicles (Girbardt, 1969), have led to notable advances over the last 15 years in an understanding of the cytology, chemistry and physiology of fungal growth. This new knowledge is now being re-defined in more precise biochemical terms and, in particular, attention is being especially focused on chitin synthesis in yeasts and moulds.

Yet great gaps still exist. For instance, a comprehensive appreciation of hyphal growth is likely to depend upon the complete definition of the composition and macromolecular organization of the hyphal wall in general, and of the tip-wall in particular. Unfortunately, although an increasing amount of information is available about the gross composition of the walls of fungi, very little is known about the tip-wall, where extension growth actually occurs. While this state of affairs is almost certainly related to the difficulties inherent in such studies, other lacunae reflect both the relatively small numbers of scientists involved at present, and the fact that the application of many techniques to the study of fungal growth, for example those genetics and mathematical analysis, are still in their infancy.

This book seeks to present a conspectus of current work and ideas. It is based upon a symposium which was organized by the Physiology Group of the British Mycological Society and held at Queen Elizabeth College, London during April 1978. The growing interest in the topic was reflected by the fact that over 200 scientists attended the meeting.

We believe that this volume adequately reflects the advances which have been made in this area of research during the last ten years, although we regret the unavoidable absence of a chapter on the cytology of hyphal growth we had originally intended to include. Nevertheless, it is our hope that the book will provide a stimulus to future work and that it will provoke many other scientists to join with the present contributors in elucidating and understanding the nature of fungal growth processes.

The layout, design and technical editing of this book are, of course, the responsibility of Cambridge University Press.

A. P. J. Trinci
Department of Microbiology
Queen Elizabeth College
London

J. H. Burnett
Department of
Agricultural Science
Oxford

1
Aspects of the structure and growth of hyphal walls

J.H.BURNETT
Department of Agricultural Science, University of Oxford, Parks Road, Oxford OX1 3PF, UK

The last two decades have been ones in which there has been a striking increase in the study of both the structure and growth of fungal walls. Initially, studies with yeasts tended to predominate, but there is now a welcome increase in the study of the walls of hyphal fungi. The other contributors to this symposium will provide detailed specialist accounts of various aspects of wall and hyphal growth. The role of this contribution, however, is to draw attention to a range of general problems and attributes of hyphal walls and their associated growth processes, especially those which are either unpopular or neglected.

Ideally, a range of information is necessary before wall structure and growth can be adequately related, namely:

(*a*) the chemical composition of the wall components, their molecular architecture and chemical connections;

(*b*) the architecture, both gross and ultrastructural, of the wall and its components;

(*c*) the biosynthetic processes involved, the transport of precursors and the transfer of wholly or partially synthesized components to their final sites of incorporation;

(*d*) the regulatory and integrating processes involved;

(*e*) variations in any of (*a*) to (*d*) due to genotype, environmental conditions, age or site in the cell in relation to differentiation.

Since wall growth is, of course, but one aspect of the total growth of the fungal cell, it is not surprising that this ideal package of information is neither available for fungi as a whole, nor for any one fungus in particular. It will emerge, I think, during this symposium, that for particular species one or more of the necessary kinds of information exist. It would, however, be dangerous to generalize from one species to another at present, since it is already clear that there is an astonishing diversity in wall composition and construction within the fungi (Bartnicki-Garcia, 1968; Hunsley & Burnett, 1970).

The topics considered here are based on experience gained with my former colleagues, Drs D. Hunsley and G. W. Gooday, using a range of fungi, although principally *Neurospora crassa* Shear & Dodge.

Wall composition and its ultrastructural architecture

Despite the considerable effort made to ascertain the chemical components of hyphal walls, detailed knowledge is still rather inadequate, although here *Schizophyllum commune* Fries is a notable exception (see Chapter 2, pp. 27–35). *Neurospora crassa* provides an excellent example of the kinds of problem encountered.

Some 25 % (w/w) of the wall in wild-type strains is composed of glucans, predominantly β-linked. Mahadevan & Tatum (1965) isolated a $\beta(1-3)$ linked glucan, and later Mishra & Tatum (1972) isolated a specific $\beta(1-3)$ glucan synthase from *Neurospora crassa*. Nevertheless, the demonstration by Potgieter & Alexander (1965, 1966) that microbial lysis of *Neurospora crassa* walls releases a little gentiobiose along with laminaribiose and laminaritriose provides evidence for the existence of some $\beta(1-6)$ linkages as well. The relative frequency of $\beta(1-3)$ and $\beta(1-6)$ linkages is unknown, as is the structure of the molecule. Presumably it is amorphous rather than microfibrillar, since the region lysed by laminarinase – an enzyme specific for such glucans – appears so in shadowed preparations viewed by electron microscopy (Hunsley & Burnett, 1970). It is not entirely clear why this should be the case since in *Schizosaccharomyces pombe* Lindner the principal microfibrillar element is said to be a branched-chain polymer of $\beta(1-3)$ linked glucose residues with $\beta(1-6)$ linked side branches some 15 to 30 glucose units long (Manners, Masson & Patterson, 1973, 1974). By contrast, in *Phytophthora cinnamomi* Rands a cell-free $\beta(1-3)$ glucan synthesis is capable of synthesizing, *in vitro*, microfibrillar material free of $\beta(1-6)$ linkages (Wang & Bartnicki-Garcia, 1976). It is evident that the numbers and location of the $\beta(1-6)$ linkages in a predominantly $\beta(1-3)$ linked glucan have important effects upon the crystallinity of the glucan, and hence upon its mechanical properties and its function in the wall.

Despite the fairly intensive examination of *Neurospora crassa* it was only in 1974 that de Vries, having developed reliable techniques for characterizing the *S*-glucan of *Schizophyllum commune*, also detected a similar, exclusively $a(1-3)$ linked glucan in its walls. Its amount and precise location in the wall of *N. crassa* is still uncertain, and there is no evidence yet that it occurs in its characteristic rodlet-form as in the outer wall region of *S. commune.*

Chitin accounts for about 10 % of the wall material and occurs as flat, ribbon-like microfibrils whose dimensions and length can vary (Potgieter & Alexander, 1965; Hunsley & Burnett, 1968, 1970). It is not entirely certain that the chitin is present in the *a*-form, although the infrared absorption spectra of *Neurospora crassa* microfibrillar material and purified crustacean *a*-chitin are very similar; however, there are some differences in the X-ray powder diagrams of the two materials. *N. crassa* microfibrillar material seems to possess a lower crystallinity than purified crustacean chitin and has an additional, weak reflection at 1.35 nm; the significance of these differences is not known (Hunsley & Kay, 1976).

Proteinaceous material (at least 6 to 8 %) also occurs in the hyphal wall, and it seems that a substantial amount of this is present as the glycoprotein first characterized by Mahadevan & Tatum (1965). A number of amino acids were first isolated from *Neurospora crassa* walls by Livingston (1969) and their amounts largely correspond with those in five peptide moieties that Wrathal & Tatum (1973) claim to have recognized in the glycoprotein (Table 1.1). These amounts differ substantially from those reported earlier by Mahadevan & Tatum (1965), by Mahadevan & Mahadkar (1970*a*) and by Manocha & Colvin (1967). The discrepancies can, however, be accounted for, since Mahadevan and coworkers only employed paper chromatography to separate and identify the amino acids, and Manocha & Colvin isolated their fraction after sonication (which can result in substantial losses of wall material; Hunsley & Burnett, 1970) and treatment with crude snail-gut enzyme, which has some proteolytic activity. The glycoprotein can be isolated in a reasonably intact form from walls and is left structurally intact within walls treated with chitinase and glusulase (Mahadevan & Tatum, 1967; Reissig & Glasgow, 1971; Wrathall & Tatum, 1973). It is likely that there are no strong covalent links between the peptide moieties and the rest of the wall. Glucose, galactose and glucuronic acid are released from the polymer, which appears to be identical with that originally described as a polygalactosamine polymer (Distler & Roseman, 1960) which bound polyphosphate (Harold, 1962). The structure of the complex carbohydrate component is not known, but it could be a branched molecule to which the peptides are covalently linked by *O*-glycosyl serine bonds (Wrathall & Tatum, 1973).

Table 1.1. *Neurospora crassa. Amino acid composition of (i) total protein of hyphal walls,[a] (ii) glycoprotein extracted by 0.5 mM NaOH and (iii) fractions I–V of glycoprotein after elution from a DEAE-cellulose column.[b] (All values expressed as a percentage of total amino acid)*

Amino acids[c]	Ala	Arg	Asp	Cys	Glu	Gly	His	Ile	Leu	Lys	Met	Phe	Pro	Hyp	Ser	Thr	Tyr	Val
Sample																		
(i) Total protein	11.1	4.9	9.7	–	8.0	8.8	1.8	4.4	6.6	6.6	–	3.1	8.0	–	8.4	9.3	1.8	7.5
(ii) 0.5 NaOH extract	10.9	3.7	12.4	0.7	10.2	13.4	2.3	3.3	6.9	6.4	1.2	2.3	0.7	–	8.6	8.2	2.9	5.8
(iii) Fraction I	13.5	0.7	15.1	0.7	10.5	11.2	1.5	4.0	6.2	4.0	0.4	3.9	0.7	–	8.7	8.2	1.7	8.7
Fraction II	12.8	1.2	14.6	–	10.8	13.7	1.4	3.7	6.6	2.7	0.9	4.6	0.6	–	8.1	7.4	3.2	7.6
Fraction III	11.3	1.5	14.0	0.8	11.2	14.7	1.6	3.7	6.7	2.6	0.9	5.0	0.7	–	7.5	7.1	3.6	7.0
Fraction IV	12.0	2.2	15.0	–	11.2	12.4	1.7	4.0	6.6	2.7	0.4	5.5	0.9	–	6.8	7.5	4.0	7.2
Fraction V	11.2	4.8	17.3	–	13.0	11.8	0.7	4.6	6.6	0.8	0.3	6.5	0.5	–	6.2	6.6	3.0	6.1

Amino acids: Ala, alanine; Arg, arginine; Asp, aspartate; Cys, cystine; Glu, glutamate; Gly, glycine; His, histidine; Ile, isoleucine; Leu, leucine; Lys, lysine; Met, methionine; Phe, phenylalanine; Pro, proline, Hyp, hydroxyproline; Ser, serine; Thr, threonine; Tyr, tyrosine; Val, valine.

[a] Data from Livingston (1969).
[b] Data from Wrathall & Tatum (1973).
[c] Tryptophane not estimated.

Although the analytical technique leading to the estimation of the peptides' molecular weight has been criticized (Gander, 1974), there is no doubt that this glycoprotein is an important wall polymer and can, in fact, be equated to a coarse, reticulated entity – the reticulum – embedded in the mature hyphal wall (Mahadevan & Tatum, 1967; Hunsley & Burnett, 1970).

The location and morphology of the reticulum was recognized by the technique of enzyme dissection, that is, the sequential or simultaneous application to intact, living hyphae of purified enzymes in appropriate buffered conditions. This technique, combined with electron microscope observations of the shadowed wall surface after successive treatments, enabled Hunsley & Burnett (1970) to make proposals concerning the ultrastructural architecture of the wall components of *Neurospora crassa*; these are summarized in Figs. 1.1 and 1.2. They recognized an outer, predominantly β-glucan layer, separated by a glycoprotein reticulum from an inner region of chitin microfibrils embedded in proteinaceous material. It should be stressed, since their description has been misunderstood by some workers, that the successive co-axially arranged regions are *not* discrete but grade into each other. These findings, using enzyme dissection, have been further substantiated (Hunsley & Kay, 1976) by the immunofluorescent localization of wall antigens to antisera developed in rabbits against the β-glucan, chitin and

Fig. 1.1. Diagram to illustrate the structure and differentiation of the wall of *Neurospora crassa* along its length to the mature primary wall.

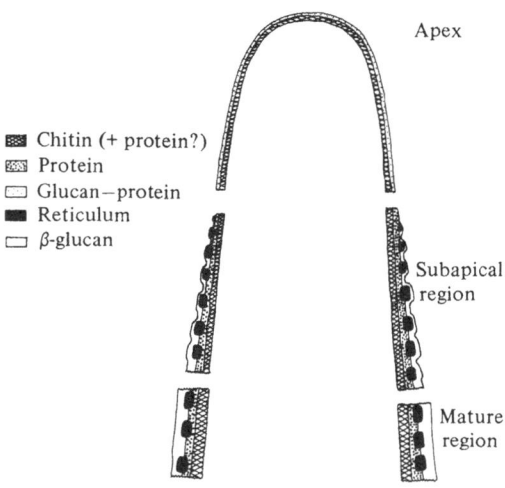

glycoprotein fractions separated according to Mahadevan & Tatum (1965). This technique has provided additional information, especially concerning the localization of wall polymers at the apex, a region notoriously difficult to study by enzyme dissection. Thus it is clear that β-glucan overlies the chitin microfibrils but some glycoprotein material occurs in the apical region, although the reticulum, as a recognizable morphological entity, has only been detected subapically. There may well be other materials. When antisera to whole walls that had then been absorbed with the three polymer fractions were applied to hyphae, both immunofluorescent and immunodiffusion methods gave evidence

Fig. 1.2. Diagram to illustrate the principal regions of the mature primary wall of *Neurospora crassa*. The relative thickness of each co-axially orientated region is to scale but the wall:lumen ratio is not. Regions, from the base up: (a) outer mixed a- and β-glucans; (b) the glycoprotein reticulum, glucans merging into proteinaceous material; (c) principally proteinaceous material; (d) innermost chitinous region with chitin microfibrils embedded principally in proteinaceous material.

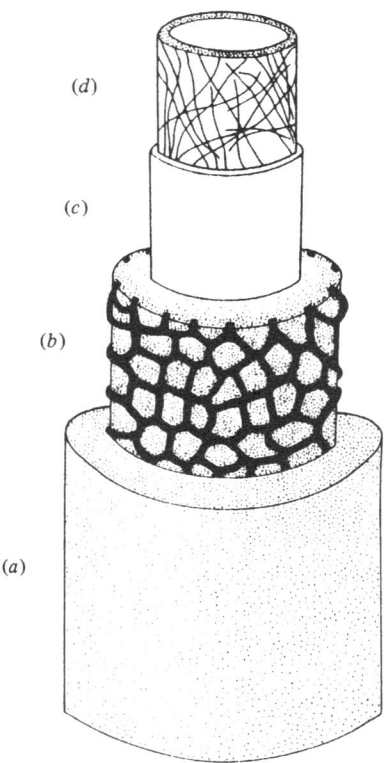

of the presence of antibodies to other, so far uncharacterized, wall antigens. Since enzyme-dissection studies had shown that wall microfibrils at the tip were exposed after treatment with pronase lacking carbohydrase activity, it may well be that the hyphal surface of the apical dome contains substantial amounts of proteinaceous material. Antigens specific to wall proteins and peptides, other than the reticulum glycoprotein, have not yet been prepared. The discrepancies between the data of Livingston (1969) and Wrathall & Tatum (1973) imply that the other wall proteins may be rich in proline (Table 1.1).

It is convenient at this point to summarize what is known concerning the composition and ultrastructural architecture of the walls of *Neurospora crassa*.

(i)   Three major polymers have been recognized; a predominantly $\beta(1-3)$ linked glucan of unknown structure with some $\beta(1-6)$ linkages in it; chitin in a microfibrillar state comparable to a-chitin; and a glycoprotein which includes glucose, galactose and glucuronate in the carbohydrate moiety. (This carbohydrate is probably linked covalently by O-glycosyl serine bonds to an uncertain number of peptides whose gross amino-acid composition is reasonably certain). In addition, $a(1-3)$ glucan also occurs (but not apparently in rodlet form), as well as unidentified proteinaceous or peptide materials.

(ii) Each polymer predominates in a co-axially arranged region in the wall. The β-glucan is outermost, then the glycoprotein and innermost are the chitin microfibrils. Proteinaceous or peptide material and, possibly, some glucan also pervade the two inner regions.

(iii) At the hyphal apex all three polymers occur; the glucan and chitin are in their usual co-axial pattern in a wall some 50 nm thick, but the glycoprotein does not form a recognizable network, other proteinaceous material may be present in some quantity. The wall thickens distally behind the apex, being c. 60 nm in the immediate subapical region and 125 nm or more in mature regions; but the inner chitin region remains more or less constant at c. 20 nm. About 4 μm behind the apex, the glycoprotein begins to form a hexagonal network whose strands are initially about 11×10 nm tangential : radial dimension) but increase up to 45×35 nm in the mature region, the gaps in the reticulum averaging 75 nm in diameter.

Clearly a good deal more needs to be learned about the chemistry of the principal polymers and the identity and location of other wall components. Moreover, nothing is known concerning linkages, if any, between wall components, although glucan–chitin links are known both in *Aspergillus niger* Van Tieghem (Stagg & Feather, 1973) and *Schizophyllum commune* (see Chapter 2 of this volume, pp. 30–32), while chitin–protein complexes are common in insects and crustacea (Ruddall, 1968) and a covalent link between the *N*-deacetylated residues of chitin and protein has been identified in *Mucor mucedo* Auct. (Datema, Wessels & van den Ende, 1977). In higher plants a hydroxyproline-rich protein provides an important linking element between carbohydrate polymers and protein (Lamport, 1965; Albersheim, 1974). So far no analogous compound has been detected in the walls of any fungus. Indeed, hydroxyproline has only been detected in the walls of fungi where chitin is replaced by a cellulose-like glucan, e.g. in Oomycetes (Crook & Johnston, 1962). Although the *Neurospora crassa* glycoprotein seems to be a distinct molecule capable of assuming a recognizable and distinctive ultrastructural morphology, it is not clear whether it is linked to other wall components *in vivo*.

It will also be obvious that these data contribute little to an understanding of the biosynthetic processes involved. None of the polymers, as such, are known to occur in the cytoplasm, and the role of vesicles or chitosomes – both of which have been recognized in *Neurospora crassa* – is unclear (see Chapter 7). Since, however, the wall thickens as it matures, and since that 2- to 3-fold thickening is predominantly in regions not adjacent to the plasmalemma, there must either be a transport system to transfer polymers from the plasmalemma to their mural site, or some synthesis must occur in the wall itself. Although there is good evidence for the presence of enzymes external to the plasmalemma and in the wall, those identified are either exclusively extracellular and hydrolytic (e.g. Trevithick & Metzenberg, 1966; Chang & Trevithick, 1972) or autolytic (e.g. Mahadevan & Mahadkar, 1970*b*; Mahadevan & Rao, 1970). The proteinaceous material in *N. crassa* walls is now being examined for synthetic activity.

As a first step, however, to studying the details of wall differentiation *in situ*, attention has been focused on the innermost chitin layer.

Differentiation in the chitinous region

There are three reasons for studying this region in some detail, namely:

(i) In all filamentous fungi investigated, the innermost region of the wall is predominantly microfibrillar – either of chitin or cellulose-like $\beta(1-4)$ glucans.

(ii) The microfibrillar region extends over the apical dome in all species examined.

(iii) One general hypothesis – the multi-net hypothesis of Roelofsen (1959, 1965) – correlates microfibrillar orientation in an important way with wall growth and, in algal and higher plant cells, microfibrillar dimensions and orientations have provided information concerning growth processes (e.g. Preston, 1974).

Thus, an understanding of the differentiation of the chitinous region in *Neurospora crassa* might well have a widespread application to problems of hyphal wall growth. Also, by chance, this region has not been extensively studied in other fungi.

Chitin microfibrils were studied after removal of the other wall components by enzyme dissection (Hunsley & Burnett, 1970; Hunsley & Kay, 1976 and unpublished), or by a treatment based on Aronson & Preston's (1960) permanganate oxidation technique (Hunsley & Burnett, 1968). This drastic preparative treatment does not seem to cause much spatial distortion of the microfibrils; their orientation is similar when the outer wall layers are progressively stripped off by the milder enzymic-dissection technique. Thus, once formed, the interwoven microfibrils apparently exist as a remarkably rigid, cylindrical, three-dimensional web, capable of retaining its shape even when the matrix materials in which it is embedded are removed. In this sense chitin and, in Oomycetes, the cellulose-like glucan microfibrils perform a genuine skeletal function.

The dimensions, numbers, spatial packing and gross configuration of the microfibrils have been examined in *Neurospora crassa* in detail and more cursorily in some other fungi. In three species examined (*N. crassa, Schizophyllum commune* and *Phytophthora parasitica* Dastur) the average dimensions of the microfibrils increased and their diameters became more variable with distance from the apex (Hunsley & Burnett, 1968, 1970; Fig. 1.3). In addition, in *N. crassa* at least, the numbers of microfibrils and their packing per unit volume increased similarly (Table 1.2, p. 13). How these changes come about is unknown, but from the very limited data available some suggestions can be made.

Increase in size could come about in three ways: through aggregation, through intussusception of existing units or through synthesis of larger microfibrils *de novo*. Aggregation and mutual cohesion of microfibrils is very evident on the surface of fungal protoplasts which are regenerating wall material, even when the density of the microfibrillar net is low, e.g. after 30 min regeneration of *Schizophyllum commune* protoplasts (van der Valk, 1976). Amorphous material is also associated at an early stage with regenerating microfibrillar aggregates and this too may facilitate the aggregation process. There is evidence from enzyme dissection that protein is closely associated with microfibrils (Hunsley & Burnett, 1970) and glucans too may promote microfibrillar aggregation (see p. 21). Aggregation is also a very characteristic feature of the microfibrillar nets formed *in vitro* by chitin synthase particles (chitosomes) of *Mucor rouxii* (Calmette) Wehm. (Ruiz-Herrera *et al.*, 1975; Bracker, Ruiz-Herrera & Bartnicki-Garcia, 1976). Indeed, if the synthetic processes *in vivo* resemble those observed *in vitro* then there is some support both for aggregation and *de novo* increase in size. Bracker *et al.* (1976) have shown that in appropriate conditions chitin microfibrils are produced *in vitro* by microvesicular chitosomes 40 to 70 nm in diameter. Within such chitosomes fine fibrils, 1 to 2 nm in diameter and possessing the insolubility properties of chitin, can be recognized. Eventually a coiled fibroid structure appearing as a 'delicate fibrillar basket' fills the chitosome, subsequently the chitosome shell opens, or is shed, and an extended, multipartite microfibril then becomes evident, apparently synthesized at its chitosomal end (Ruiz-Herrera *et al.*, 1975). Bracker *et al.* (1976) have summarized the development of such microfibrils as follows:

> Chitin microfibrils are ribbon-shaped, kinked, and of various dimensions. Thick fibrils are composed of thinner fibrils. Fine fibrils are seen merging into thicker ones, and occasionally a

Fig. 1.3. Frequency distribution of diameters of microfibrils (*a*) in the apical zone and (*b*) in the mature primary wall of *Neurospora crassa*.

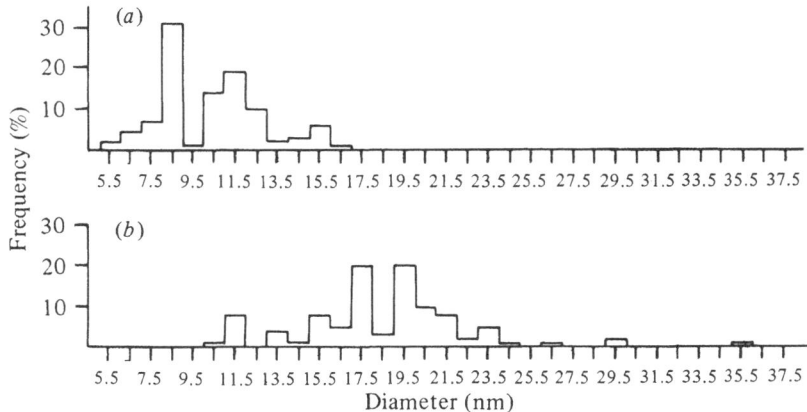

microfibril arising from a fibroid particle can be seen as a lateral tributary to another microfibril. The microfibrils synthesized during short (5 min) incubations are much thinner on average than those observed after long (3 to 11 h) incubations.

Thus, both at the time of synthesis (*in vitro*) and even when larger, as on regenerating protoplasts, fibrillar chitin moieties can aggregate. It is not clear how this is achieved. Hydrogen bonding could presumably occur between CO and NH originating from the *N*-acetylglucosamine side groups on adjacent chitin chains:

$$
\begin{array}{cc}
| & | \\
N-H \cdots O=C \\
| & | \\
C=O \cdots H-N \\
| & |
\end{array}
$$

Presumably, by analogy with cellulose, the crystalline regions of chitin are surrounded by less-ordered paracrystalline regions and here lateral hydrogen bonding could occur between adjacent chitin moieties. The question also arises as to whether the minimal size of a unit which can become aggregated is a chitin chain or something larger, e.g. the 'elementary fibrils', 3.5×3.5 nm, postulated by Mühlethaler (1967) for chitin by analogy with similar chains for bacterial and plant cellulose (Mühlethaler, 1960; Frey-Wyssling & Mühlethaler, 1963). Although Preston (1974) has vigorously denied the existence of such subunits, claiming that their appearance is due to an optical artefact, measurements made after careful application of negative-staining techniques to isolated microfibrils have been substantiated by shadowing techniques (Ohad & Danon, 1964). Fig. 1.4*a* illustrates the frequency distribution of the diameters of subunits observed in negatively stained (1 % w/v aqueous uranyl acetate; Harris & Westwood, 1964), isolated microfibrils of *Neurospora crassa* (Hunsley, unpublished). It can be seen that they show a wide scatter but most of them fall in the range 3 to 4 nm, although some are narrower. The diameters of the narrowest, indeed, are comparable with those of the 'fine fibrils' of *Mucor rouxii* (Bracker *et al.*, 1976). Chitosomes have now, incidentally, been demonstrated to occur in *N. crassa* (Bracker *et al.*, 1976; see also Chapter 7). It is possible, therefore, that the chitin microfibrils can increase in size through aggregation mediated by hydrogen bonding of subunits *c.* 3 to 4 nm in diameter. Fig. 1.4*b* compares, diagrammatically, an average apical and an average subapical microfibril on such an assumption. (In terms of numbers of aggregated chitin chains, the difference can be expressed as *c.* 90 compared with *c.* 260.) In the intact hyphal wall, of course, the matrix materials in which the microfibrils are embedded may also be involved in the process. There is no evidence available on this aspect.

The observations of Bracker *et al.* (1976) do, however, suggest that the size of a chitin microfibril is determined at an early stage in its development. They also noted that the diameters varied, being broader after longer incubation. Thus there is a further possibility that, *in vivo*, conditions alter subapically so as to favour the synthesis of larger microfibrils than at the apex. There is little doubt that microfibrils are, in fact, synthesized behind the apex. The diameter of *Neurospora crassa* hyphae increases dramatically by a factor of 6 to 15 (Steele & Trinci, 1975) throughout the extension zone, i.e. from the tip to the point where the diameter becomes constant; yet the width of the chitinous region remains constant or slightly decreases in thickness (p. 7). In *N. crassa*, in addition, the numbers and packing of the microfibrils remain constant or, more usually, increases distally, i.e. the numbers of microfibrils per unit area of the chitinous zone increases with the distance from the apex as do their average dimensions, and hence their packing (Table 1.2). It is evident that if new microfibrils were not

Fig. 1.4. (*a*) Frequency distribution of the diameters of the subunits revealed in isolated, negatively stained microfibrils of *Neurospora crassa*. (*b*) Reconstruction of an average microfibril from the apical zone (*left*) and of the mature primary wall (*right*) of *Neurospora crassa* on the assumption that each is built up from subunits of about 3–4 nm × 3–4 nm.

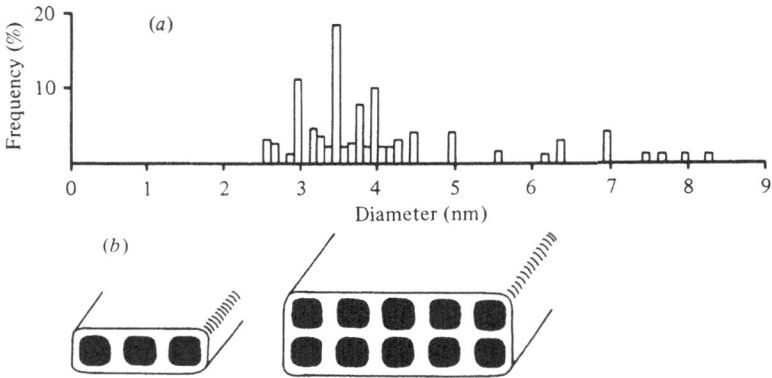

synthesized, the numbers per unit area in the chitinous zone behind the apex would decline. This is not to say that apically synthesized microfibrils may not extend so far back, conceivably increasing in size by aggregation as already discussed; but this seems unlikely. There are, incidentally, few data concerning the maximum length of chitin microfibrils. *In vitro* Ruiz-Herrera *et al.* (1975) found individual microfibrils 12 to 18 nm wide and up to 2 μm long in these preparations. In a favourable preparation of *Schizophyllum commune*, prepared by the oxidation technique, I have traced a chitin microfibril arising at about 0.5 μm from the hyphal tip back to over 3 μm, and others, arising further back, for distances between 0.8 and 3.5 μm. These observed upper limits equal the shortest extension zones measured in *N. crassa* but fall far short of the longest (2.2–29.2 μm (Steele & Trinci, 1975)). Finally, of course, observations of autoradiographs of a number of species have shown that radioactively labelled *N*-acetylglucosamine continues to be incorporated in wall chitin throughout the extension zone and even further back, e.g. *Mucor rouxii* (Bartnicki-Garcia & Lippman, 1969), *N. crassa, S. commune* (Gooday, 1971).

In summary, then, the regulation of chitin synthesis from the apex to the base of the cylindrical, tapered extension zone appears to be crucial to wall growth. Microfibrils are synthesized throughout the extension zone, to maintain the numbers per unit volume established initially in the chitinous region of the apex although, more usually, their numbers increase somewhat. At the same time, the microfibrils come to occupy more of the region, by virtue of the fact that on average the more distal

Table 1.2. Neurospora crassa. *Average numbers, dimensions and relative packing-factors of chitin microfibrils in the apical and mature region of a vegetative hypha*

Region	Number[a] (μm^{-1})	Dimension[b] (nm)	Packing[c] (%)
Apex (0 to 5 μm)	44.2	10.4 × 4	18
Mature primary wall (15 to 20 μm)	52.7	18.9 × 7	35

[a] Average numbers on surface per 1 μm transect across a hypha treated to reveal microfibrils.
[b] Mean diameters from Fig. 1.3
[c] Relative packing per unit volume of chitinous region, assuming all microfibrils perpendicular to long axis of hypha and region 20 nm thick.

they are, the larger their dimensions. Aggregation, mediated by hydrogen bonding, plays an important part in the development of the microfibril and possibly promotes the cohesion of already synthesized microfibrils. Matrix materials, conceivably proteins or glucans, may enhance this process.

Aggregation, together with the characteristic interlacing of the microfibrils, must also be responsible both for establishing and providing the strength of the three-dimensional chitin wall which occupies the innermost region of the wall. Since the growing hypha expands radially throughout the extension zone, the wall is exposed both to longitudinal and, especially, transverse stresses (Castle, 1937). Thus the mechanical integrity and response to deformation of the chitinous region – which accounts for about half the wall thickness at the tip (p. 7) – is of the utmost importance if wall rupture is not to occur. The strength of the microfibrillar web through interlacing must contribute greatly to these requirements, but it is not clear how interlacing arises. However, it may be suggested that the characteristic 'kinks' observed by Bracker *et al.* (1976) in newly synthesized microfibrils, coupled with the probability of their end-synthesis from a terminal chitosome (Ruiz-Herrera *et al.*, 1975), would increase the probability of the interweaving of microfibrils synthesized in close juxtaposition – as, indeed, in-vitro preparations exhibit. It is possible that chitosomes occur in ordered arrays in the living cell, perhaps in a hexagonal arrangement similar to that of the particles supposedly synthesizing microfibrils on the outer surface of the plasmalemma of yeast (Moor & Mühlethaler, 1963). In such a case the outwardly directed microfibrils would tend to be synthesized parallel to each other. 'Kinks' would then be of especial importance in deflecting fibrils, so promoting their interweaving. Subsequently, aggregation would enhance the mechanical strength of the microfibrillar web by binding its interlocked components together. Interlacing would arise even more readily if chitosomes were not in ordered arrays, as might be the case if some were located in the chitinous region of the wall itself. It will be recalled in this context that about 85 % of the recoverable chitin synthase activity of *Mucor rouxii* is associated with cell walls (McMurrough, Flores-Carreón & Bartnicki-Garcia, 1971).

The importance of interlocking and binding bears also on the question of whether or not lytic enzymes are essential for wall growth. If the interlacing and bonding of microfibrils of determinate length are the basis of the chitin web then their slippage, combined with the simultaneous synthesis of new interlocking microfibrils, could provide a basis for wall expansion without loss of mechanical integrity, or the invocation of lytic enzymes. I do not, of course, deny the presence of lytic enzymes in fungal cell walls, the evidence for which has been summarized effectively by Rosenberger (Chapter 12 pp. 265–77). Clearly, wall lysis is necessary when lateral hyphae develop, but there is not yet convincing direct evidence for the occurrence of such enzymes at the growing apex. Indeed, teleologically speaking, the apex would seem to be a most dangerous location for a lytic entity!

One further aspect of the chitinous region in relation to growth will be considered, namely the orientation of the microfibrils. Roelofsen (1950*a, b,* 1958) examined the microfibrillar orientation of rapidly expanding Phase I sporangiophores of *Phycomyces blakesleeanus* Burgeff. He found that the microfibrils on the inner surface of the chitinous region exhibited a predominantly transverse orientation (with respect to the long axis of the sporangiophore), whereas on the outer surface of the region the orientation was predominantly axial. He supposed that this came about through the following sequence:

(i)   Microfibrils are deposited on the inner surface of the wall during growth and, in a tubular cell, they will be deposited either in a transverse orientation, or in a helix with a low pitch.

(ii) When a new set of microfibrils is deposited, the first layer will be shifted outwards and stretched in an axial direction as a consequence of elongation.

(iii) Consequently, as growth proceeds the microfibrils will become re-oriented within the wall, slipping along each other until, eventually, they adopt a predominantly axial orientation with a wider mesh on the outer surface of the chitinous zone (Fig. 1.5).

This process was described by Roelofsen as 'multinet growth'. It is characteristic of growing tubular cells with or without tip growth. In tubular cells where growth is confined exclusively to the tip, multinet growth is not found. Instead, the outer aspect of the microfibrillar region usually exhibits no preferred orientation and the arrangement is described as random or isotropic. Here the inner aspect is usually similar to the outer. Sometimes both aspects exhibit axial orientation.

It might be supposed that multinet growth would not, therefore, occur at the fungal apex, but its demonstration in *Phycomyces blakesleeanus* suggests that other fungi should be examined. In fact, in the species examined no evidence for multinet growth has been found either in germ or vegetative hyphae. Typical results for *Neurospora crassa* are illustrated in Fig. 1.6. They were obtained by using the angular frequency method of Probine & Preston (1961). In essence the angles made between a random sample of microfibrils and the long axis of hypha are measured on electron micrographs of chemically or enzymically treated hyphae and the angular distribution plotted as a histogram. Axial orientation is indicated by a bias to low angular distributions, transverse orientation by high angles and isotropic orientation either by the lack of bias, or by a peak at mid-values. Fig. 1.7. reconstructs the findings diagrammatically.

Figs. 1.6 *a, b* and *c* represent the angular frequency distributions of the microfibrils of the outer surface of the chitinous zone at the apex of chemically treated 8-h-old germ tubes (Figs. 1.6*a* and *b*) and in the region behind the zone of extension (Fig. 1.6*c*) of the same germ tube illustrated in Fig. 1.6*b*. When a conidium germinates, the germ tube is at first oval and almost as broad as it is long; from this a narrower, cylindrical germ hypha develops. Isotropic orientation is shown both by the apex initially produced (Figs. 1.6*a* and 1.7*a* and by its homologue, the swollen mid-region of a germ tube that has progressed to the later stage (Figs. 1.6*c* and 1.7*c*). In contrast the narrow elongating germ hypha (Figs. 1.6*b* and 1.7*b*), exhibits axial orientation, with 50 % or more of the microfibrils lying between 0° and 30° to the long axis of the

Fig. 1.5. Diagram to illustrate the change in orientation of microfibrils in successive layers of the microfibrillar zone in the primary wall of an elongating cylindrical cell (After Roelofsen).

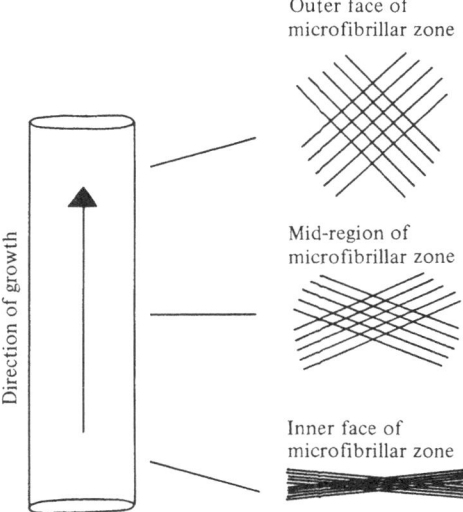

Outer face of
microfibrillar zone

Mid-region of
microfibrillar zone

Inner face of
microfibrillar zone

Direction of growth

hypha. In all cases the microfibrils at the inner surface of the chitinous zone showed an identical orientation.

Figs. 1.6*d, e* and *f* represent a similar situation at the apex of a vegetative hypha after chemical treatment (Figs. 1.6*d* and 1.7*a*) and enzymic dissection (Figs. 1.6*e* and 1.7*a*) respectively, and behind the zone of elongation after chemical treatment (Figs. 1.6*f* and 1.7*c*). In all three cases the orientation is isotropic, and that at the inner surface of the chitinous zone is similar. Very rarely, vegetative hyphae exhibit a tendency to transverse orientation but in such cases the orientations at the inner and outer surfaces are identical. Thus in *Neurospora crassa* there is no evidence for multinet growth. Indeed, this is also the case in three other species (*Schizophyllum commune, Coriolus versicolor* (Linnaeus ex Fries) Fries and *Phytophthora parasitica*), although they, in fact, show different patterns of microfibrillar orientation from that of *N. crassa*. These will not be considered further here.

Fig. 1.6. Frequency distributions of the angle between the long axis of the cell and microfibrils exposed on the surface of the chitinous region of *Neurospora crassa* by various treatments. Each distribution is based on measurements on at least one hundred randomly selected microfibrils. (*a*) Apex of the germ tube of a just-germinated conidium (see insert); isotropic orientation. (Outer surface, oxidative treatment.) (*b*) Apex of germ hypha developed from germ tube similar to (*a*) (see insert); axial orientation. (Outer surface, oxidative treatment.) (*c*) Swollen mid-part of germ hypha illustrated in (*b*) and homologous to (*a*) (see insert); isotropic orientation. (Outer surface, oxidative treatment.) (*d*) Apex of vegetative hypha; isotropic orientation. (Inner surface, oxidative treatment.) (*e*) Apex of vegetative hypha; isotropic orientation. (Outer surface, enzyme dissection.) (*f*) Mature primary wall of vegetative hypha; isotropic orientation. (Outer surface, oxidative treatment.)

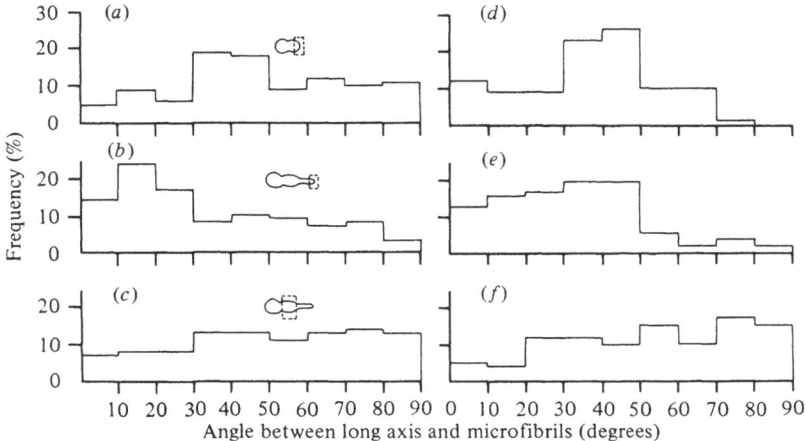

The occurrence of both isotropic and axial orientations in the same species has been seen in other organisms exhibiting tip growth, e.g. *Raphanus* root hairs (Dawes & Bowler, 1959; Scott, Bystrom & Bowler, 1963). It has been suggested by Roelofsen (1965) that the difference reflects the shape of the tip. He suggests that the more elongated and ellipsoidal the tip, the more likely is the bias to axial orientation. This difference certainly applies to the germ tube and germ hypha of *Neurospora crassa* (Fig. 1.6a and b), but there is no evidence that this is how the different orientations are determined. The matter is being further investigated. The occurrence of multinet growth in *Phycomyces blakesleeanus* may well be exceptional as, indeed, is its remarkable sporangiophore, with its long and complex growth zone (Trinci & Halford, 1975).

Coordination and regulation in wall development

The examination of the ways in which a single chemical component of one region of the wall differentiates has indicated how complex must be differentiation as a whole, involving, as it does, all the wall components. Evidently the various processes need to be precisely

Fig. 1.7. Diagrams to illustrate the principal types of orientation of microfibrils in the chitinous region at the apex and in the mature primary wall of *Neurospora crassa*.

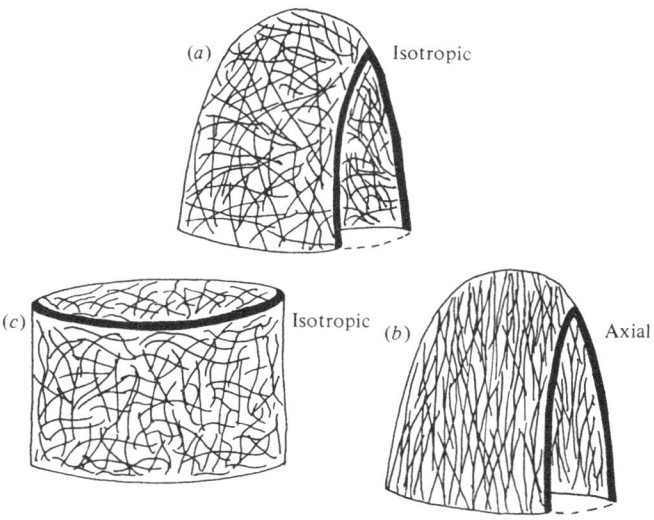

regulated and coordinated. To illustrate this the disruptive effect of replacing most of the sucrose in the medium by 4 % (w/v) sorbose, or the addition of griseofulvin to the normal medium will be considered. Fig. 1.8, based on unpublished data of Dr D. Hunsley (1971), summarizes the principal changes in wall architecture and compares them with a normal hyphal wall of *Neurospora crassa*.

Sorbose is of particular interest since its site of action is probably outside the plasmalemma. Thus aerial hyphae, free of sorbose medium, are not affected, and hyphae injected with sorbose are said to grow completely normally (Tatum, Barratt & Cutter, 1949; de Terra & Tatum, 1961). Sorbose is known to shorten cell length, increase branching – even at the hyphal tip – and induce localized bulges in the walls, although the maximum specific growth rate may only be slightly reduced (Crocker & Tatum, 1968; Trinci & Collinge, 1973). On solid medium, the width of the peripheral growth zone is reduced as are the numbers and size of apical vesicles at the hyphal tip (Trinci & Collinge, 1973). Biochemically, sorbose cultures have walls with approximately twice as much extractable glucosamine and less β-glucan than do normal walls, and the activity of $\beta(1-3)$ glucan synthase is reduced (de Terra & Tatum, 1961; Mahadevan & Tatum, 1965; Mishra & Tatum, 1972). The only published electron-micrograph of an affected wall of *Neurospora crassa* shows a greatly thickened cell wall with a dense outer fibrous-like layer (Shatkin & Tatum, 1959).

The wall is, indeed, almost three times thicker than the normal wall and even thicker opposite septa (Fig. 1.8a). Almost the only component that does not seem to be morphologically affected is the reticulum, though in places its strands may be far thicker (up to 80 nm) than in a normal wall. The region outside the reticulum can, with permanganate fixation, be seen to be made up of dense, fibrillar-like material, while with glutaraldehyde fixation it appears amorphous, apparently enclosing numerous empty vesicles. The whole outer region is susceptible to enzymic attack whether by laminarinase, pronase or chitinase. Indeed, microfibrils of the chitin region, which is much thicker than usual, can readily be revealed by any of these enzyme treatments. This contrasts with the normal situation, where only sequential treatments with laminarinase and pronase are effective (Hunsley & Burnett, 1970). The normal isotropic orientation of the microfibrils is retained, but the diameter of the microfibrils is very variable, ranging from 4.5 to 20.5 nm at the apex to 5.5 to 21.5 nm in the mature wall region, with a mean value throughout of about 11.4 nm. In addition, localized patches

Fig. 1.8. Diagrams to illustrate the effects of 4% L-sorbose (*a*) and of up to 30 μg ml^{-1} griseofulvin (*c*) on the ultrastructural architecture of the walls of *Neurospora crassa*. (*b*) is a normal wall for comparison.

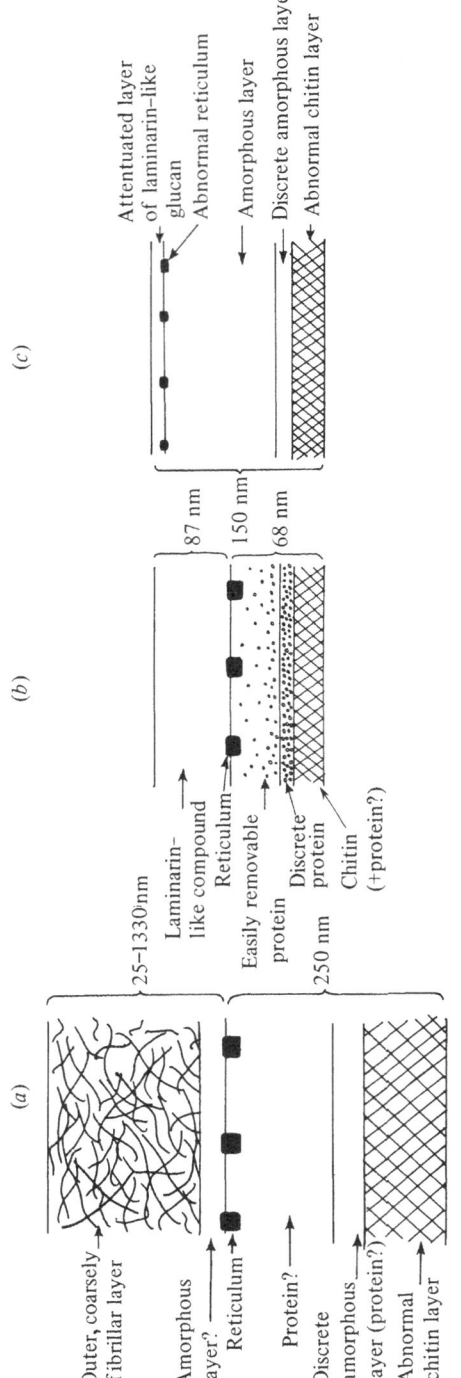

occur with mean diameters of *c.* 8 nm and others of 15.6 nm, as well as areas where the microfibrils are poorly aggregated.

The impression given is that the normal balance of wall structure and chemistry is disturbed although certain features, the reticulum and the orientation of the chitin web, appear to be relatively little affected. The existing biochemical evidence supports the observed major disturbance of the predominantly β-glucan layer but there is no obvious reason for the disturbance of the normal differentiation pattern of chitin microfibrils, or the matrix materials within the reticulum region. It may be hypothesized that β-glucan synthesis may well occur in the wall and that sorbose interferes both with this and, probably, the transport of it or its precursors to their normal location. It may also be suggested that, in the normal wall, β-glucan–chitin linkages occur but are disrupted by sorbose, resulting in some of the differences observed in the chitin region. This implies that normally β-glucan might be involved in microfibril aggregation phenomena in addition to the bonding between chitin chains. The striking feature of these largely inexplicable effects of sorbose is that coordinated wall differentiation is almost entirely lost.

Griseofulvin has similar but not quite such dramatic effects (Fig. 1.8c). In a useful review Bent & Moore (1966) summarized the effects of griseofulvin. It causes wall thickening, distortion, irregular swelling and curling of the hyphae of most chitinous fungi but not in those where chitin is absent, or only present in small amounts, e.g. Oomycetes and yeasts. In addition, it has apparently unrelated, antimitotic, colchicine-like effects. This aspect has subsequently received more attention in fungi (Heath, 1978).

Different concentrations of griseofulvin have different effects on *Neurospora crassa* hyphae and their walls, but a number of ultrastructural features are common. The reticulum and the region which normally lies outside it are most affected. The latter is very thick and consists of an irregular mass of amorphous material, including β-glucan, which is apparently sloughing off. An aberrant reticulum may be seen either at the surface, or just within this layer. The roughly hexagonal shape of its meshes is lost and the strands appear loose and disorganized. Sometimes bundles of several strands varying from 90 to 225 nm thick are apparent, or they may be replaced by single isolated strands randomly distributed, especially at higher concentrations of griseofulvin (30 μg ml^{-1}). Inside this abortive reticulum is a wide amorphous layer which is not present in a normal wall but, within it, the chitinous region is about the same thickness as usual. However, as in sorbose cultures, the microfibrils show no increase, on average, in their dimensions from the apex backwards. Moreover, in localized regions they appear to be less distinct than usual but lightly aggregated, giving a fasciculate appearance, often enhanced by the bundles appearing to be dislocated at their ends. This latter appearance is not seen after enzyme dissection and could, therefore, be an artefact induced by the chemical treatment. Even so, it does suggest that the microfibrils are more fragile than normal. These microfibrillar aggregates are often associated with amorphous material causing loss of definition: another abnormal feature. Orientation is more or less normal but axially orientated appearances are seen more often than in untreated hyphae.

Thus, like sorbose, griseofulvin affects more than one region of the hyphal wall. The outer glucan region is suppressed, the reticulum disorganized and the physical state of the chitin microfibrils disorganized. The balanced development of the wall is completely disturbed.

Certain effects are common to both agents. Localized swellings arise in both cases and, ultrastructurally, there is an absence of increasingly thicker microfibrils distal to the apex, while microfibrillar aggregation is abnormal. These observations may be held to support the view expressed earlier (pp. 9–14) that the change in microfibrillar dimensions and their aggregation are associated with the increasing rigidification of the hyphal wall behind the apex.

The observations raise more problems than they solve, but they demonstrate that the coordinated regulation of wall synthesis and the incorporation of wall components is an important and relatively unexplored aspect of wall growth.

The study of alterations imposed on the ultrastructural architecture of walls either of mutants, or of normal hyphae under different environmental conditions in *Neurospora crassa* is being continued. It is an essential complement to studies on their growth or biochemistry since it is the only technique which provides adequate spatial information bearing on coordinated and regulated growth.

There is little in this article about some of the topics defined at the outset, notably biosynthetic and transport processes. But these will be dealt with adequately and in detail, by others. In concentrating on wall architecture, whether at the chemical, ultrastructural or macroscopic level, it will be evident that there are many problems yet to be solved. In particular, I suspect that there is much diversity in wall composition that can be related either to the functional role of the wall, or to a particular growth requirement during development.

References

Albersheim, P. (1974). The primary cell wall and control of elongation growth. In *Plant Carbohydrate Biochemistry*, ed. J. B. Pridham, pp. 154–64. New York, London: Academic Press.

Aronson, J. M. & Preston, R. D. (1960). An electron microscopic and X-ray analysis of the walls of selected lower Phycomycetes. *Proceedings of the Royal Society of London*, **B152**, 346–53.

Bartnicki-Garcia, S. (1968). Cell wall chemistry, morphogenesis and taxonomy. *Annual Review of Microbiology*, **22**, 87–108.

Bartnicki-Garcia, S. & Lippman, E. (1969). Fungal morphogenesis: cell wall construction in *Mucor rouxii. Science*, **165**, 302–4.

Bent, K. J. & Moore, R. H. (1966). The mode of action of griseofulvin. In *Biochemical Studies of Antimicrobial Drugs* (16th Symposium of the Society for General Microbiology), eds. B. A. Newton & P. E. Reynolds, pp. 82–110. Cambridge University Press.

Bracker, C. E., Ruiz-Herrera, J. & Bartnicki-Garcia, S. (1976). Structure and transformation of chitin synthetase particles (chitosomes) during microfibril synthesis *in vitro. Proceedings of the National Academy of Sciences, USA*, **73**, 4570–4.

Castle, E. S. (1937). The distribution of velocities of elongation and of twist in the growth zone of Phycomyces in relation to spiral growth. *Journal of Cellular and Comparative Physiology*, **9**, 477–89.

Chang, P. Y. L. & Trevithick, J. R. (1972). Distribution of wall-bound invertase during the asexual life-cycle of *Neurospora crassa. Journal of General Microbiology*, **70**, 23–9.

Crocker, B. & Tatum, E. L. (1968). The effect of sorbose on metabolism and morphology of *Neurospora. Biochimica et Biophysica Acta*, **156**, 1–8.

Crook, E. M. & Johnston, I. R. (1962). The qualitative analysis of the cell walls of selected species of fungus. *Biochemical Journal*, **83**, 325–31.

Datema, R., Wessels, J. G. H. & van den Ende, H. (1977). The hyphal wall of *Mucor mucedo* 2. Hexosamine containing polymers. *European Journal of Biochemistry*, **80**, 621–6.

Dawes, C. J. & Bowler, E. (1959). Light and electron microscope studies of the cell wall structure of the root hairs of *Raphanus sativus. American Journal of Botany*, **46**, 561–5.

de Terra, N. & Tatum, E. L. (1961). Colonial growth of *Neurospora. Science*, **134**, 1066–8.

de Vries, O. M. H. Formation and cell wall regeneration of protoplasts from *Schizophyllum commune*. Doctoral Thesis. Rijksuniversiteit te Groningen, The Netherlands.

Distler, J. J. & Roseman, S. (1960). Galactosamine polymers produced by *Aspergillus parasiticus. Journal of Biological Chemistry*, **235**, 2534–41.

Frey-Wyssling, A. & Mühlethaler, K. (1963). Die elementarfibrillen der Cellulose. *Die Makromolekulare Chemie*, **62**, 25–30.

Gander, J. E. (1974). Fungal cell wall glycoproteins and peptido-polysaccharides. *Annual Review of Microbiology*, **28**, 103–19.

Gooday, G. W. (1971). An autoradiographic study of hyphal growth of some fungi. *Journal of General Microbiology*, **67**, 125–33.

Harold, F. M. (1962). Binding of inorganic polyphosphate to the cell wall of *Neurospora crassa. Biochimica et Biophysica Acta*, **57**, 59–66.

Harris, W. J. & Westwood, J. C. N. (1964). Phosphotungstate staining of *Vaccinia* virus. *Journal of General Microbiology*, **34**, 491–5.

Heath, I. B. (1978). *Nuclear Division in the Fungi.* New York, London: Academic Press.

Hunsley, D. (1971). Wall structure in some hyphal fungi. Ph.D Thesis, University of Newcastle upon Tyne, UK.

Hunsley, D. & Burnett, J. H. (1968). Dimensions of microfibrillar elements in fungal walls. *Nature (London)*, **218**, 462–3.

Hunsley, D. & Burnett, J. H. (1970). The ultrastructural architecture of the walls of some hyphal fungi. *Journal of General Microbiology*, **62**, 203–218.

Hunsley, D. & Kay, D. (1976). Wall structure of the *Neurospora* hyphal apex: immunofluorescent localization of wall surface antigens. *Journal of General Microbiology*, **95**, 233–48.

Lamport, D. T. A. (1965). The protein component of primary cell walls. *Advances in Botanical Research*, vol. 2, ed. R. D. Preston, pp. 151–218. New York, London: Academic Press.

Livingston, L. R. (1969). Locus-specific changes in cell wall composition characteristic of osmotic mutants of *Neurospora crassa*. *Journal of Bacteriology*, **99**, 85–90.

Mahadevan, P. R. & Mahadkar, U. R. (1970a). Major constituents of the conidial wall of *Neurospora crassa*. *Indian Journal of Experimental Biology*, **8**, 207–10.

Mahadevan, P. R. & Mahadkar, U. R. (1970b). Role of enzymes in growth and morphology of *Neurospora crassa*: cell-wall bound enzymes and their possible role in branching. *Journal of Bacteriology*, **101**, 941–7.

Mahadevan, P. R. & Rao, S. R. (1970). Enzyme degradation of conidial wall during germination of *Neurospora crassa*. *Indian Journal of Experimental Biology*, **8**, 293–7.

Mahadevan, P. R. & Tatum, E. L. (1965). Relationship of the major constituents of the *Neurospora crassa* cell wall to wild-type and colonial morphology. *Journal of Bacteriology*, **90**, 1073–81.

Mahadevan, P. R. & Tatum, E. L. (1967). The localization of structural polymers in the cell wall of *Neurospora crassa*. *Journal of Cell Biology*, **35**, 295–302.

Manners, D. J., Masson, A. J. & Patterson, J. C. (1973). The structure of a β-(1→3)-D-glucan from yeast cell walls. *Biochemical Journal*, **135**, 31–6.

Manners, D. J., Masson, A. J. & Patterson, J. C. (1974). The heterogeneity of glucan preparations from the walls of various yeasts. *Journal of General Microbiology*, **80**, 411–17.

McMurrough, I., Flores-Carreón, A. & Bartnicki-Garcia, S. (1971). Pathway of chitin synthesis and cellular localization of chitin synthetase in *Mucor rouxii*. *Journal of Biological Chemistry*, **246**, 3999–4007.

Manocha, M. S. & Colvin, J. R. (1967). Structure and composition of the cell wall of *Neurospora crassa*. *Journal of Bacteriology*, **94**, 202–12.

Mishra, N. C. & Tatum, E. L. (1972). Effect of L-sorbose on polysaccharide synthetases of *Neurospora crassa*. *Proceedings of the National Academy of Sciences, USA*, **69**, 313–17.

Moor, H. & Mühlethaler, K. (1963). Fine structure of frozen-etched yeast cells. *Journal of Cell Biology*, **17**, 609–28.

Mühlethaler, K. (1960). Die Feinstruktur der Zellulose mikrofibrillen. *Biehaft zu den Zeitschriften des Schweizerischen Forstvereins*, **30**, 55–64 (*Festschrift Albert Frey-Wyssling*).

Mühlethaler, K. (1967). Ultrastructure and formation of plant cell walls. *Annual Review of Plant Physiology*, **18**, 1–24.

Ohad, I. & Darro, D. (1964). On the dimensions of cellulose microfibrils. *Journal of Cell Biology*, **22**, 302–5.

Potgieter, H. J. & Alexander, M. (1965). Polysaccharide components of *Neurospora crassa* hyphal walls. *Canadian Journal of Microbiology*, **11**, 122–5.

Potgieter, H. J. & Alexander, M. (1966). Susceptibility and resistance of several fungi to microbial lysis. *Journal of Bacteriology*, **91**, 1526–32.

Preston, R. D. (1974). *The Physical Biology of Plant Cell Walls*. London: Chapman & Hall.

Probine, M. C. & Preston, R. D. (1961). Cell growth and the structure and mechanical properties of the wall in internodal cells of *Nitella opaca*. I. Wall structure and growth. *Journal of Experimental Botany*, **12**, 261–82.

Reissig, J. L. & Glasgow, J. E. (1971). Mucopolysaccharide which regulates growth in *Neurospora*. *Journal of Bacteriology*, **106**, 882–9.

Roelofsen, P. A. (1950a). The origin of spiral growth in *Phycomyces* sporangiophores. *Recueil des Travaux Botaniques Néerlandais*, **42**, 72–110.

Roelofsen, P. A. (1950b). Cell wall structure in the growth zone of *Phycomyces* sporangiophores. I. Model experiments and microscopical observations. *Biochimica et Biophysica Acta*, **6**, 340–56.

Roelofsen, P. A. (1958). Cell wall structure as related to surface growth. *Acta Botanica Neerlandica*, **7**, 77–89.

Roelofsen, P. A. (1959). *The Plant Cell Wall* (Encyclopedia of Plant Anatomy, vol. III part 4, eds. W. Zimmermann & P. G. Ozenda, 2nd edition) Berlin: Gebruder Borntraeger.

Roelofsen, P. A. (1965). Ultrastructure of the wall in growing cells and its relation to the direction of growth. *Advances in Botanical Research*, vol. 2 ed. R. D. Preston, pp. 67–149. New York, London: Academic Press.

Ruddall, K. M. (1968). Chitin and its association with other molecules. *Journal of Polymer Science*, **C28**, 83–102.

Ruiz-Herrera, J., Sing, V. O., Van der Woude, W. J. & Bartnicki-Garcia, S. (1975). Microfibril assembly by granules of chitin synthetase. *Proceedings of the National Academy of Sciences, USA*, **72**, 2706–10.

Scott, F. M., Bystrom, B. G. & Bowler, E. (1963). Root hairs, cuticle and pits. *Science*, **140**, 63–4.

Shatkin, A. J. & Tatum, E. L. (1959). Electron microscopy of *Neurospora crassa* mycelia. *Journal of Biophysical and Biochemical Cytology*, **6**, 423–6.

Stagg, C. M. & Feather, M. S. (1973). The characterisation of a chitin-associated D-glucan from the cell walls of *Aspergillus niger*. *Biochimica et Biophysica Acta*, **320**, 64–72.

Steele, G. C. & Trinci, A. P. J. (1975). The extension zone of mycelial hyphae. *New Phytologist*, **75**, 583–7.

Tatum, E. L., Barratt, R. W. & Cutter, V. M. (1949). Chemical induction of colonial paramorphs in *Neurospora* and *Syncephalastrum*. *Science*, **109**, 509–11.

Trevethick, J. R. & Metzenberg, R. L. (1966). Molecular sieving by *Neurospora* cell walls during secretion of invertase isozymes. *Journal of Bacteriology*, **92**, 1010–15.

Trinci, A. P. J. & Collinge, A. (1973). Influence of L-sorbose on the growth and morphology of *Neurospora crassa*. *Journal of General Microbiology*, **78**, 179–92.

Trinci, A. P. J. & Halford, E. A. (1975). The extension zone of Stage 1 sporangiophores of *Phycomyces blakesleeanus*. *New Phytologist*, **74**, 81–3.

van der Valk, P. (1976). Light and electron microscopy of cell-wall regeneration by *Schizophyllum commune* protoplasts. Doctoral Thesis. Rijksuniversiteit te Groningen, The Netherlands.

Wang, M. C. & Bartnicki-Garcia, S. (1976). Synthesis of β-1,3 glucan microfibrils by a cell-free extract from *Phytophthora cinnamomi*. *Archives of Biochemistry and Biophysics*, **175**, 351–4.

Wrathall, C. R. & Tatum, E. L. (1973). The peptides of the cell wall of *Neurospora crassa*. *Journal of General Microbiology*, **78**, 139–53.

2
Wall structure and growth in *Schizophyllum commune*

J.G.H.WESSELS AND J.H.SIETSMA
Department of Developmental Plant Biology, Biological Centre, University of Groningen, Haren, The Netherlands

Introduction

Schizophyllum commune Fries is a wood-rotting basidiomycete with 'gilled' fruit-bodies belonging to the Aphyllophorales (Donk, 1964). Among Hymenomycetes this species is the best studied from a genetic point of view, with emphasis on the incompatibility system that controls sexual morphogenesis (Raper, 1966; Raper & Raper, 1973). At the same time, interest has developed in the structural and chemical aspects of hyphal and fruit-body morphogenesis, particularly in relation to the incompatibility system (Niederpruem & Wessels, 1969; Wessels, 1978). From these studies the morphogenetic significance of degradation and synthesis of wall components has become apparent and this paper attempts to review what is known about this aspect in *S. commune*.

Chemical structure of the hyphal wall

In the water-insoluble portion of the hyphal wall of *Schizophyllum commune* three distinct homopolymers, S-glucan, R-glucan and chitin, can be recognized, together comprising about 70 % of the dry weight (Table 2.1). In addition to S-glucan, the alkali-soluble portion contains glucose, mannose, xylose and amino acids linked in an unknown way. In addition to R-glucan and chitin, the alkali-insoluble portion contains an appreciable amount of amino acids and (acetyl)glucosamine, possibly involved in covalent linkages between R-glucan and chitin (see below). The whole alkali-insoluble fraction will therefore be referred to as the R-glucan–chitin complex.

Apart from the water-insoluble components listed in Table 2.1, *Schizophyllum commune* produces variable amounts of an extracellular water-soluble polysaccharide. This can produce a high viscosity in the culture fluid after growth and it adheres to the walls as a jelly-like substance (slime, mucilage). It is, therefore, a matter of opinion to consider this substance a wall component or not. Kikumoto *et al.* (1970), who called the substance schizophyllan, showed that it is a glucan with $\beta(1-3)$ and $\beta(1-6)$ glycosidic linkages. A detailed study employing enzymolysis, periodate oxidation and methylation data (Wessels *et al.*, 1972; Sietsma & Wessels, 1977) showed that it is a glucan consisting of $\beta(1-3)$-linked chains with branches of single glucose units attached by $\beta(1-6)$ linkages at every third unit, on average, along the chain (see Fig. 2.1). Similar glucans, varying in the number of branches, are produced by a variety of fungi (Gorin & Spencer, 1968).

S-glucan was isolated from the walls of *Schizophyllum commune* as an alkali-soluble glucan that could be precipitated by acidification (Wessels, 1965). It displayed a typical X-ray diffraction pattern, both in the native wall and in the precipitated glucan. Comparison of these

Table 2.1. *Composition of the water-insoluble portion of the hyphal wall of* Schizophyllum commune[a]

Component		Proportion of dry weight (%)
Alkali-soluble polymers		
S–glucan (as glucose $-H_2O$)		21.8
Other components containing:		
Glucose $(-H_2O)$	6.8	
Mannose $(-H_2O)$	3.4	
Xylose $(-H_2O)$	0.2	
Amino acids	1.1	11.5
Alkali-insoluble polymers		
R-glucan (as glucose $- H_2O$)		39.0
Chitin (as *N*-acetylglucosamine $- H_2O$)		10.0
Other components containing:		
(*N*-acetyl)glucosamine $(-H_2O)$	2.5	
Amino acids	6.1	8.6
Lipids		3.0
		93.9

[a] Analysis of hyphal walls prepared from a 5-day-old culture of the monokaryon 699. Data calculated from Sietsma & Wessels (1977), except for the value of amino acids in the alkali-insoluble polymers which is based on the sum of individual amino acids in this fraction instead of a total ninhydrin value (5.3%) with leucine as a standard.

X-ray data with those obtained with $a(1-3)$glucan (pseudo-nigeran) from other fungi indicated the identity of these glucans (Bacon *et al.*, 1968). This was later confirmed by periodate and methylation data showing that S-glucan is an exclusively $(1-3)$-linked a-glucan (Wessels *et al.*, 1972; Sietsma & Wessels, 1977). In contrast, Siehr (1976) has reported on the occurrence of about 10 % $a(1-6)$ linkages in S-glucan. But when comparing the data of Siehr with ours it should be kept in mind that the cell-wall preparation and the S-glucan fraction were prepared in a different way. With regard to the apparent identity of the crystal spacings in the native and precipitated S-glucan it should be noted that a small difference has been observed in the diffraction patterns of native and precipitated S-glucan from *Laetiporus sulphureus* (Bull. ex Fr.) Murr. and *Piptoporus betulinus* (Bull.) Karst. although such a shift could not be detected in S-glucan from *S. commune* (Jelsma & Kreger, personal communication).

The polymeric organization of the other monomers in the hydrolysate of the alkali-soluble fraction is unknown. Xylose, mannose and the amino acids are found in the acid-precipitable fraction and may be derived from a xylan, a mannan and a protein, respectively. The glucose found in the extract after acid precipitation is predominantly $(1-3)$ linked and thus may represent low molecular weight $a(1-3)$ glucan.

The main constituents of the alkali-insoluble portion of the wall of *Schizophyllum commune* are R-glucan and chitin. Since these two polymers cannot be isolated as such without the use of agents that break covalent bonds, they must be carefully defined. Quantitatively, R-glucan refers to the total amount of glucan, as measured with the anthrone reagent; it can be extracted from the alkali-insoluble fraction by successive treatments with 0.5 M HCl (1 h, 100 °C) and 1 M KOH (20 min, 60 °C). These treatments also solubilize nearly all amino acids and some of the (acetyl)glucosamine from the fraction (see Table 2.1). The (anhydro)acetylglucosamine in the residue is referred to as chitin. This seems justified because treatment of the alkali-insoluble R-glucan–chitin complex with β-glucanases free of chitinase leave approximately the same amount of chitin as the acid–alkali treatment. Both the chemical and enzymatic extraction methods result in residues which exhibit sharp X-ray diffraction lines of chitin and which have an identical microfibrillar appearance in the electron microscope (Sietsma & Wessels, 1977; van der Valk, Marchant & Wessels, 1977).

The structure of the glucan in the R-glucan–chitin complex was examined by fractionation of the products obtained by digestion with exo-β(1–3)glucanase, periodate oxidation and methylation analysis (Sietsma & Wessels, 1977). The data indicate a highly branched β-glucan with (1–3) and (1–6) linkages. The R-glucans from different strains and even from different structures within the same strain do vary in the proportions of linkage types, chain lengths and degree of cross-linking. In general, with the R-glucan from fruit-bodies as an exception, there are more (1–6) than (1–3) linkages in the glucan. Again our results deviate from those of Siehr (1976) who determined a higher proportion of (1–3) linkages. In using the same strain (699) we found a ratio of end groups: (1–3) linkages: (1–6) linkages: branching points of 12.2:34.9:40.0:12.4, whereas Siehr observed 12.2:48.8:24.4:14.6. The high proportion of (1–3) linkages reported by Siehr is possibly because of contamination of the R-glucan with a(1–3)glucan due to his use of cold alkali to extract the S-glucan. Even our more drastic treatment with hot alkali was not completely effective in extracting S-glucan, as judged from the X-ray diffraction pattern of the residue obtained after digestion of the alkali-insoluble fraction with exo-β(1–3)glucanase.

Figure 2.1 summarizes the structure of the fragments obtained from R-glucan. Fragment A is resistant to exo-β(1–3)glucanase and comprises 16 % of the glucan. Although it is rather short (DP averages 10) it remains insoluble in close association with chitin. In the rest of the glucan, blocks of uninterrupted (1–3)-linked chains (8.5 % of the glucan, DP averages 30) and (1–6)-linked chains (29 % of the glucan, DP averages 8) occur. The other uninterrupted chains are much shorter, ranging from two to four glucose units in (1–3) linkage and two to six glucose units in (1–6) linkage. Because of the ratio of end groups to branching points, as determined by methylation analysis, these fragments must all be part of a highly branched structure (Fig. 2.1). A conspicuous feature of this structure is that many of the (1–6)-linked branches on the (1–3)-linked chains consist of only one glucose residue. This part of the R-glucan is thus very similar to the water-soluble glucan excreted by the fungus (Fig. 2.1).

In addition to glucan fragments, purified exo-β(1–3)glucanase (free of chitinase) released about 16 % of the total hexosamine in the R-glucan–chitin complex as N-acetylglucosamine (Sietsma & Wessels, 1977). This suggests that this N-acetylglucosamine does not originate from chitin but occurs in the R-glucan–chitin complex linked in an unknown way, in an abundance of 1 mol of N-acetylglucosamine to 24 mol of glucose. Probably this N-acetylglucosamine is also part of the hexosamine fraction that can be extracted from the R-glucan–chitin complex by dilute acid (Table 2.1). A similar observation was made by Troy & Koffler (1969) who found that N-acetylglucosamine was released when walls of *Penicillium chrysogenum* were treated with $\beta(1\text{–}3)$glucanase free of chitinase.

There is some evidence that covalent linkages exist between the β-glucan and chitin in the R-glucan–chitin complex. Such linkages are suggested because the structure of the β-glucan would not explain its extreme insolubility. At least the short glucan chains remaining after digestion with exo-$\beta(1\text{–}3)$glucanase would be expected to be released if not linked to an insoluble residue. To test this possibility, the chitin in the R-glucan–chitin complex was selectively degraded by deacetylation

Fig. 2.1. Structure of polysaccharides in the wall of *Schizophyllum commune*. The partial structure given for R-glucan refers to R-glucan obtained from strain 699. Symbols: ●, glucose; ○, N-acetylglucosamine. *A* represents the structure of a mixed-linked glucan fragment, possibly covalently bonded to chitin. *B* represents a branched $\beta(1\text{–}3)$ glucan fragment, the length of the branches varies from one to eight units; *B* may be linked to *A* at the points indicated by dashed lines. *C* represents short $\beta(1\text{–}3)$ linked chains (one to four units) attached to the $\beta(1\text{–}6)$ linked chains of structure *B*; *C* may also be attached to *A*.

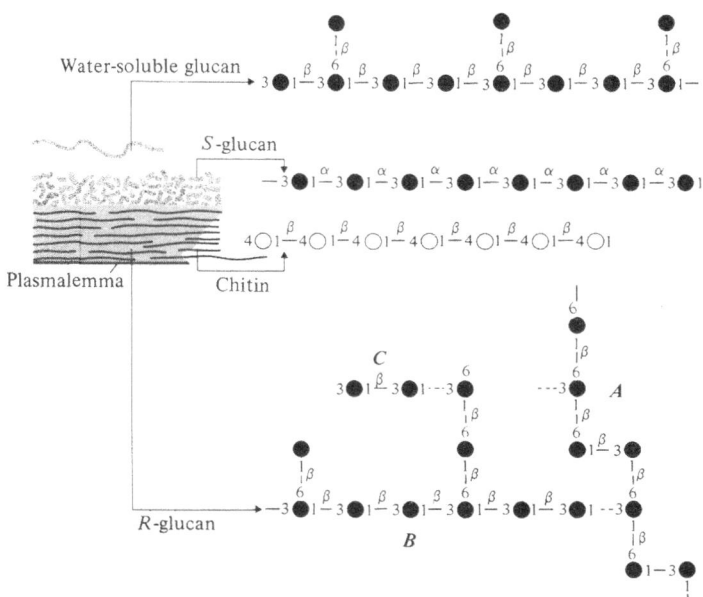

(Horton & Lineback, 1965) followed by nitrous acid treatment (Datema, Wessels & van den Ende, 1977) or by treatment with chitinase scrupulously freed from β-glucanases. The results (Sietsma & Wessels, to be published) show that after nitrous acid treatment about 90 % of the glucan becomes soluble in alkali and about 50 % even in water. Up to 10 d was required to completely digest the chitin with chitinase but, again, nearly 90 % of the glucan was released in an alkali-soluble form.

The nature of the linkages between the β-glucan and chitin remains a matter of conjecture, but the amino acids in the R-glucan–chitin complex may be involved. Among these, lysine is prominent, comprising 49 % of the amino acids on a mole basis, whereas citrulline and glutamic acid account for 20 % and 12.5 % respectively. During prolonged digestion of the R-glucan–chitin complex with a crude R-glucanase preparation, a small fragment was released which consisted of lysine and/or citrulline linked to (acetyl)glucosamine. Presumably such a fragment could be part of the bridge between the glucan and chitin. Interestingly, Wang & Bartnicki-Garcia (1970) have found that the chitinous residue remaining after successive extractions with alkali–acid–alkali of the wall of *Verticillium albo-atrum* Reinke & Berthold contained only two amino acids, lysine and histidine. They proposed that these amino acids formed the linkage between chitin and protein in the wall.

The location of polysaccharides in the wall

Knowledge of the chemical and physical identity of the various wall components makes it possible to selectively remove or stain these components and to determine their location in the wall by electron microscopy. In an earlier study, Hunsley & Burnett (1970) arrived at a model of the wall of *Schizophyllum commune* in which the various components are layered from the inside to the outside as follows: a layer of chitin microfibrils, possibly intermixed with protein; a discrete layer of protein; a layer of amorphous β-glucan; and a layer of a-glucan. A more recent study from our laboratory (van der Valk & Wessels, 1977) challenges this model; our model proposes only one inner layer, composed of an R-glucan–chitin complex containing amino acids, and it provides a number of details not observed earlier. This study arrives at a model of wall architecture (given in Fig. 2.2) which will be discussed in some detail.

The water-soluble $\beta(1–3)$, $\beta(1–6)$glucan (mucilage), when present, covers the hyphae with a gelatinous mass. This polysaccharide may infiltrate the whole wall and the large amounts that are sometimes secreted can gelatinize the whole culture medium. Repeated washing of hyphal wall preparations removes this glucan.

The outer layer of the water-insoluble portion of the wall consists of $a(1-3)$glucan (S-glucan). The abundance of S-glucan in this layer is indicated by the lack of staining by the Thiéry reagents (no periodate-sensitive sites), the removal of this outer layer by alkali and the high proportion of S-glucan in the alkaline extract (Table 2.1). The fact, however, that the β-glucan and chitin become more susceptible to enzymatic degradation after alkaline extraction of the wall (Wessels & De Vries, 1973; van der Valk *et al.*, 1977) indicates that S-glucan also inter-mixes with these components in the inner wall layer. In any case very little S-glucan is present at the inner surface of the wall since alkali does not cause any detectable morphological change of this surface.

At least part of the S-glucan in the wall occurs in a microcrystalline condition (Wessels *et al.*, 1972). No fibre structure was discernible in the compact S-glucan layer of the hyphal wall but irregular fibres (diameter 20–30 nm) with regular cross-striations were seen in well-spread preparations of reprecipitated S-glucan (van der Valk, 1976). It is possible that such fibres also occur in the native S-glucan of the hyphal wall but are not seen because they are densely packed. This is suggested by the

Fig. 2.2. Architecture of lateral wall and cross-wall of *Schizophyllum commune.*

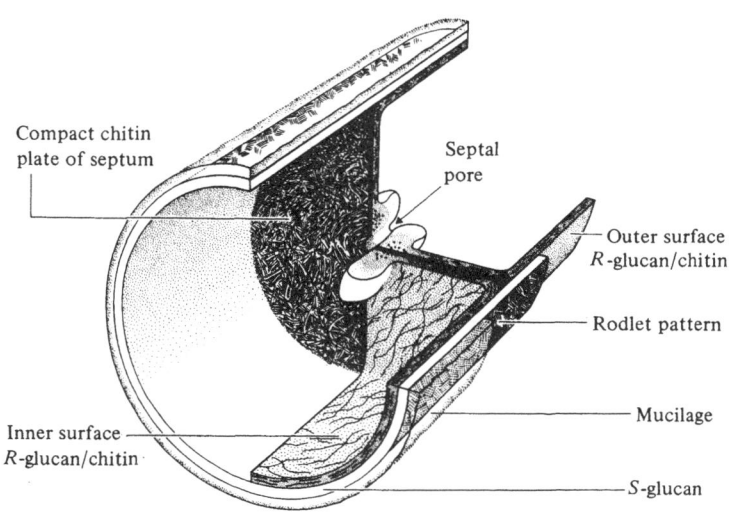

fact that similar fibres are seen in the loose mass of native *S*-glucan formed by regenerating protoplasts (van der Valk & Wessels, 1976). What can be seen on the freeze-etched outer surface of the hyphal wall is a complex pattern of small parallel arrays of rodlets with a periodicity of about 10 nm (Wessels *et al.*, 1972; Fig. 2.2). Since a similar pattern was seen on the surface of reprecipitated *S*-glucan, the pattern was regarded as reflecting the crystalline nature of *S*-glucan. However, a quite similar structure in the wall of microconidia of *Trichophyton mentagrophytes* (Robin) Blanchard has recently been attributed to a distinct glycoprotein layer which is very resistant to proteolytic digestion (Hashimoto, Wu-Yuan & Blumenthal, 1976). Thus there remains uncertainty with respect to the chemical nature of the rodlet pattern as seen in the walls of hyphae and spores of various fungi (see Burnett, 1976).

The whole inner wall which contains the *R*-glucan–chitin complex stains heavily with the Thiéry method. Since the amount of periodate consumed closely agrees with the number of (1–6) linkages in *R*-glucan, the Thiéry staining can be regarded as specific for *R*-glucan. Together with results obtained by examining shadowed preparations of the inner wall before and after selective removal of *R*-glucan with chemical and enzymatic methods, this suggests that *R*-glucan occurs throughout the inner wall and embeds the chitin microfibrils. At the outer surface of the inner wall the chitin microfibrils are thickly covered with this *R*-glucan, generating a smooth surface, whereas at the inner surface some microfibrils can be seen protruding through the *R*-glucan matrix (Fig. 2.2).

At least in the monokaryon, the septum does not seem to contain *S*-glucan. This is concluded from the facts that enzyme preparations devoid of $a(1–3)$glucanase can completely dissolve the septum (Janszen & Wessels, 1970; Wessels & Marchant, 1974) and that alkali treatment does not change the ultrastructural appearance of the septum except for a thinning of the septal swelling (van der Valk & Wessels, 1977). Essentially the septum consists of a central plate of densely interwoven chitin microfibrils anchored in the *R*-glucan–chitin complex of the lateral wall. The absence of *R*-glucan in this central plate is suggested by the absence of Thiéry staining. On both sides this chitin plate is covered by a layer similar to, and continuous with, the inner *R*-glucan–chitin layer of the lateral wall.

Around the septal pore the chitin microfibrils run circularly and are completely covered by a thick ring-shaped deposit of *R*-glucan, which forms the central part of the septal swelling. Towards the periphery of the septal swelling this *R*-glucan seems to merge into apparently highly hydrated material that constitutes the bulk of the septal swelling. Both its susceptibility to β-glucanase and its positive reaction with the Thiéry reagents suggests that this material is similar to the highly hydrated $\beta(1-3)$, $\beta(1-6)$glucan that covers the surface of the hyphae.

Enzymes involved in wall degradation

Schizophyllum commune makes several hydrolases potentially active against its wall components. Among these, $\beta(1-3)$glucanase and β-glucosidase were apparent (Schneberger & Luchsinger, 1967; Wilson & Niederpruem, 1967). Comparison of enzyme activities at different stages against laminarin, pustulan and the R-glucan–chitin complex revealed the additional existence of a $\beta(1-6)$glucanase and of an enzyme activity acting specifically on R-glucan (Wessels & Niederpruem, 1967). Sephadex filtration of an extracellular enzyme preparation (Wessels, 1969*a*) permitted the separation of a β-glucosidase, an exo-$\beta(1-3)$glucanase ($\beta(1-3)$glucan glucohydrolase), an endo-$\beta(1-3)$glucanase ($\beta(1-3,4)$glucan glucanohydrolase), a $\beta(1-6)$glucanase, and R-glucanase. This R-glucanase, an enzyme of low molecular weight (about 15 500) was tentatively classified as a $\beta(1-6)$glucan glucanohydrolase because of its activity on pustulan. Whereas all glucanases in the enzyme preparation effected degradation of yeast glucan, only the R-glucanase effected degradation of R-glucan. The reaction products were mainly large, alcohol-precipitable, water-soluble glucan fragments. Unpublished methylation data revealed that after 30 % enzymatic solubilization of the R-glucan, the percentage of (1–3) linkages and of branching points in the whole reaction mixture was hardly changed. However, there was a decrease (from 40.0 % to 36.2 %) in the proportion of (1–6) linkages and an increase (from 12.2 % to 17.0 %) in the proportion of end-groups. Since all linkage types were still represented in the released glucan fragments, it appeared that R-glucanase selectively hydrolysed a few internal (1–6) linkages in the R-glucan–chitin complex, which resulted in the release of water-soluble branched glucan fragments.

The relative resistance of R-glucan to $\beta(1-3)$glucanases, as compared with yeast glucan, can be explained on the basis of the scarcity of non-substituted $\beta(1-3)$-linked glucan chains in R-glucan. These occur abundantly in yeast glucan, which is a much less branched structure (Manners, Masson & Patterson, 1973*a*; Manners *et al.*, 1973*b*). However, the water-soluble branched glucan fragments released by R-glucanase from R-glucan are more susceptible to $\beta(1-3)$glucanase, particularly exo-β(1–3)glucanase (Wessels, 1969a), suggesting that β(1–3)glucanase-sensitive sites are opened by the R-glucanase. Such a sequential action of enzymes in degrading polymers has also been suggested in other cases (see Reese, 1977). With respect to the control of R-glucan degradation, this means that changes in R-glucanase are of primary importance, unless the activity of exo-β(1–3)glucanase becomes very high. At high activities, this enzyme has actually been used to degrade the branched β(1–3)glucan chains in R-glucan (B in Fig. 2.1) for structural studies (Sietsma & Wessels, 1977). The same enzyme is also quite active in degrading the extracellular water-soluble glucan (Fig. 2.1) giving glucose and gentiobiose (Wessels *et al.*, 1972). As expected from the branched nature of these glucans, endo-β(1–3)glucanase is ineffective (see Perlin, 1963).

At this point it should be stressed that henceforth the term R-glucanase will be used to denote any enzyme activity releasing anthrone-positive materials from the R-glucan–chitin complex. Therefore, R-glucanase activities as measured in different strains and at different times during growth are not always necessarily due to the β(1–6)glucan glucanohydrolase referred to above. For instance, in those cases where a high exo-β(1–3)glucanase develops, this enzyme may contribute somewhat to the measured R-glucanase activity.

With regard to actual degradation of R-glucan by R-glucanase *in vivo*, it should be borne in mind that R-glucan is not a defined structure but that variations exist in the proportions of linkage types in the R-glucans from different strains and structures of *Schizophyllum commune* (Sietsma & Wessels, 1977). Such differences, which may even exist in different parts of a hypha, e.g. lateral wall and septum, may influence the susceptibility of the R-glucan to R-glucanase. Growth conditions may also profoundly influence the type of R-glucan produced. For instance, it has been observed that R-glucan produced in a dikaryon grown in carbon dioxide-enriched air is rather resistant to degradation both by R-glucanase and exo-β(1–3)glucanase (Sietsma, Rast & Wessels, 1977). Another important point concerns possible physical protection of R-glucan against degradation. For instance, it was found that in a particular mutant dikaryon, the R-glucan was inaccessible to R-glucanase although the isolated R-glucan was normally susceptible (Wessels, 1966). But, in addition, in strains in which the R-glucan in the wall is accessible to the enzyme, alkali extraction of the wall greatly enhances the enzymatic degradation of R-glucan (Wessels & de Vries, 1973; van der Valk *et al.*, 1977). Similarly, R-glucan protects the chitin in the wall against degradation by chitinase, this is in agreement with the chitin microfibrils being embedded in the *R*-glucan matrix.

It is clear that any of these wall-hydrolysing enzymes may be involved in hyphal extension growth, branch initiation, hyphal fusion, and hook-cell formation and fusion during the construction of clamp connection, but no specific information has been obtained in *Schizophyllum commune*. However, evidence has accumulated to show that *R*-glucanase and chitinase are instrumental in septal dissolution during nuclear migration and that, in the dikaryon, *R*-glucanase serves the breakdown of *R*-glucan during growth of the fruit-bodies. The latter process occurs after exhaustion of the exogenous carbon supply, this exhaustion also induces the degradation of the water-soluble glucan.

Dissolution of cross-walls

Septal degradation in *Schizophyllum commune* occurs as a prerequisite to nuclear migration and is controlled by the incompatibility factors (Raper, 1966; Raper & Raper, 1973). The transient process of septal dissolution in fully compatible matings has been studied with the electron microscope (Raudaskoski, 1973; Mayfield, 1974) but most observations concern common-*A* heterokaryons (Jersild, Mishkin & Niederpruem, 1967) and homokaryons with a mutation in the *B*-incompatibility factor (Koltin & Flexer, 1969; Marchant & Wessels, 1973, 1974; Raudaskoski & Koltin, 1973), in which septal dissolution occurs continuously. This means that in the latter cases septa are normally synthesized in the apical cells but degradation may start within the hour (Niederpruem, 1971). When continuous septal degradation occurs an abundance of vesicles is observed in the cytoplasm (Raudaskoski & Koltin, 1973) at least some of which carry acid phosphatase (Raudaskoski, 1976). During septal degradation such vesicles are apparently undergoing fusion with the plasma membrane covering the cross walls, suggesting the extrusion of hydrolytic enzymes into the cross walls (Marchant & Wessels, 1974).

Of the hydrolytic enzymes possibly involved in cross-wall dissolution, the activity of *R*-glucanase only is sharply increased when septal dissolution occurs (Wessels & Niederpruem, 1967; Wessels 1969*b*; Wessels & Koltin, 1972). Little if any increase is found in the activities of $\beta(1-3)$glucanases and of chitinase.

By incubating mycelial wall preparations, which still contain the cross-walls, with hydrolytic enzymes it could be shown that a combination of *R*-glucanase and chitinase dissolves the cross-walls while leaving the lateral walls structurally intact (Janszen & Wessels, 1970; Wessels & Marchant, 1974). Chitinase alone was quite ineffective, and R-glucanase alone effected dissolution of the R-glucan component of the cross-wall but left the central chitinous plate intact. Simultaneous determination of the degree of cross-wall dissolution and of the break-down of wall polymers in the whole preparation revealed that the R-glucan–chitin in the cross-walls must be much more susceptible to R-glucanase and chitinase than the R-glucan–chitin in the lateral wall. This is probably due to the absence of S-glucan in the cross-walls. After prolonged incubation of the wall preparation with R-glucanase and chitinase, only the outer S-glucan layer of the lateral walls and the chitinous rings that anchor the cross-walls in the lateral walls remain (Fig. 2.3).

The foregoing results suggest that in the presence of a constant chitinase activity an increase in R-glucanase activity would suffice to effect cross-wall dissolution without much affecting the lateral walls. Yet in strains with continuous septal degradation the hyphal walls are apparently weakened, resulting in the formation of irregularly shaped hyphae and frequent bursting of hyphae with the extrusion of cytoplasm

Fig. 2.3. Scheme illustrating dissolution of cross-walls in hyphal wall preparations of *Schizophyllum commune* by R-glucanase and chitinase. (*a*) Before treatment with enzymes. Black corresponds to R-glucan–chitin, the hatched area represents S-glucan and the central chitinous plate of the cross-wall is indicated as white. The dotted area indicates the septal swelling round the pore. (*b*) After moderate treatment with the enzymes the cross-wall is dissolved and some R-glucan–chitin has disappeared from the lateral wall. This situation probably corresponds to that obtained *in vivo*. (*c*) Prolonged enzyme treatment only leaves the S-glucan layer and a chitinous ring from the septum.

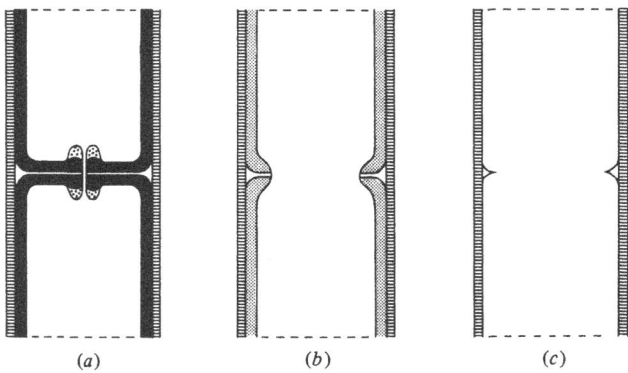

(*a*)   (*b*)   (*c*)

(Raper, 1966). Indeed we observed (Wessels, 1969*b*) that such strains accumulate only half the amount of *R*-glucan when compared with strains without septal dissolution, whereas the accumulation of the other wall components, *S*-glucan and chitin, was not affected. Whether this decreased accumulation of *R*-glucan is actually due to an increased rate of degradation by the high *R*-glucanase activity in these growing mycelia remains to be established.

In contrast to the high susceptibility of cross-walls of the monokaryon towards *R*-glucanase and chitinase, the cross-walls in wall preparations of a dikaryon were found to be quite resistant in this respect (Wessels & Marchant, 1974). This notwithstanding the fact that these enzymes were just as effective in removing *R*-glucan and chitin from the lateral walls. This indicates that the *R*-glucan in the cross-walls of the dikaryon is less susceptible to *R*-glucanase, or is somehow physically protected against degradation. Whatever the mechanism of cross-wall stability in the dikaryon, this phenomenon may have a bearing on the turning off of septal dissolution after the establishment of dikaryotic cells in a fully compatible mating. Resistance of cross-walls in the dikaryon towards enzymatic degradation would also explain why entire cross-walls can be seen in dikaryons after depletion of the external carbon supply, a condition that leads to a high *R*-glucanase activity in the dikaryon.

Wall degradation during carbon starvation

The amount of water-soluble $\beta(1-3)$, $\beta(1-6)$glucan excreted into the medium by *S. commune* decreases in older cultures of both monokaryons and dikaryons (Wang & Miles, 1964; Sietsma *et al.*, 1977; Niederpruem, Marshall & Speth, 1978). This is not surprising since both monokaryons and dikaryons produce $\beta(1-3)$glucanases (Wessels & Niederpruem, 1967) of which the exo-$\beta(1-3)$glucanase is able to degrade this branched glucan to glucose and gentiobiose (Wessels *et al.*, 1972). Part of the breakdown products may be re-utilized for the synthesis of wall polysaccharides, since *R*-glucan and *S*-glucan can increase somewhat after depletion of the glucose in the medium (Wessels, 1969*b*). Substantial activities of $\beta(1-3)$glucanases are found during the growth period, but these increase somewhat after glucose depletion. $\beta(1-3)$glucanases are also found in strains with no detectable production of water-soluble $\beta(1-3)$, $\beta(1-6)$glucan but, in strains which do produce this glucan, we have consistently found higher activities.

The net breakdown of *R*-glucan is characteristic for the dikaryon where it occurs after exhaustion of the external carbon supply. During this period a limited number of fruit-body primordia develop full grown pilei and the evidence indicates that the breakdown of *R*-glucan in stunted primordia and mycelium provides both energy and building materials for the construction of these pilei (Fig. 2.4).

Table 2.2 compares two different dikaryons with respect to the degradation of wall polymers after the external carbon supply (glucose) in the medium has become depleted. At the time the glucose is just exhausted, both strains have accumulated approximately 25 % of the added glucose in wall materials. The dikaryons differ, however, in that dikaryon K8 does not make water-soluble glucan but makes much more *S*-glucan than dikaryon 699 × 845. During the 5 to 6 days of carbon starvation that follow, both dikaryons develop pilei and degrade approximately 65 % of their *R*-glucan. During this period dikaryon 699 × 845 also consumes 68 % of its water-soluble glucan. Both dikaryons degrade little if any of their *S*-glucan and chitin.

As a response to glucose depletion, *R*-glucanase activity appears in these dikaryons but not in monokaryons under similar conditions

Fig. 2.4. Increase in the *S*-glucan:*R*-glucan ratio in the cell walls of primordia and stroma during development of pilei in *Schizophyllum commune* (strain K8) in the absence of an external carbon supply. (*a*) The initial state refers to the moment that the glucose in the medium is just depleted. (*b*) The final state is reached after 7 d of carbon starvation: a large amount of *R*-glucan in stroma and stunted primordia has been degraded, supporting the development of pilei.

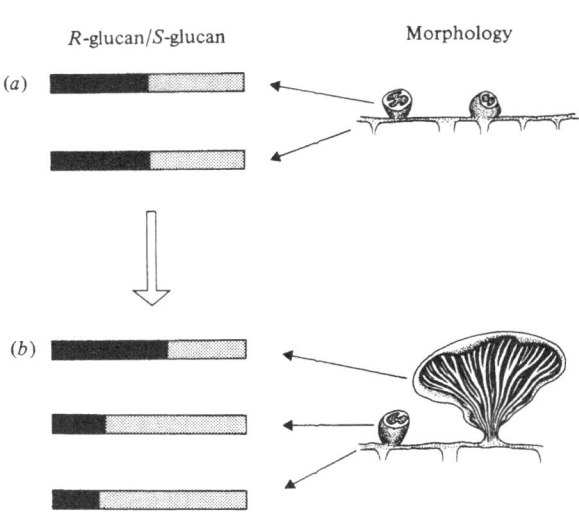

(Wessels 1966; Wessels & Niederpruem, 1967). *R*-glucanase is necessary for *R*-glucan degradation to occur, but instances are found in which *R*-glucan degradation does not ensue. Even though the enzyme activity normally develops, a mutation (Wessels, 1965, 1966) or environmental conditions prevailing during growth of the dikaryon, such as high temperature (Wessels, 1965) or high carbon dioxide concentrations (Sietsma *et al.*, 1977), may prevent *R*-glucan from being degraded *in vivo*. In the case of the mutant, evidence was found for physical protection of *R*-glucan in the wall; the isolated *R*-glucan was normally susceptible to *R*-glucanase. In the case of inhibition by high carbon dioxide concentration in the atmosphere, a decreased susceptibility of isolated *R*-glucan to *R*-glucanase was indicated as the mechanism. In none of these cases of impaired breakdown of *R*-glucan were pilei produced.

This raises the question of the causal relationship between *R*-glucan degradation in abortive primordia and mycelium and the development of pilei. It was found that the development of pilei is suppressed by glucose in the medium in the absence of an exogenous nitrogen supply. On the other hand, the maintenance of a steady but low concentration glucose in the medium was conducive to the formation of pilei in a mutant unable to degrade its *R*-glucan or to produce fruit-bodies developed beyond the cup stage. Therefore, the theory was advanced (Wessels, 1965) that the slow degradation of a wall polymer, such as *R*-glucan, in the vegetative mycelium and in stunted fruit-body primordia provides for a steady flow of carbon compound at low concentration

Table 2.2. *Decrease in wall polymers of two different dikaryons of* Schizophyllum commune *during carbon starvation and pileus growth*

	K8		699 × 845	
Time (d)	7	12	6	12
Water-soluble glucan	0	0	77.9	24.4
S-glucan	80.1	70.6	27.3	27.2
R-glucan	55.6	17.9	57.6	19.2
Chitin	5.2	4.1	4.8	5.0
S-glucan:R-glucan ratio	1.44	3.94	0.47	1.42

The glucans are expressed as mg anhydroglucose per culture and the chitin as mg anhydro-*N*-acetylglucosamine per culture; cultures containing 30 ml of medium with initially 600 mg of glucose as a carbon source. The first column for each dikaryon represents an analysis on the day the glucose in the medium just became depleted. Data from Wessels (1965) and Sietsma *et al.* (1977).

towards the developing pilei in normal situations. On the one hand, this would provide for both the necessary building material and the energy requirement of the developing pilei; on the other hand, suppression of pileus development by too-high concentrations of a carbon compound, as generated by an exogenous supply of glucose, would not occur. On this view, R-glucan would simply serve as an endogenous reserve polysaccharide, the breakdown of which ensures the maintenance of a critical concentration of metabolites at the site of pileus formation.

In addition, R-glucan degradation may fulfil a more specific role, being part of a general degradative process initiated in the vegetative mycelium and stunted fruit-body primordia. In the absence of an exogenous nitrogen source the developing pilei not only receive carbohydrates but also nitrogenous compounds from pre-existing cells, as evidenced by measurements of total nitrogen (Wessels, 1965). In fact, after pileus development the hyphae of vegetative mycelium and stunted fruit-body primordia have lost much of their cytoplasmic contents. Therefore, R-glucan degradation appears to be part of a controlled, general degradative process in certain hyphae coupled with the transport of the breakdown products to the sites of synthesis in developing pilei. In this respect the breakdown of R-glucan in certain cells, coupled to immediate absorption and translocation of the breakdown product, would probably provide a much more efficient system than re-utilization of the breakdown products of an exogenous polysaccharide.

The re-utilization of previously formed cell constituents, including wall materials, has long been suspected from cytological work (Lohwag, 1941). But only recently have quantitative analyses of cell components during carpophore construction in other basidiomycetes (Kitamoto & Gruen, 1976; Robert, 1977*a*, *b*) and cleistothecium formation in *Aspergillus nidulans* (Eidam) Winter (Zonneveld, 1972, 1974; Polacheck & Rosenberger, 1977) shown the generality of the processes first demonstrated in *S. commune*. The latter case is particularly analogous to *S. commune* because here, too, the breakdown of a wall polysaccharide is essential. Failure to perform this degradation prevents the formation of cleistothecia. Unlike *S. commune*, however, in this case the *S*-glucan component is the major wall reserve polysaccharide.

Spore formation in carpophores may also draw heavily on pre-existing materials. Corner (1932) has shown 'corrosion' of walls in detached fruit-bodies of a polypore and Bromberg & Schwalb (1976) found in *S. commune* that, even in the presence of glucose in the medium, about

30 % of the materials of discharged spores derives from previously synthesized materials.

Synthesis of wall components

Table 2.2 shows that considerable quantitative variations in wall polysaccharides can occur within a definable morphological phenotype of *S. commune*. Such quantitative differences in wall polysaccharides in dikaryons have also been noted by Schwalb (1977). Apparently, genetic differences not related to the expression of morphological traits are responsible for this variation. Conversely, it becomes very difficult to relate such quantitative differences as found in morphological mutants (Wang, Schwalb & Miles, 1968) to the expressed phenotype. By comparing co-isogenic strains, the expression of the B-factor on phenotype (dissolved septa and gnarled knobby hyphae) has been related to the synthesis of walls with a decreased amount of R-glucan (Wessels, 1969b; Wessels & Koltin, 1972), but, as discussed earlier, this was probably due to increased degradation rather than decreased synthesis of this glucan. Environmental conditions may also drastically change the proportions of the various wall polysaccharides. For instance, growth of a dikaryon in air supplemented with 5% carbon dioxide causes a shift in the S-glucan : R-glucan ratio from 0.47 to 0.97, while twice the amount of water-soluble glucan is excreted by the hyphae (Sietsma *et al.*, 1977).

Yet there is little doubt that many processes related to hyphal morphogenesis are intimately associated with the synthesis of wall components. The main obstacle to unravelling this relationship is our ignorance concerning the enzymic machinery that fabricates the wall, and the cytoplasmic structures that are involved in the localization of wall synthesis in time and space. As in other mycelial fungi, the incorporation of wall precursors during extension growth of mycelial hyphae of *Schizophyllum commune* mainly occurs at the hyphal apex (Gooday, 1971). In these hyphae, little synthesis can be detected in subapical regions except that synthesis is resumed at the sites of septum formation and lateral branch initiation (van der Valk & Wessels, 1977). In the dikaryon, the backward-growing hook cell that forms the clamp is also initiated as a new apical growth centre at a specific site and at a specific time in the cell cycle. Girbardt (1973) has suggested that the lateral position of the apical assembly of vesicles (his '*Spitzenkörper*') is of prime importance in determining the direction of growth of the hook cell, because it changes the pattern of wall synthesis at the apex. In accordance with such a view of cytoplasmic organization determining the direction of growth we have succeeded in forcing the hook cells to grow in a 'forward' direction by growing the dikaryon in a centrifugal field (unpublished). There is also the problem of secondary wall thickening as it occurs in fruit-bodies and stroma layers (Niederpruem & Wessels, 1969; van der Valk & Marchant, 1978) in addition to the many other examples of cellular morphogenesis that occur in this and other systems in which wall synthesis plays a crucial role.

To begin to understand these processes of wall synthesis, particularly in relation to growth and morphogenesis of *Schizophyllum commune*, we have done a study on wall synthesis by protoplasts of this organism. Since the wall polymers are relatively well characterized we hoped to visualize the synthesis of individual components on the naked plasma membrane. In addition, the use of protoplasts would permit the gentle isolation of subcellular structures functioning in wall synthesis. Since this work has been reviewed recently (Wessels, van der Valk & de Vries, 1976) and is also the subject of another paper in this symposium (by Peberdy, Chapter 3), a very brief summary will suffice here.

A method was developed for the efficient production of protoplasts from *Schizophyllum commune* mycelium, employing an enzyme cocktail prepared from the cultural fluid of *Trichoderma harzianum* Rifai. This cocktail is different from most other enzyme mixtures used, in that it contains $a(1-3)$glucanase (de Vries & Wessels, 1972; 1973*a*). This enzyme cocktail is also very effective with other fungi (de Vries & Wessels, 1973*b*) and is now used by many other workers. By using magnesium sulphate as an osmotic stabilizer, nucleated protoplasts without any mycelial debris could be obtained by flotation. All of these protoplasts regenerated a wall, and about 50 % reverted to hyphal growth in a liquid medium. The course of wall synthesis was followed by chemical analysis (de Vries & Wessels, 1975), electron microscopy (van der Valk & Wessels, 1976), and autoradiography (van der Valk & Wessels, 1977). The first components detected on the naked plasma membrane are chitin microfibrils and *S*-glucan, and this occurs even if protein synthesis is completely inhibited. A few hours later, however, only the chitin microfibrils are formed, together with the embedding *R*-glucan. In hyphal tubes that form from the regenerated protoplasts, the synthesis of these two components is coupled from the start. This accounts for the fact that, after removal of the *S*-glucan layer that covers both regenerated protoplasts and emerging hyphae, shadowed preparations reveal non-embedded chitin microfibrils at the outside of the protoplast walls, whereas in the hyphal initials the microfibrils are completely buried in the R-glucan matrix. Polyoxin D, regarded as a specific inhibitor of chitin synthesis, prevents both the synthesis of chitin and R-glucan. This suggests an intimate relationship between the synthesis of these two components, possibly based on the existence of covalent linkages between them, as discussed earlier.

Electron microscopic autoradiography shows that the synthesis of chitin, even in very thick-walled cells, occurs in close association with the plasma membrane (van der Valk & Wessels, 1977). Recent experiments (to be published) have shown that membrane preparations from mycelium and protoplasts are very active in synthesizing crystalline microfibrillar chitin from UDP-N-acetylglucosamine. Experiments employing membranes from Concanavalin A coated protoplasts have shown that the plasma membrane does indeed contain chitin synthase, and that an inactive form of the enzyme is present in other subcellular fractions.

References

Bacon, J. S. D., Jones, D., Farmer, V. C. & Webley, D. M. (1968). The occurrence of $a(1-3)$glucan in *Cryptococcus, Schizosaccharomyces* and *Polyporus* species, and its hydrolysis by a *Streptomyces* culture filtrate lysing cell walls of *Cryptococcus*. *Biochimica et Biophysica Acta*, **158**, 313–15.

Bromberg, S. K. & Schwalb, M. N. (1976). Studies on basidiospore development in *Schizophyllum commune*. *Journal of General Microbiology*, **96**, 409–13.

Burnett, J. H. (1976). *Fundamentals of Mycology*, 2nd edition. London: Edward Arnold.

Corner, E. J. H. (1932). The fruit body of *Polystictus xanthopus* Fr. *Annals of Botany*, **46**, 71–111.

Datema, R., Wessels, J. G. H. & van den Ende, H. (1977). The hyphal wall of *Mucor mucedo*. 2. Hexosamine-containing polymers. *European Journal of Biochemistry*, **80**, 621–7.

Donk, M. A. (1964). A conspectus of the families of Aphyllophorales. *Persoonia*, **3**, 199–324.

Girbardt, M. (1973). Die Pilzzelle. In *Grundlagen der Cytologie*, eds. G. C. Hirsch, H. Ruska & P. Sitte, pp. 441–60. Jena: Fischer-Verlag.

Gooday, G. W. (1971). An autoradiographic study of hyphal growth of some fungi. *Journal of General Microbiology*, **67**, 125–33.

Gorin, P. A. J. & Spencer, J. F. T. (1968). Structural chemistry of fungal polysaccharides. *Advances in Carbohydrate Chemistry*, **23**, 367–417.

Hashimoto, T., Wu-Yuan & Blumenthal, H. J. (1976). Isolation and characterization of the rodlet layer of *Trichophyton mentagrophytes* microconidial wall. *Journal of Bacteriology*, **127**, 1543–9.

Horton, D. & Lineback, D. R. (1965). *N*-deacetylation. Chitosan from chitin. In *Methods in Carbohydrate Chemistry*, vol. 5, ed. R. L. Whistler, pp. 403–6. New York, London: Academic Press.

Hunsley, D. & Burnett, J. H. (1970). The ultrastructural architecture of the walls of some hyphal fungi. *Journal of General Microbiology*, **62**, 203–18.

Janszen, F. H. A. & Wessels J. G. H. (1970). Enzymic dissolution of hyphal septa in a basidiomycete. *Antonie van Leeuwenhoek. Journal of Microbiology and Serology*, **36**, 255–7.

Jersild, R., Mishkin, S. & Niederpruem, D. J. (1967). Origin and ultrastructure of complex septa in *Schizophyllum commune* development. *Archiv für Mikrobiologie*, **57**, 20–32.

Kikumoto, S., Miyajima, T., Yoshizumi, S., Fujimoto, S. & Kimura, K. (1970). Polysaccharide produced by *Schizophyllum commune*. Part I. Formation and some properties of an extracellular polysaccharide. *Journal of the Agricultural and Chemical Society (Japan)*, **44**, 337–42.

Kitamoto, Y. & Gruen, H. E. (1976). Distribution of cellular carbohydrates during development of the mycelium and fruit bodies of *Flammulina velutipes*. *Plant Physiology*, **58**, 485–91.

Koltin, Y. & Flexer, A. S. (1969). Alteration of nuclear distribution in *B*-mutant strains of *Schizophyllum commune*. *Journal of Cell Science*, **4**, 739–49.

Lohwag, H. (1941). Anatomie der Asco- und Basidiomyceten. In *Handbuch der Pflanzenanatomie*, vol 6 (II, 3c), Ed. K. Linsbauer, pp. 309–58. Berlin, Stuttgart: Borntraeger.

Manners, D. J., Masson, A. J. & Patterson, J. C. (1973a). The structure of a β-(1–3)-D-glucan from yeast cell walls. *Biochemical Journal*, **135**, 19–30.

Manners, D. J., Masson, A. J., Patterson, J. C., Björndal, H. & Lindberg, B. (1973b). The structure of a β(1–6)-D-glucan from yeast cell walls. *Biochemical Journal*, **135**, 31–6.

Marchant, R. & Wessels, J. G. H. (1973). Septal structure in normal and modified strains of *Schizophyllum commune* carrying mutations affecting septal dissolution. *Archiv für Mikrobiologie*, **90**, 35–45.

Marchant, R. & Wessels, J. G. H. (1974). An ultrastructural study of septal dissolution in *Schizophyllum commune*. *Archives of Microbiology*, **96**, 175–82.

Mayfield, J. (1974). Septal involvement in nuclear migration in *Schizophyllum commune*. *Archives of Microbiology*, **95**, 115–24.

Niederpruem, D. J. (1971). Kinetic studies of septum synthesis, erosion and nuclear migration in a growing *B*-mutant of *Schizophyllum commune*. *Archiv für Mikrobiologie*, **75**, 189–96.

Niederpruem, D. J. & Wessels, J. G. H. (1969). Cytodifferentiation and morphogenesis in *Schizophyllum commune*. *Bacteriological Reviews*, **33**, 505–35.

Niederpruem, D. J., Marshall, C. & Speth, J. L. (1978). Control of extracellular slime accumulation in monokaryons and resultant dikaryons of *Schizophyllum commune*. *Sabouraudia*, **15**, 283–95.

Perlin, A. S. (1963). The action of β-glucanases on β-glucans of mixed linkage. In *Advances in Enzymatic Hydrolysis of Cellulose and Related Materials*, ed. E. T. Reese, pp. 185–95. Oxford, New York: Pergamon Press.

Polacheck, I. & Rosenberger, R. F. (1977). *Aspergillus nidulans* mutant lacking a-(1,3)-glucan, melanin, and cleistothecia. *Journal of Bacteriology*, **132**, 650–6.

Raper, J. R. (1966). *Genetics of Sexuality in Higher Fungi*. New York: The Ronald Press.

Raper, J. R. & Raper, C. A. (1973). Incompatibility factors: regulatory genes for sexual morphogenesis in higher fungi. *Brookhaven Symposia in Biology*, **25**, 19–38.

Raudaskoski, M. (1973). Light and electron microscope study of unilateral mating between a secondary mutant and a wild-type strain of *Schizophyllum commune*. *Protoplasma*, **76**, 35–48.

Raudaskoski, M. (1976). Acid phosphatase activity in the wild-type and *B*-mutant hyphae of *Schizophyllum commune*. *Journal of General Microbiology*, **94**, 373–9.

Raudaskoski, M. & Koltin, Y. (1973). Ultrastructural aspects of a mutant of *Schizophyllum commune* with continuous nuclear migration. *Journal of Bacteriology*, **116**, 981–8.

Reese, E. T. (1977). Degradation of polymeric carbohydrates by microbial enzymes. In

Recent Advances in Phytochemistry, vol. 11, eds. F. Loewus & V. C. Runeckles, pp. 311–67. New York: Plenum Press.

Robert, J. C. (1977*a*). Fruiting of *Coprinus congregatus*: biochemical changes in fruit-bodies during morphogenesis. *Transactions of the British Mycological Society,* **68,** 379–387.

Robert, J. C. (1977*b*). Fruiting of *Coprinus congregatus:* relationships to biochemical changes in the whole culture. *Transactions of the British Mycological Society,* **68,** 389–95.

Schneberger, G. L. & Luchsinger, W. W. (1967). Beta-D-glucanases of *Schizophyllum commune. Canadian Journal of Microbiology,* **13,** 969–78.

Schwalb, M. (1977). Cell wall metabolism during fruiting of the basidiomycete *Schizophyllum commune. Archives of Microbiology,* **114,** 9–12.

Siehr, D. J. (1976). Studies on the cell wall of *Schizophyllum commune.* Permethylation and enzymatic hydrolysis. *Canadian Journal of Biochemistry,* **54,** 130–6.

Sietsma, J. H. & Wessels, J. G. H. (1977). Chemical analysis of the hyphal wall of *Schizophyllum commune. Biochimica et Biophysica Acta,* **496,** 225–39.

Sietsma, J. H., Rast, D. and Wessels, J. G. H. (1977). The effect of carbon dioxide on fruiting and on the degradation of a cell-wall glucan in *Schizophyllum commune. Journal of General Microbiology,* **102,** 385–9.

Troy, F. A. & Koffler, H. (1969). The chemistry and molecular structure of the cell walls of *Penicillium chrysogenum. Journal of Biological Chemistry,* **244,** 5563–76.

Valk, P. van der (1976). Light and electron microscopy of cell-wall regeneration by *Schizophyllum commune* protoplasts. Thesis, University of Groningen.

Valk, P. van der & Marchant, R. (1978). Hyphal ultrastructure in fruit-body primordia of the basidiomycetes *Schizophyllum commune* and *Coprinus cinereus. Protoplasma,* **95,** 57–72.

Valk, P. van der & Wessels, J. G. H. (1976). Ultrastructure and localization of wall polymers during regeneration and reversion of protoplasts of *Schizophyllum commune. Protoplasma,* **90,** 65–87.

Valk, P. van der & Wessels, J. G. H. (1977). Light and electron microscopic autoradiography of cell-wall regeneration by *Schizophyllum commune* protoplasts. *Acta Botanica Neerlandica,* **26,** 43–52.

Valk, P. van der, Marchant, R. & Wessels, J. G. H. (1977). Ultrastructural localization of polysaccharides in the wall and septum of the basidiomycete *Schizophyllum commune. Experimental Mycology,* **1,** 69–82.

Vries, O. M. H. de & Wessels, J. G. H. (1972). Release of protoplasts from *Schizophyllum commune* by a lytic preparation from *Trichoderma viride. Journal of General Microbiology,* **73,** 13–22.

Vries, O. M. H. de & Wessels, J. G. H. (1973*a*). Release of protoplasts from *Schizophyllum commune* by combined action of purified *a*-1,3-glucanase and chitinase derived from *Trichoderma viride. Journal of General Microbiology,* **76,** 319–30.

Vries, O. M. H. de & Wessels, J. G. H. (1973*b*). Effectiveness of a lytic enzyme preparation from *Trichoderma viride* in releasing spheroplasts from fungi, particularly basidiomycetes. *Antonie van Leeuwenhoek. Journal of Microbiology and Serology,* **39,** 397–400.

Vries, O. M. H. de & Wessels, J. G. H. (1975). Chemical analysis of cell wall regeneration and reversion of protoplasts from *Schizophyllum commune. Archives of Microbiology,* **102,** 209–18.

Wang, C. S. & Miles, P. G. (1964). The physiological characterization of dikaryotic mycelia of *Schizophyllum commune. Physiologia Plantarum,* **17,** 573–88.

Wang, C. S., Schwalb, M. N. & Miles, P. G. (1968). A relationship between cell wall composition and mutant morphology in the basidiomycete *Schizophyllum commune. Canadian Journal of Microbiology,* **14,** 809–811.

Wang, M. C. & Bartnicki-Garcia, S. (1970). Structure and composition of walls of the yeast form of *Verticillium albo-atrum. Journal of General Microbiology,* **64,** 41–54.

Wessels, J. G. H. (1965). Morphogenesis and biochemical processes in *Schizophyllum commune* Fr. *Wentia,* **13,** 1–113.

Wessels, J. G. H. (1966). Control of cell-wall glucan degradation in *Schizophyllum commune. Antonie van Leeuwenhoek. Journal of Microbiology and Serology,* **32,** 341–55.

Wessels, J. G. H. (1969*a*). A *β*-1,6-glucan glucanohydrolase involved in hydrolysis of cell-wall glucan in *Schizophyllum commune. Biochimica et Biophysica Acta,* **178,** 191–3.

Wessels, J. G. H. (1969*b*). Biochemistry of sexual morphogenesis in *Schizophyllum commune*: Effect of mutations affecting the incompatibility system on cell-wall metabolism. *Journal of Bacteriology,* **98,** 697–704.

Wessels, J. G. H. (1978). Incompatibility factors and the control of biochemical processes. In *Genetics and Morphogenesis of Higher Basidiomycetes*, eds M. Schwalb & P. G. Miles, pp. 81–104. New York, London: Academic Press.

Wessels, J. G. H. & Koltin, Y. (1972). *R*-glucanase activity and susceptibility of hyphal walls to degradation in mutants of *Schizophyllum* with disrupted nuclear migration. *Journal of General Microbiology,* **71,** 471–5.

Wessels, J. G. H. and Marchant, R. (1974). Enzymic degradation of septa in hyphal wall preparations from a monokaryon and a dikaryon of *Schizophyllum commune. Journal of General Microbiology,* **83,** 359–68.

Wessels, J. G. H. & Niederpruem, D. J. (1967). Role of a cell-wall glucan-degrading enzyme in mating of *Schizophyllum commune. Journal of Bacteriology,* **94,** 1594–1602.

Wessels, J. G. H., & Vries, O. M. H. de (1973). Wall structure, wall degradation, protoplast liberation and wall regeneration in *Schizophyllum commune*. In *Yeast, Mould and Plant Protoplasts*, eds J. R. Villanueva, I. Garcia Acha, S. Gascón & F. Uruburo, pp. 295–306. New York, London: Academic Press.

Wessels, J. G. H., Kreger, D. R., Marchant, R., Regensburg, B. A. & Vries, O. M. H. de (1972). Chemical and morphological characterization of the hyphal wall surface of the Basidiomycete *Schizophyllum commune. Biochimica et Biophysica Acta,* **273,** 346–58.

Wessels, J. G. H., Valk, P. van der & Vries, O. M. H. de (1976). Wall synthesis by fungal protoplasts. In *Microbial and Plant Protoplasts*, eds J. F. Peberdy, A. H. Rose, H. J. Rogers & E. C. Cocking, pp. 267–81. New York, London: Academic Press.

Wilson, R. W. & Niederpruem, D. J. (1967). Control of *β*-glucosidases in *Schizophyllum commune. Canadian Journal of Microbiology,* **13,** 1009–20.

Zonneveld, B. J. M. (1972). Morphogenesis in *Aspergillus nidulans. The significance of a*(1–3)glucan of the cell wall and *a*(1–3)glucanase for cleistothecium development. *Biochimica et Biophysica Acta,* **273,** 174–87.

Zonneveld, B. J. M. (1974). *a*(1–3)glucan synthesis correlated with *a*(1–3)glucan synthesis, conidiation and fructification in morphogenetic mutants of *Aspergillus nidulans. Journal of General Microbiology,* **81,** 445–51.

3
Wall biogenesis by protoplasts

J.F.PEBERDY

Department of Botany, University of Nottingham, Nottingham, UK

Introduction

In the two decades since the first reports of the isolation and cultural characteristics of fungal protoplasts (Emerson & Emerson, 1958; Bachmann & Bonner, 1959; Eddy & Williamson, 1959) an extensive literature has accumulated about their regenerative properties. The patterns of development exhibited by protoplasts in culture have provided interesting systems for investigations of wall biogenesis and the role of the wall as a determinant of form and shape.

The literature is most extensive in the fields of light and electron microscopy of protoplasts, reflecting, at least in the early period, the technical problems associated with the production of large quantities of protoplasts. In recent years, methods have been scaled up and more effective enzyme cocktails have been developed, providing the opportunity to produce large yields of protoplasts. Thus it is now possible to investigate the chemistry of the regenerated wall, and other biochemical events associated with regeneration and reversion.

An advantage of using protoplasts in wall research is the relative rapidity of wall regeneration and reversion to normal morphology. There are, however, certain disadvantages: firstly, protoplast preparations are very heterogeneous, especially in their biochemical activities; secondly the frequency of reversion in protoplast preparations is highly variable; and finally protoplast regeneration and reversion proceed asynchronously. The latter two features could be expressions of the biochemical heterogeneity. Nevertheless, studies on protoplast regeneration and reversion have provided an essential background for recent studies in which protoplasts have been used in the genetic modification of fungi. The success of protoplast fusion or transformation of protoplasts depends ultimately on the recovery of viable organisms by reversion.

Several reviews on regeneration of protoplasts from yeasts and filamentous fungi have been published (Necas, 1971; Villanueva & Garcia Acha, 1971; Peberdy, 1972; Wessels, van der Valk & de Vries, 1976).

Frequency of protoplast reversion

The frequency of reversion in protoplasts from a particular fungus is never 100 %. Bachmann & Bonner (1959) reported reversion of preparations of *Neurospora crassa* Shear & Dodge protoplasts varying between 20 and 80 %. Garcia Acha, Lopez-Belmonte & Villanueva (1966) showed that regeneration and reversion of *Fusarium culmorum* (W. G. Smith) Saccardo protoplasts was influenced by the nature of the carbon source; at best, reversion (based on colonies developing from protoplasts plated in agar media) was 82 %, and at worst 5 %. Fukui *et al.* (1969) estimated regeneration of protoplasts of *Geotrichum candidum* Link ex Persoon by counting the number of osmotically stable cells; they observed 100 % regeneration. A high level (85–100 %) of wall regeneration was also reported for protoplasts of *Schizophyllum commune* Fries (de Vries & Wessels, 1975) which floated after centrifugation; of these floating protoplasts, 50 % were capable of reversion to the normal hyphal form. Reversion frequencies obtained for *Aspergillus nidulans* (Eidam) Winter were generally lower (10–30 %) than reported for other fungi (Isaac, 1978).

The inability of some protoplasts from filamentous fungi to revert to their normal morphology is not understood. The non-reverting protoplasts may lack a nucleus (Garcia Acha *et al.*, 1966) but other factors may also be important in determining whether or not regeneration occurs. In *Aspergillus nidulans* the proportion of enucleate protoplasts in a preparation can be quite high (Fig. 3.1), e.g. up to 50 % of protoplasts released in the first hour lack a nucleus (Isaac, 1978). Gibson & Peberdy (1972) suggested that the protoplasts released during this early period probably arise from the tips of the hyphae. Calculations based on the data of Fiddy & Trinci (1976) (Fig. 3.2) indicate that the nuclear-free zone at the tips of leading hyphae of *A. nidulans* could yield up to ten enucleate protoplasts. After longer periods of lytic digestion, protoplasts are released from the distal regions of hyphae and it is possible that many of these may lack the organellar and biochemical requirements for regeneration and reversion.

Fig. 3.1. Number of nuclei per protoplast in *Aspergillus nidulans*. Protoplasts produced after 1 h (white bars) 3 h (dark bars) lytic digestion using (*a*) KCl and (*b*) MgSO₄ as osmotic stabilizers were fixed and stained in Giemsa stain. The estimations are based on counts of 250 protoplasts. (S. Isaac, unpublished results.)

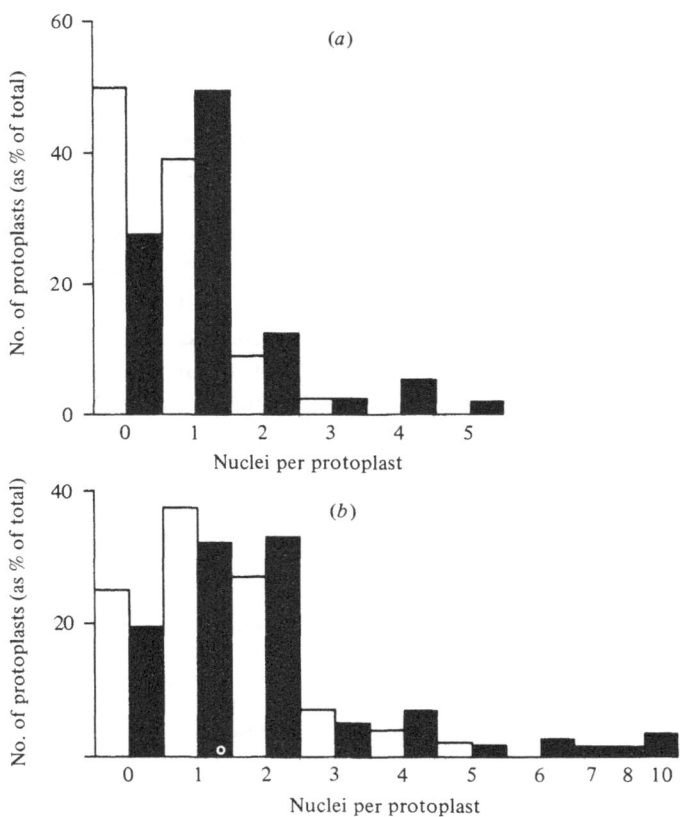

Fig. 3.2. Diagrammatic representation of the nuclear-free zone at the hyphal tip in *Aspergillus nidulans*. Dimensions are based on data from Fiddy & Trinci (1976). Volume of nuclear-free zone = 141 μm³; protoplast volumes: 3 μm diameter = 14 μm³; 5 μm diameter = 65 μm³.

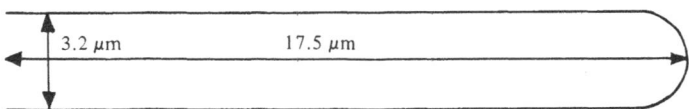

There are only a few reports on the reversion frequency of yeast protoplasts. Necas (1962) reported the mass reversion of *Saccharomyces cerevisiae* Hansen protoplasts in a medium containing gelatin, and Svoboda (1966) demonstrated 50 to 70 % reversion for the same organism in agar medium. In *Schizosaccharomyces pombe* Lindner the frequency of reversion to normal cells was 50 % in liquid medium and 90 % in gelatin medium (Havelkova, 1969).

Morphology and cytology of protoplast reversion
Yeasts

Reversion of protoplasts from *Saccharomyces* spp. occurs only on solid media containing agar or gelatin (Necas, 1961; Svoboda, 1966; Svoboda & Necas, 1966). The reversion of *Saccharomyces cerevisiae* protoplasts has been described by Necas (1971). The wall synthesized by the protoplasts retains the spherical shape of the protoplast and not the characteristic ellipsoidal shape of normal yeast cells (Fig. 3.3). During the period of wall regeneration the protoplasts increase in size

Fig. 3.3. Regeneration and reversion of protoplast of *Saccharomyces cerevisiae*. The regenerated protoplast (arrowed) retains a roughly spherical shape and forms irregularly shaped buds at the first generation. (From Necas, 1971.)

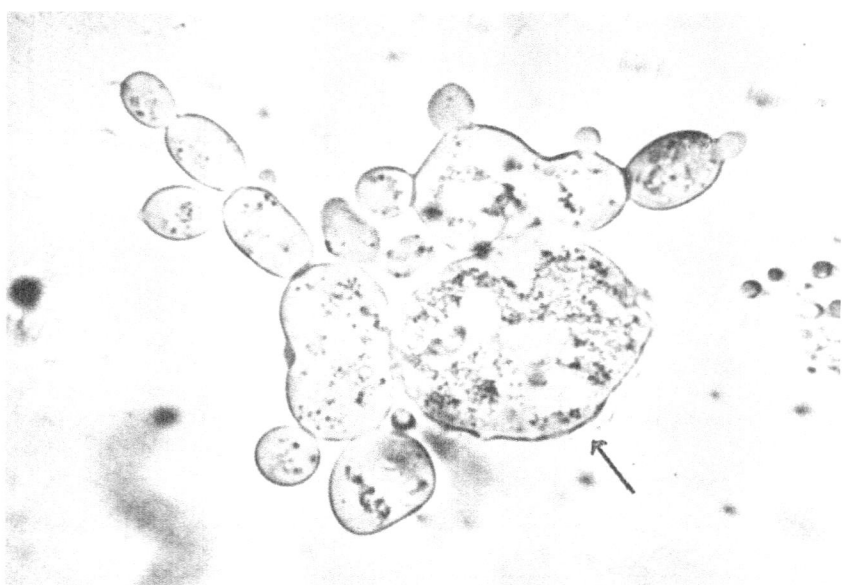

and nuclear division occurs. The formation of multinucleate protoplasts is assumed to be related to the absence of the normal cell wall. As synthesis by the protoplasts is completed, budding occurs. The first bud formed after reversion is atypical in shape, is frequently multinucleate and exhibits multipolar budding. The second-generation buds are usually normal in appearance and behaviour.

In contrast, protoplasts of the fission yeast *Schizosaccharomyces pombe* regenerate a new wall in both liquid and solid media, and nuclear division occurs only at the time of division of the regenerated protoplast. The two hemispherical cells produced at division give rise to a normal elongated cell by a budding process. Detailed ultrastructural studies of protoplasts during regeneration showed that extensive dictyosome formation occurred at the start of regeneration but decreased as the wall was being laid down (Havelkova, 1969).

Fig. 3.4. Protoplasts of *Aspergillus nidulans* showing pleomorphic development (arrowed). (Peberdy, unpublished results.)

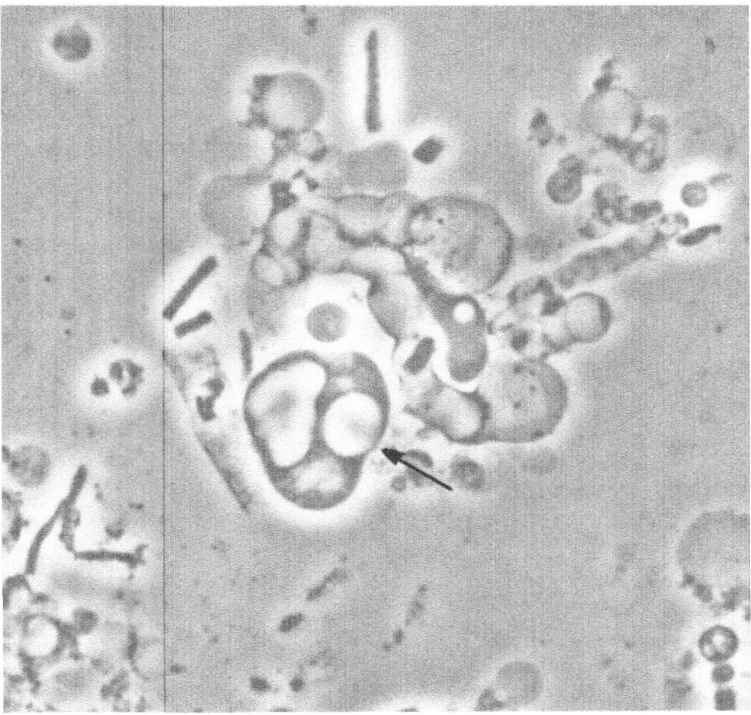

Filamentous fungi

Protoplasts from filamentous fungi will regenerate new walls and undergo reversion in either liquid or solid media. As stated above, not all protoplasts revert to the organism's normal morphology; some remain as naked protoplasts or give rise to pleomorphic forms which probably have a wall of sorts (Fig. 3.4). The protoplasts capable of reversion may do so through several patterns of development. In many species one pattern tends to dominate, but in others a variety of developmental forms may be seen. The reason for this diversity is unknown.

One type of regeneration and reversion found in several species commences with the outgrowth from the protoplast of an abnormally

Fig. 3.5. Reversion of protoplasts of *Aspergillus nidulans*. The protoplasts were cultured in osmotically stabilized Vogel's salts medium (Vogel, 1957), containing *N*-acetylglucosamine as the carbon source, for 16 h. The material was stained with the optical brightener Tinopal BOPT and viewed by fluorescence microscopy. The aberrant (*a*) and normal (*n*) hyphae are readily distinguished.

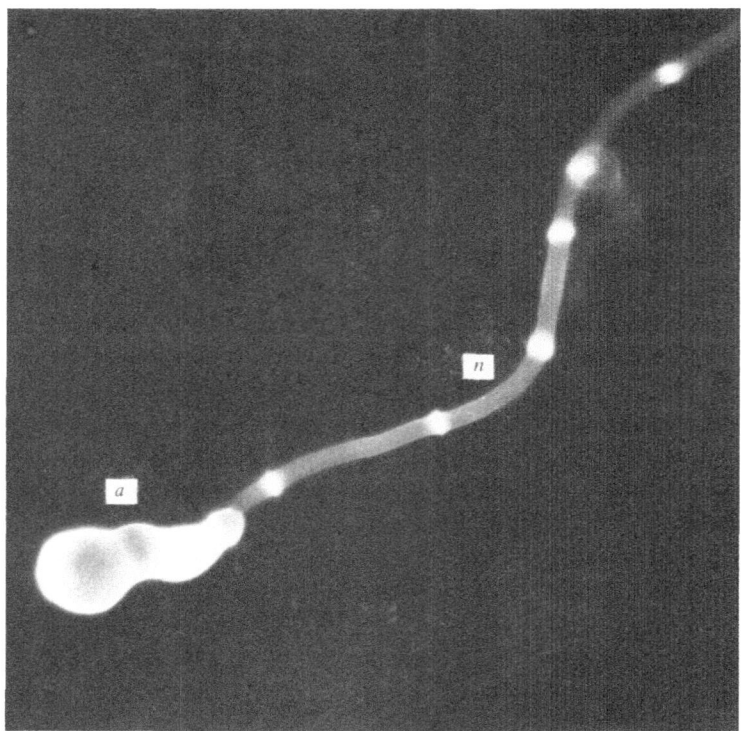

shaped hypha resembling chains of budding cells. Ultimately a normal hypha arises from the aberrant outgrowth, and reversion is complete. This pattern of development is found in *Aspergillus nidulans* (Fig. 3.5; Peberdy & Gibson, 1971), *Penicillium chrysogenum* Thom (Anné, Eyssen & De Somer, 1974) and *Geotrichum candidum* (Dooijewaard-Kloosterziel, Sietsma & Wouters, 1973). In *Trichoderma viride* Persoon ex S. F. Gray a similar outgrowth develops from the protoplast, but its tip eventually lyses and a normal hypha grows from the original protoplast (Benitez, Ramos & Garcia Acha, 1975). Reverting protoplasts of *Schizophyllum commune* make a new primary cell from which hyphal outgrowths occur (de Vries & Wessels, 1975). Non-reverting protoplasts of this fungus frequently develop as a budding chain of cells. Species in which both reversion patterns have been described include *Pythium* (Sietsma & De Boer, 1973) and *Fusarium culmorum* (Garcia Acha *et al.*, 1966).

Ultrastructure and chemical composition of cell-wall material formed by protoplasts

Yeasts

When cultured in liquid medium, protoplasts of *Saccharomyces* spp. fail to revert to normal vegetative cells, however they do form an incomplete wall which consists of a network of microfibrils (Fig. 3.6; Eddy & Williamson, 1959; Necas, 1965; Kopecka, Ctvrtnicek & Necas, 1967). Deposition of the microfibrils continues for 6 h and then is repressed (Necas *et al.*, 1970). The lack of normal wall development in protoplasts in liquid media has been attributed to the loss of the amorphous wall component or its precursors, or synthesizing enzymes, through the microfibrillar net and into the medium (Necas & Svoboda, 1967). Analysis of the microfibrillar net produced by *Saccharomyces carlesbergensis* Hansen protoplasts revealed the presence of mannose, glucose, *N*-acetylglucosamine and traces of amino acids (Eddy & Williamson, 1959). The hexosamine accounted for 27 % of the total dry weight and indicated a high chitin content. The solubility of the regenerated microfibrillar net from *Saccharomyces cerevisiae* in dilute alkali suggested it was composed of glucan (Kopecka *et al.*, 1967) and this was confirmed by an analysis of purified nets, which yielded only glucose as the hydrolysis product (Svoboda & Necas, 1970).

The microfibrillar nets have been studied by Kreger & Kopecka (1976) using preparations shown to be free of contamination by normal wall and bud scar residues. Dilute alkali removed about 40 % of the net material, leaving a residue of clusters of microfibrils. X-ray diffraction patterns showed two microcrystalline components – an unbranched $\beta(1-3)$glucan and chitin. The glucan formed the alkali-soluble fraction and differed from the normal wall β-glucan in that it lacked $\beta(1-6)$ linkages. The chitin component accounted for 15 % of the dry weight of the net material, but could not be demonstrated in electron microscope preparations. Further evidence that chitin is produced by protoplasts is found in experiments in which ^{14}C-labelled glucose was supplied during regeneration (Farkas & Svoboda, personal communication). Isotopic labelling of both β-glucan and chitin was found. In normal yeast cells chitin is found only in the primary septum of the bud scar (see Chapter 9); however, regenerating protoplasts may produce chitin for one of two reasons. Removal of the wall during protoplast isolation might uncouple the normal pattern of chitin deposition. Alternatively, if chitin production is normally linked to nuclear division during budding, and if

Fig. 3.6. Shadow-cast preparation of the microfibrillar network synthesized by *Saccharomyces cerevisiae* protoplasts cultured in a liquid medium. From Kreger & Kopecka (1976) reproduced by courtesy of Drs M. Kreger and M. Kopecka.

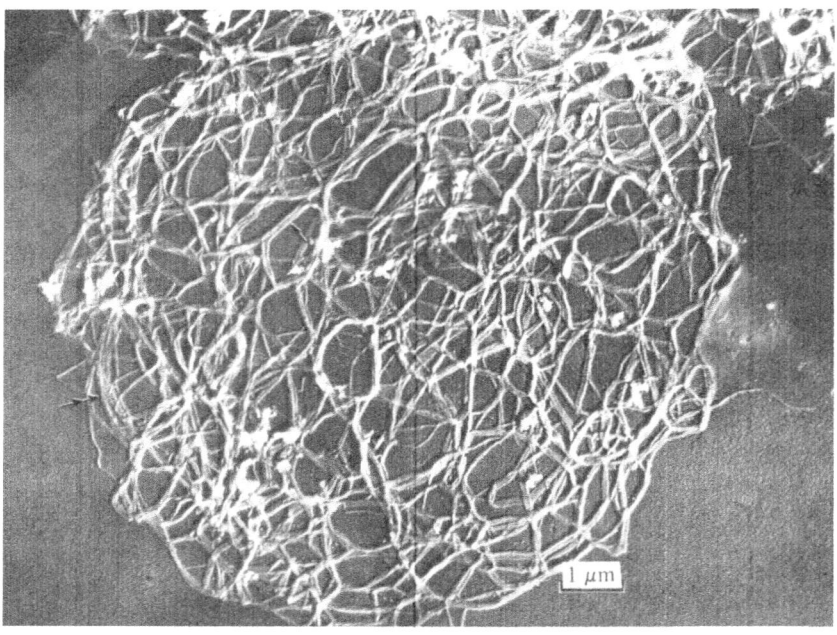

multiple mitoses also occur during regeneration in liquid medium, then pulses of chitin synthesis could occur in synchrony with these division events.

The biosynthetic activity of *Saccharomyces cerevisiae* protoplasts in liquid medium is not restricted to β-glucan and chitin. Other polymers normally found in the cell wall, e.g. mannan and glycoproteins, are formed and secreted into the culture medium (Lampen, 1968).

The wall produced by yeast protoplasts cultured in solidified media contains both fibrillar and amorphous components. After incubation for 1 h a wall layer 15 nm thick is formed, this is made up of a fibrillar component partially embedded in an amorphous matrix. As regeneration continues the wall becomes progressively thicker, up to 200–300 nm. The wall now has two distinct layers, a thick outer, discontinuous, amorphous layer covering the inner fibrillar layer. The deposition of the amorphous material is markedly affected by the concentration of gelatin in the culture medium as well as the incubation temperature. This suggests that gelatin has a role of containment, preventing diffusion of the amorphous mannan component, and/or the enzymes associated with its incorporation into the wall fabric, into the culture medium (Necas, 1961, 1962).

Protoplasts of *Schizosaccharomyces pombe* also commence wall regeneration with the formation of a microfibrillar network. As the protoplasts age, the net becomes denser and amorphous material is found associated with it (Necas, Svoboda & Havelkova, 1968); the retention of the amorphous material by the wall has been attributed to the broad and dense microfibrils produced by this organism (Necas, 1971). Support for this view is provided by the synchronous deposition of fibrillar and matrix components by protoplasts embedded in solidified media (Havelkova, 1969). Wall regeneration has been studied in several other yeast species (Havelkova, 1969, Svoboda, 1966), and in general the patterns of development are similar to *Schizosaccharomyces pombe*.

Chemical analyses have been made on the regenerated wall synthesized by *Candida utilis* (Henneberg) Lodder ex Kreger-van Rij in liquid medium (Novaes-Ledieu & Garcia Mendoza, 1970). The protoplasts first formed tubular structures which later changed to ellipsoid yeasts. A chemical difference was found in the walls of the two morphological forms. The normal cell and the ellipsoid reversion stage contained glucan (42 to 48 %), mannan (25 to 31 %) and protein (20 %). The tubular forms had a high chitin (12 to 18 %) and protein (20 %) content. The differences were correlated with the two morphological changes which occur during reversion, but no further explanation of the high chitin content was given.

Filamentous fungi

Ultrastructural aspects of wall biogenesis by protoplasts have been investigated in several filamentous fungi. Despite the variation in morphology of the regenerating protoplasts, there is a marked similarity in the ultrastructure of the newly formed wall. Wall biogenesis commences with the deposition of a microfibrillar net with the subsequent deposition of amorphous material (Gabriel, 1968; Uruburu, Garcia Acha & Villanueva, 1970; Ramos & Garcia Acha, 1975; Benitez, Villa & Garcia Acha, 1975; Sietsma *et al.*, 1975; Gibson, Buckley & Peberdy, 1976; van der Valk & Wessels, 1976) and thus is very similar to wall regeneration in yeasts.

Detailed studies have been made on wall regeneration in the basidiomycete *Schizophyllum commune* (de Vries & Wessels, 1975; van der Valk & Wessels, 1976). Chemical analyses showed an initial production of chitin and S-$a(1$–$3)$glucan followed after a lag of 3 h by the formation of R-$\beta(1$–$3)$, $\beta(1$–$6)$glucan. Hyphal emergence occurred after 8 h, when synthesis of all polymers was in progress (Fig. 3.7). Electron microscopic investigations confirmed these analytical studies. The wall produced after 3 h was loosely organized, consisting mainly of chitin fibrils closely bound to the plasma membrane with aggregates of alkali soluble S-glucan present as a fluffy covering. After this time further amorphous material, including the R-glucan, was laid down. The completely regenerated wall showed a different wall architecture from the hyphal wall produced at reversion (Figs. 3.8 and 3.9).

Using the antibiotics polyoxin D and cycloheximide it proved possible to uncouple the biogenesis of the three wall polymers. Polyoxin D, in increasing concentrations up to $100~\mu g~ml^{-1}$ blocked the formation of chitin and R-glucan, resulting in walls containing predominantly S-glucan (Table 3.1) in the form of thick loose fibres (van der Valk & Wessels, 1976). In contrast cycloheximide (an inhibitor of protein synthesis) had no immediate effect upon the formation of chitin and S-glucan. However, after 5 h the synthesis of these polymers was inhibited: no R-glucan was synthesized in the presence of this antibiotic. These results suggest that the chitin and S-glucan synthases were present in the protoplasts but the R-glucan synthase required *de novo* synthesis. Protoplast reversion did not occur in the presence of these antibiotics (de Vries & Wessels, 1975).

Observations on *Pythium* (Sietsma *et al.*, 1975) were similar to those described above, with the same pattern for regeneration and reversion. Wall regeneration by the protoplasts involved the formation of a microfibrillar net which over a 12 h period became denser and the individual microfibrils thicker. The hypha which emerged from the protoplast at this time had the granular surface texture typical of a normal mature hypha.

The situation in *Aspergillus nidulans* is rather different, the microfibrillar net produced by the protoplast is composed of chitin (Fig. 3.10) and is the only visible component for the first 6 h of regeneration. As the aberrant hypha arising from the protoplast grows, the net becomes denser over its surface and gradually becomes packed with amorphous material (Fig. 3.11). There were no obvious ultrastructural differences between the surfaces of the aberrant and normal hyphae.

Fig. 3.7. The deposition of wall polymers by protoplasts from *Schizophyllum commune* during regeneration and reversion. Symbols: ●, *S*-glucan; ▲, *R*-glucan; ■, chitin. (After de Vries & Wessels, 1975.)

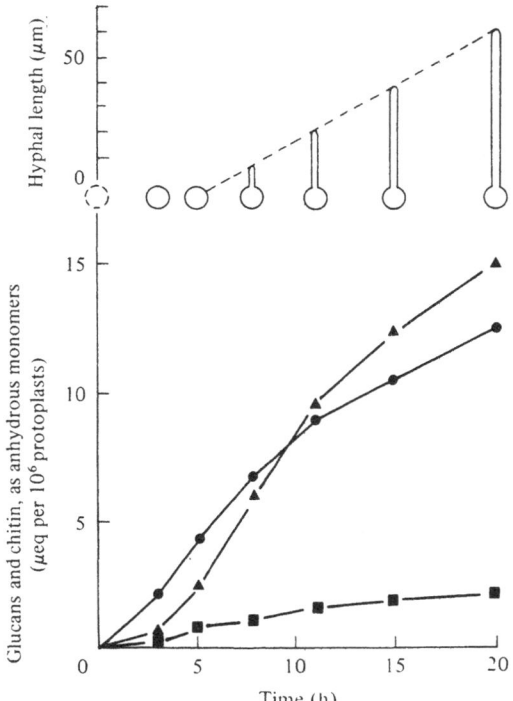

Fig. 3.8. Shadow-cast preparation of protoplast reversion in
Schizophyllum commune showing the transition in surface texture at
the point of hyphal emergence. The wall surface of the hypha (H) is
smoother than that of the protoplast regenerated wall (PC). From van
der Valk & Wessels (1976) reproduced by courtesy of Professor J. G.
H. Wessels.

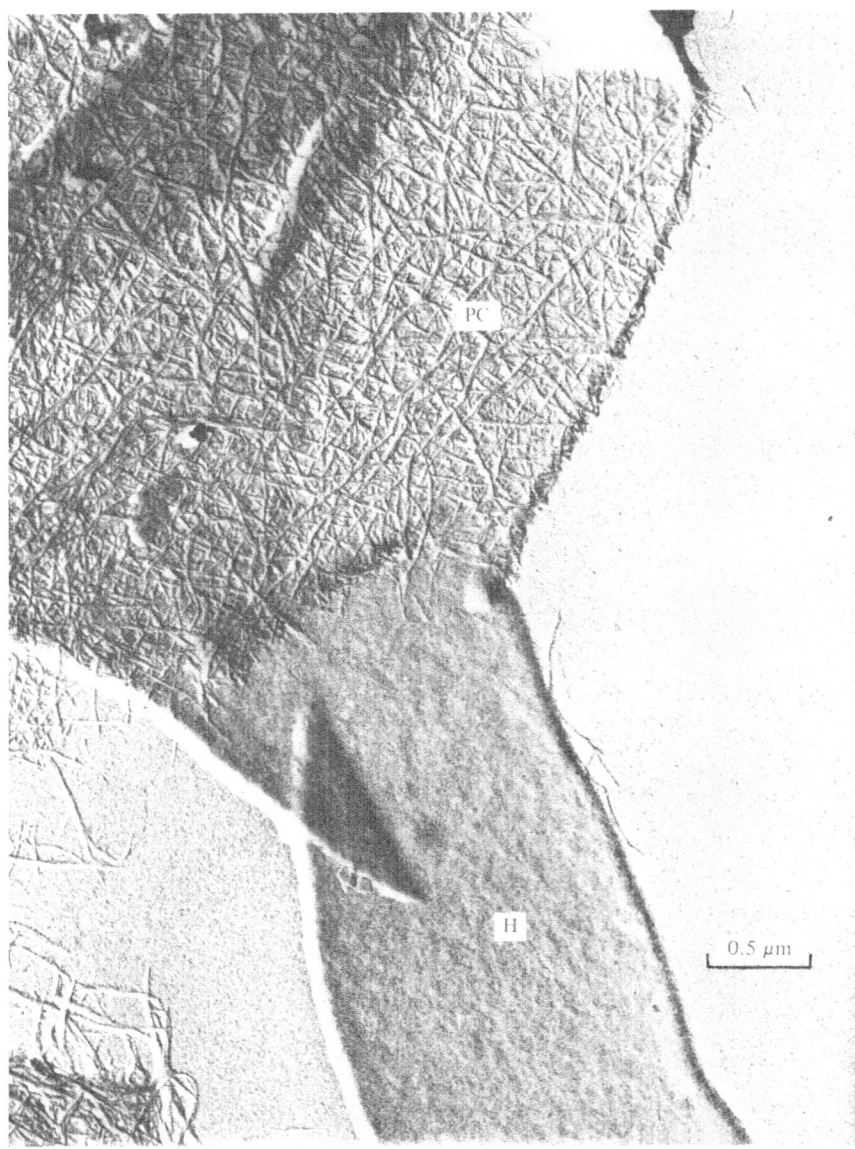

Chemical analyses on the regenerated wall confirmed the ultrastructural observations (Peberdy, unpublished results). The experimental protocol is shown in Table 3.2. The distribution of the isotope into the three major wall polymers is shown in Fig. 3.12. The high level of isotope in the hexosamine (chitin) can be correlated with the formation of the fibrillar net; the subsequent deposition of glucans probably equates with the development of the amorphous material. The build-up of polymers by the protoplasts contrasts markedly with the events in hyphae (Table 3.3). For comparison, hyphae from exponential cultures

Table 3.1. *The effect of polyoxin D on polymer formation during wall regeneration by* Schizophyllum commune *protoplasts. (After de Vries & Wessels, 1975)*

Inhibitor concentration (μg ml^{-1})	Polymer fraction after 7 h regeneration (%)		
	Chitin	R-glucan	S-glucan
0	100	100	100
10	14	2	133
25	7	1	139
50	0	2	132
100	0	1	125

Fig. 3.9. Model of wall architecture in the primary cell (protoplast) and hypha of reverted protoplasts of *Schizophyllum commune*. (After van der Valk & Wessels, 1976.)

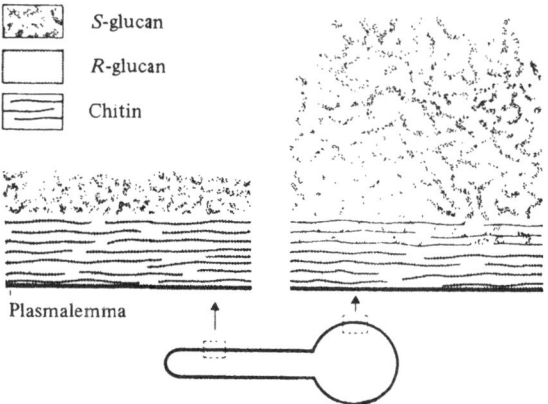

S-glucan

R-glucan

Chitin

Plasmalemma

were labelled for several hours with ^{14}C-labelled sugars and the isolated wall material was treated in a similar manner to that from the protoplasts. With either labelled glucose or *N*-acetylglucosamine, the incorporation of isotope into the chitin and glucan fraction was very similar to the amounts of these polymers found in hyphal walls by chemical analysis (Bull, 1970).

In fungi so far referred to, the skeletal and matrix wall components produced during regeneration are in general the same as are found in the normal hyphal wall. An exception to this situation was reported in *Trichoderma viride* (Benitez, Villa & Garcia Acha, 1975). The fibrillar component of the regenerated wall of this fungus is not chitin but $\beta(1-3)$, $\beta(1-6)$glucan. The absence of chitin has not been explained and

Fig. 3.10. Wall regeneration in *Aspergillus nidulans* protoplasts. Shadowed preparation of a protoplast at an early stage of regeneration showing the loose fibrillar network. (From Gibson *et al.*, 1976.)

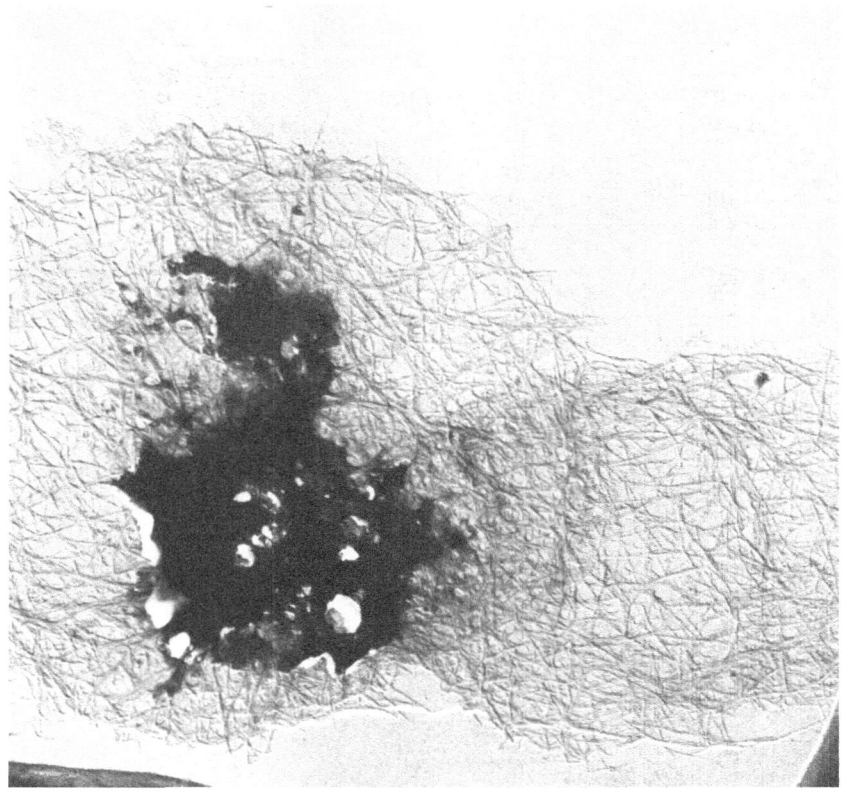

Fig. 3.11. Wall regeneration in *Aspergillus nidulans* protoplasts.
Shadowed preparation of a protoplast near to the reversion stage. The
loose fibrillar net synthesized by the protoplast remains uncemented by
amorphous material (A). As the density of net increases the
amorphous material becomes associated with it (B). (From Gibson *et
al.*, 1976.)

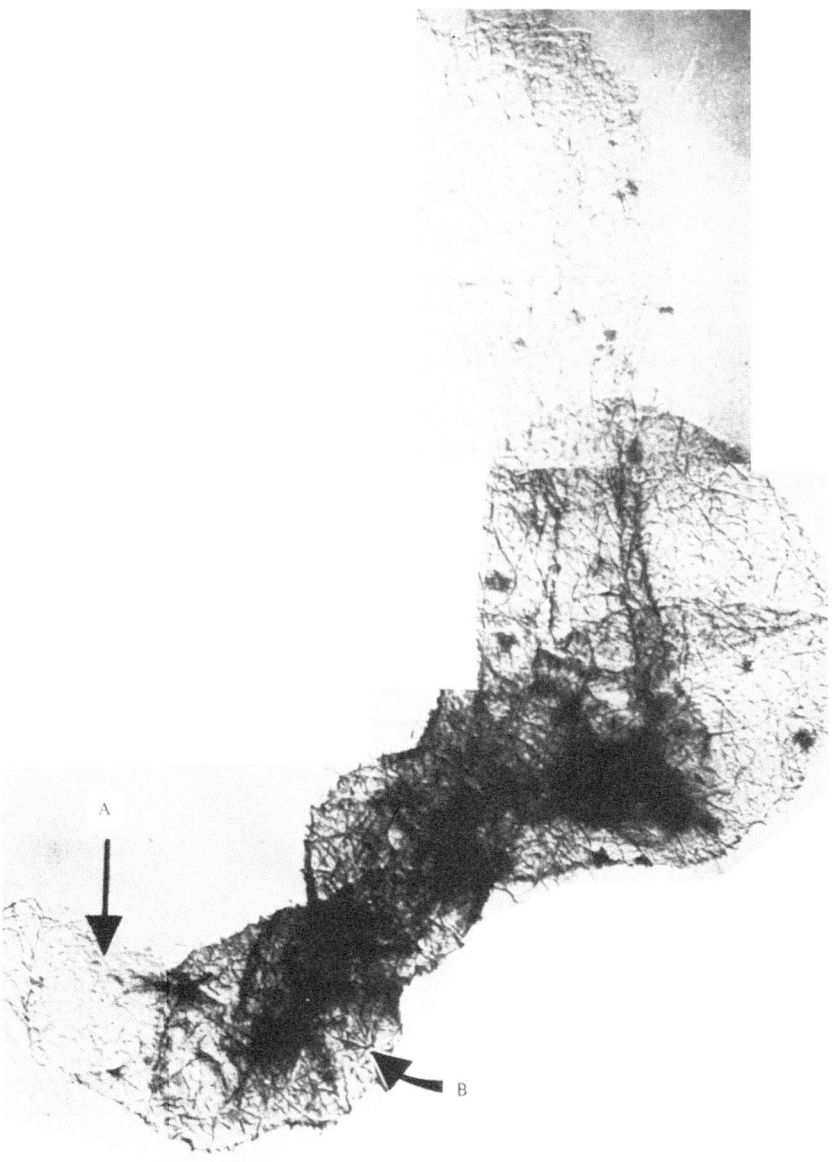

these results clearly warrant a study of the wall synthases present in the protoplasts.

Chitin synthase in protoplasts

Chitin synthase is readily detected in protoplasts (Moore & Peberdy, 1976; Archer, 1977; Isaac, Ryder & Peberdy, 1978). The distribution of the enzyme in protoplast fractions corresponded with findings for mycelium homogenates in *Aspergillus flavus* Link ex Fries

Table 3.2. *Experimental protocol for* ^{14}C*-labelled glucose incorporation into the regenerated wall of* Aspergillus nidulans *protoplasts*

Washed protoplasts suspended in glucose–salts medium (Vogel, 1957) osmotically stabilized with 0.6 M NH$_4$Cl. Culture (1 ml containing 10^8 protoplasts) was added to each tube together with 20 μCi [U−^{14}C]glucose.

$\downarrow$

Protoplast cultures incubated under stationary conditions at 20°C and harvested at various time intervals over a period of 15 h.

$\downarrow$

The regenerated protoplasts collected by centrifugation and after three washings in 0.2 M phosphase buffer, pH 5.8, subjected to mechanical breakage on a vortex mixture.

$\downarrow$

The recovered wall material was treated with a cocktail of ribonuclease, deoxyribonuclease and pronase (1 mg ml^{-1} of each) at 90°C overnight in the presence of thiomersal.

$\downarrow$

The wall residue was treated with cold KOH for 16 h.

Extract dried on glass-fibre filter for counting.

Residue treated with *Trichoderma* lytic enzyme for 16 h at 37°C. Hydrolysate chromatographed on paper (butanol:pyridine:water, 6:4:3) and products determined using standards. Counts made on portions of the chromatogram in the equivalent area to the standard spot.

(Table 3.4). The highest specific activity for the enzyme was found in the 200 000 g pellet and was thus similar to *Aspergillus nidulans* protoplast material. Archer (1977) also reported a high level of the enzyme in the microsomal fraction of protoplasts of *Aspergillus fumigatus* Frisenius.

Suspensions of protoplasts derived from lengthy lytic digestions are clearly heterogeneous in relation to their original hyphal location and their biochemical capabilities. Following up earlier observations on the

Table 3.3. *Incorporation of labelled sugars into hyphal walls of* Aspergillus nidulans

Labelled substrate	Distribution of isotope[a] (%)			
	Cold KOH hydrosylate	Lytic enzyme digest		
		Glc	GlcN	GlcNAc
Glucose	30.1	34.6	1.2	14.1
GlcNAc	28.0	37.2	1.7	13.1

[a] Average values from two analyses.
Abbreviations: Glc, glucose; GlcN, glucosamine; GlcNAc, N-acetylglucosamine.

Table 3.4. *Distribution of chitin synthase in fractions prepared from* Aspergillus flavus *protoplast and mycelium preparations. (From Moore & Peberdy, 1976)*

Fraction	Enzyme activity			
	Protoplast		Mycelium	
	Total	Specific[a]	Total	Specific[a]
2 000 g	7.9	0.075	8.8	0.078
10 000 g	11.8	0.105	11.6	0.237
100 000 g	19.4	0.340	14.3	0.345
100 000 g supernatant	60.9	0.242	65.3	0.238

[a] Specific activity expressed as nmol substrate incorporated per min per mg protein.

pattern of protoplast release in *Aspergillus nidulans* (Gibson & Peberdy, 1972) further studies have been made on a number of biochemical and physiological parameters of protoplasts released at different times during lytic digestion (Isaac, 1978). Experiments were designed to measure the levels of active and trypsin-activatable chitin synthase in protoplasts released at 1, 2 and 3 h of lytic digestion (Isaac *et al.*, 1978). With KCl as osmotic stabilizer, a definite pattern in the levels of these two components of the enzyme system was observed (Table 3.5). Protoplasts released after 1 h had a high level of active enzyme but none of the activatable component. In contrast later fractions had less of the active form and very high levels of activatable enzyme. The distribution of the enzyme components in $MgSO_4$-stabilized protoplasts was not so precise, but a similar trend could be detected. If the protoplasts released after 1 h in the lytic digestion were from the hyphal apices and later

Fig. 3.12. Incorporation of ^{14}C labelled glucose into polymers during wall biogenesis in *Aspergillus nidulans* protoplasts. Details of the experimental procedure are given in Table 3.2. Symbols: ▲, *N*-acetylglucosamine released by lytic enzyme (chitin); ●, glucose released by lytic enzyme (β-glucan); ■, alkali-soluble extract (α-glucan). (Peberdy, unpublished results.)

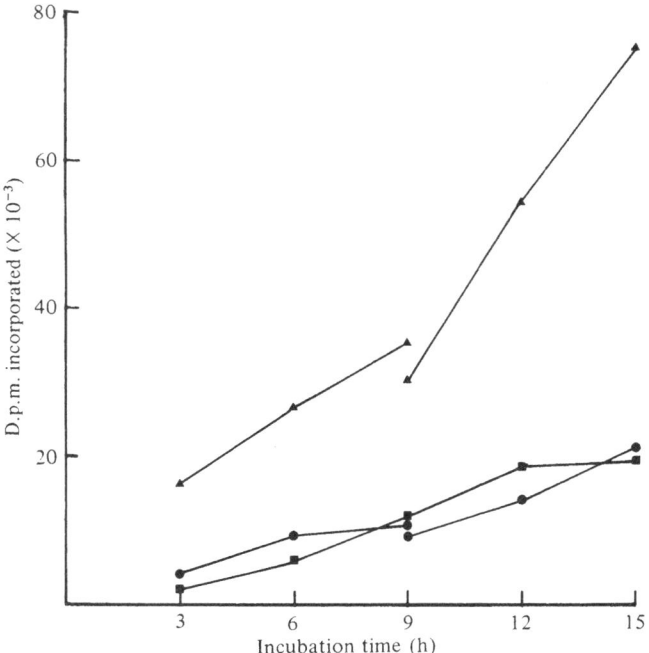

Table 3.5. *Chitin synthase activities of protoplasts produced in KCl- and MgSO$_4$-stabilized lytic digestions.* *(From Isaac et al., 1978)*

Digestion period (h)	Protoplast yield (10^7 protoplasts per mg dry mycelium)	Chitin synthase activities				Increase in activity after trypsin digestion (%)
		Untreated protoplasts		Protoplasts digested with trypsin		
		Specific activity	Total activity	Specific activity	Total activity	
(a) Protoplasts produced in KCl						
1	0.14	6.13±0.18	237.52±7.24	4.14±0.12	168.31±4.10	–
2	4.25	1.73±0.04	28.04±0.67	2.70±0.03	44.39±0.56	56.07
3	7.55	0.69±0.01	17.42±0.14	1.05±0.01	26.24±0.28	66.57
(b) Protoplasts produced in MgSO$_4$						
1	0.42	0.70±0.03	33.73±1.29	0.37±0.05	17.65±2.34	–
2	7.28	1.05±0.06	6.55±0.44	1.84±0.08	27.19±1.19	75.2
3	7.45	1.70±0.01	16.98±0.13	1.51±0.08	15.02±0.84	–

Specific chitin synthase activities are expressed as nmol *N*-acetylglucosamine incorporated per min per mg dry protein; total activities as nmol *N*-acetylglucosamine incorporated per min per 10^{10} protoplasts. Values quoted are from representative experiments and are the means of at least three determinations.

protoplasts from distal regions then these observations might provide a clue as to the mechanism of control of chitin synthesis in the growing hypha.

Conclusion

The study of wall regeneration by fungal protoplasts has contributed to our understanding of the hyphal wall in a variety of ways. The rebuilding of a new wall on the naked plasma membrane has provided an understanding of the ultrastructural relationships between the skeletal and amorphous polymers, as well as the kinetics of deposition. Another useful attribute of the protoplast system has been the ability to uncouple the biosynthesis of the different polymers. Expansion of this field of investigation should prove to be highly profitable as new inhibitory compounds specific for the wall synthase systems are discovered. Temperature-sensitive mutants with biochemical lesions associated with particular polymers (Bainbridge, Chapter 4) could also prove to be very useful for experimentation on uncoupled polymer synthesis.

A continuing interest in protoplasts has been in their use as a gentle procedure for producing cell-free systems. Refinement in techniques for the fractionation of protoplasts respective of their hyphal location would be of great benefit. By this means it would be possible to break down the highly integrated functions of the hypha, in particular those involved in the control of the localized biosynthesis of the cell wall.

References

Anné, J., Eyssen, H. & De Somer, P. (1974). Formation and regeneration of *Penicillium chrysogenum* protoplasts. *Archives of Microbiology*, **98**, 159–66.

Archer, D. B. (1977). Chitin biosynthesis in protoplasts and sub-cellular fractions of *Aspergillus flavus*. *Biochemical Journal*, **164**, 653–8.

Bachmann, B. J. & Bonner, D. M. (1959). Protoplasts from *Neurospora crassa*. *Journal of Bacteriology*, **78**, 550–6.

Benitez, T., Ramos, S. & Garcia Acha, I. (1975). Protoplasts from *Trichoderma viride*: Formation and regeneration. *Archives of Microbiology*, **103**, 199–203.

Benitez, T., Villa, T. G. & Garcia Acha, I. (1975). Chemical and structural differences in mycelial and regeneration walls of *Trichoderma viride*. *Archives of Microbiology*, **105**, 277–82.

Bull, A. T. (1970). Chemical composition of wild-type and mutant *Aspergillus nidulans* cell walls. The nature of the polysaccharide and melanin constituents. *Journal of General Microbiology*, **63**, 75–94.

Dooijewaard-Kloosterziel, A. M. P., Sietsma, J. H. & Wouters, J. T. M. (1973). Formation and regeneration of *Geotrichum candidum* protoplasts. *Journal of General Microbiology*, **74**, 205–9.

Eddy, A. A. & Williamson, D. H. (1959). Formation of aberrant cell walls and of spores by the growing of yeast protoplasts. *Nature (London)*, **183**, 1101–4.

Emerson, S. & Emerson, M. R. (1958). Production, reproduction and reversion of protoplast-like structures in the osmotic strains of *Neurospora crassa*. *Proceedings of the National Academy of Sciences, USA*, **44**, 668–71.

Fiddy, C. & Trinci, A. P. J. (1976). Mitosis, septation, branching and the duplication cycle in *Aspergillus nidulans*. *Journal of General Microbiology*, **97**, 169–84.

Fukui, K., Sagara, Y., Yoshida, N. & Matsuoka, T. (1969). Analytical studies on regeneration of protoplasts of *Geotrichum candidum* by quantitative thin layer plating. *Journal of Bacteriology*, **98**, 256–63.

Gabriel, M. (1968). Formation and regeneration of protoplasts in the mold *Rhizopus nigricans*. *Folia Microbiologica (Praha)*, **13**, 231–4.

Garcia Acha, I., Lopez-Belmonte, F. & Villanueva, J. R. (1966). Regeneration of mycelial protoplasts of *Fusarium culmorum*. *Journal of General Microbiology*, **45**, 515–23.

Gibson, R. K. & Peberdy, J. F. (1972). Fine structure of protoplasts of *Aspergillus nidulans*. *Journal of General Microbiology*, **72**, 529–38.

Gibson, R. K., Buckley, C. E. & Peberdy, J. F. (1976). Wall ultrastructure in regenerating protoplasts of *Aspergillus nidulans*. *Protoplasma*, **89**, 381–7.

Havelkova, M. (1969). Electron microscopy study of cell structures and their changes during growth and regeneration in *Schizosaccharomyces pombe* protoplasts. *Folia Microbiologica (Praha)*, **14**, 155–64.

Isaac, S. (1978). Physiological and biochemical properties of protoplasts of *Aspergillus nidulans*. Ph.D Thesis, University of Nottingham, UK.

Isaac, S., Ryder, N. S. & Peberdy, J. F. (1978). Distribution and activation of chitin synthase in protoplast fractions released during the lytic digestion of *Aspergillus nidulans* hyphae. *Journal of General Microbiology*, **105**, 45–50.

Kopecka, M., Ctvrtnicek, O. & Necas, O. (1967). Formation and properties of fibrillar network formed in yeast protoplasts as the first step of biosynthesis of cell wall. In *Symposium uber Hefeprotoplasten*, ed. R. Muller, pp. 73–6. Berlin: Akademie-Verlag.

Kreger, D. R. & Kopecka, M. (1976). On the nature and formation of the fibrillar nets produced by protoplasts of *Saccharomyces cerevisiae* in liquid media; an electron microscopic, X-ray diffraction and chemical study. *Journal of General Microbiology*, **92**, 207–20.

Lampen, J. O. (1968). External enzymes of yeast: their nature and formation. *Antonie van Leeuwenhoek. Journal of Microbiology and Serology*, **34**, 1–18.

Moore, P. M. & Peberdy, J. F. (1976). A particulate chitin synthase from *Aspergillus flavus* Link: the properties, location and levels of activity in mycelium and regenerating protoplast preparations. *Canadian Journal of Microbiology*, **22**, 915–21.

Necas, O. (1961). Physical conditions as important factors for the regeneration of naked yeast protoplasts. *Nature (London)*, **192**, 580–1.

Necas, O. (1962). The mechanism of regeneration of yeast protoplasts. I. Physical conditions. *Folia Biologica (Praha)*, **8**, 256–62.

Necas, O. (1965). Mechanism of regeneration of yeast protoplasts. II. Formation of cell wall *de novo*. *Folia Biologica (Praha)*, **11**, 97–102.

Necas, O. (1971). Cell wall synthesis in yeast protoplasts. *Bacteriological Reviews*, **35**, 149–70.

Necas, O. & Svoboda, A. (1967). Mechanism of regeneration of yeast protoplasts. IV. Electron microscopy on regenerating protoplasts. *Folia Biologica (Praha)*, **13**, 379–85.

Necas, O., Svoboda, A. & Havelkova, M. (1968). Mechanism of regeneration of yeast protoplasts. V. Formation of the cell wall in *Schizosaccharomyces pombe*. *Folia Biologica (Praha)*, **14**, 80–5.

Necas, O., Soska, I., Kopecka, M. & Reich, J. (1970). Application of ^{3}H-glucose in the study of cell wall synthesis in yeast protoplasts. In *Yeast Protoplasts. Proceedings of the 2nd International Symposium, Brno*, eds. O. Necas & A. Svoboda, pp. 209–31. Brno, Czechoslovakia: J. E. Purkyne University Press.

Novaes-Ledieu, M. & Garcia Mendoza, C. (1970). Biochemical studies on walls synthesised by *Candida utilis* protoplasts. *Journal of General Microbiology*, **61**, 335–42.

Peberdy, J. F. (1972). Protoplasts from fungi. *Science Progress (Oxford)*, **60**, 73–86.

Peberdy, J. F. & Gibson, R. K. (1971). Regeneration of *Aspergillus nidulans* protoplasts. *Journal of General Microbiology*, **69**, 325–30.

Ramos, S. & Garcia Acha, I. (1975). Cell wall enzymatic lysis of the yeast form of *Pullularia pullulans* and wall regeneration by protoplasts. *Archives of Microbiology*, **104**, 271–7.

Sietsma, J. H. & De Boer, W. R. (1973). Formation and regeneration of protoplasts from *Pythium* PRL 2142. *Journal of General Microbiology*, **74**, 211–17.

Sietsma, J. H., Child, J. J., Nesbitt, L. R. & Haskins, R. H. (1975). Ultrastructural aspects of wall regeneration by *Pythium* protoplasts. *Antonie van Leeuwenhoek, Journal of Microbiology and Serology*, **41**, 17–23.

Svoboda, A. (1966). Regeneration of yeast protoplasts in agar gels. *Experimental Cell Research*, **44**, 640–2.

Svoboda, A. & Necas, O. (1966). Regeneration of yeast protoplasts prepared by snail enzyme. *Nature (London)*, **210**, 845.

Svoboda, A. & Necas, O. (1970). Experimental decoupling in the synthesis of fibrillar and amorphous components during regeneration of the cell wall in *Saccharomyces cerevisiae* protoplasts. In *Yeast Protoplasts. Proceedings of the Second International Symposium on Yeast Protoplasts*, eds. O. Necas & A. Svoboda, pp. 211–15. Brno, Czechoslovakia: J. E. Purkyne University Press.

Uruburu, F., Garcia Acha, I. & Villanueva, J. R. (1970). Electron microscopy of *Fusarium culmorum* regenerating protoplasts. In *Yeast Protoplasts. Proceedings of the Second International Symposium on Yeast Protoplasts*, eds. O. Necas & A. Svoboda, pp. 143–6. Brno, Czechoslovakia: J. E. Purkyne University Press.

Valk, P. van der & Wessels, J. G. H. (1976). Ultrastructure and localisation of wall polymers during regeneration and reversion of protoplasts of *Schizophyllum commune*. *Protoplasma*, **90**, 65–87.

Villanueva, J. R. & Garcia Acha, I. (1971). Production and use of fungal protoplasts. In *Methods in Microbiology* vol. 4, ed. C. Booth, pp. 665–718. London, New York: Academic Press.

Vogel, H. J. (1957). A convenient medium for *Neurospora*. *Microbial Genetics Bulletin*, **13**, 42–3.

Vries, O. M. H. de & Wessels, J. G. H. (1975). Chemical analysis of cell wall regeneration and reversion of protoplasts of *Schizophyllum commune*. *Archives of Microbiology*, **102**, 209–18.

Wessels, J. G. H., van der Valk, P. & Vries, O. M. H. de (1976). Wall synthesis by fungal protoplasts. In *Microbial and Plant Protoplasts*, eds. J. F. Peberdy, A. H. Rose, H. J. Rogers & E. C. Cocking, pp. 267–81. London, New York: Academic Press.

4

The use of temperature-sensitive mutants to study wall growth

B. W. BAINBRIDGE, B. P. VALENTINE* AND
P.MARKHAM

*Microbiology Department, Queen Elizabeth College,
Campden Hill Road, London W8 7AH, UK*

* Beecham Pharmaceuticals, Clarendon Road, Worthing, Sussex BN14 8QH, UK

Introduction

The last forty years have seen an increasing use of genetics in the study of the biochemistry of development. This has now reached its peak with the availability of sophisticated techniques for the extracellular manipulation and cloning of specific genes permitting the purification of individual gene products. Since, under normal conditions, some gene products may be present as only one or two molecules per cell, this is a considerable achievement. An important technical advance in such studies of development has been the discovery of conditional lethal mutants. These strains exhibit a normal phenotype under one set of conditions but are lethal under another. A particular class of conditional lethal mutants are temperature sensitive, that is they are normal at one temperature, called the permissive temperature, but abnormal at another, the non-permissive temperature. The reason for the importance of these mutants is that in principle any protein, even presumably essential enzymes such as chitin synthase, may become thermolabile following mutation and therefore a wide range of developmental mutants can be isolated. Although such mutants have been used extensively to study bacteriophages and bacteria (Hayes, 1968) and yeast (Hartwell, 1967), systematic attempts have been made only recently to isolate a wide range of temperature-sensitive mutants in filamentous fungi (Unrau & Holliday, 1970; Waldron & Roberts, 1974; Inoue & Ishikawa, 1975; Morris, 1976; Orr & Rosenberger, 1976; Valentine & Bainbridge, (1978). A few of these mutants have been used to study wall synthesis, but most of our knowledge of wall synthesis stems from the use of the electron microscope and biochemical analysis of purified walls. In *Neurospora crassa* Shear & Dodge, conventional morphological mutants have provided an interesting insight into wall composition, growth and development, but as this work has recently been reviewed (Mishra, 1977) it will not be discussed in detail here. The systematic use of temperature-sensitive mutants should provide information on the enzymes and structural polymers needed for normal wall synthesis, as their use should permit the controlled elimination of specific wall components. Thus, temperature-sensitive mutants can be analysed at a non-permissive temperature which prevents growth, at intermediate temperatures where growth is abnormal, and at permissive temperatures where growth is normal. A correlation can then be made between wall composition and hyphal growth.

In *Aspergillus nidulans* (Eidam) Winter a limited range of temperature-sensitive mutants have been reported which have reduced amounts of wall monomers such as glucosamine (Cohen, Katz & Rosenberger, 1969) and mannose (Valentine & Bainbridge, 1975, 1978). The lack of these sugars resulted in abnormal wall growth and this provided clues to the mechanism of normal polarized wall growth. Provision in the medium of the sugar lacking in the wall eliminated the developmental defects. This opened up the possibility of labelling sugar components in the wall, and the distribution of these labelled sugars could then be followed during purification and fractionation of the wall. This approach should enable a detailed analysis to be made of the distribution of glucosamine and mannose in wall polymers. The composition of the wall could also be followed over a range of temperatures and in different morphological phenotypes, and this could provide information about the mechanism of polarized wall growth.

A range of temperature-sensitive mutations affecting nuclear division of *Aspergillus nidulans* has been reported (Orr & Rosenberger, 1976; Morris, 1976) and it is possible that a genetic analysis could be made of nuclear division, septation, branching and wall synthesis. *A. nidulans* has been shown to have a duplication cycle in which nuclear division and septation are normally coordinated (Clutterbuck, 1970; Fiddy & Trinci, 1976; Bainbridge, 1976) and, as the genetics of this organism is well studied, it would appear to be an ideal organism for the genetic study of the link between septation and nuclear division, particularly in septation-deficient mutants (Morris, 1976). Branching could also be studied, and a step in this direction has been made by the study of a temperature-sensitive colonial mutant of *Neurospora crassa, cot-1,* in which extensive branching occurs when the temperature is raised (Steele & Trinci, 1977). Electron microscopists and fungal biochemists should also be able to exploit such mutants to provide material enriched in hyphal apices, septa or vesicles containing synthetic or lytic enzymes involved in wall synthesis.

Range of temperature-sensitive mutants isolated

The initial isolation of mutants was achieved by treating spores with a mutagen followed by plating, with or without filtration enrichment, to give separate colonies at the permissive temperature of 25 °C (Valentine & Bainbridge, 1978). These were then replica-plated on to fresh plates which were incubated at 42–45 °C, the non-permissive temperature. Any colonies which failed to grow or grew poorly were selected as potential temperature-sensitive mutants. Another approach has been to select for cold-sensitive mutants where the non-permissive temperature is 20 °C (Waldron & Roberts, 1974). The phenotypes of the mutants have been classified under the microscope after 48 h, and a range of types is shown in Fig. 4.1*a* (Valentine, 1975). An alternative method was to pre-incubate spores at 25 °C for 24 h and then to transfer the plates to 45 °C for a further 24 h. Further classes could then be identified as those affecting hyphal growth rather than the earlier stages of spore germination (Fig. 4.1*b*). One class of mutants showed either distorted swollen hyphae or regular spherical swellings called 'balloons' and this group of mutants will be discussed in greater detail later.

Influence of incubation temperature on linear growth

Seventy mutants of *Aspergillus nidulans* have been screened for their growth at a range of temperatures between 25 °C and 45 °C (Valentine, 1975). Mutant and parental strains were grown on minimal agar plates until the parental strain reached a diameter of 8 mm. The mutant colonies were then fixed, measured and the final colony diameter expressed as a percentage of the control value. The mutants fell into three classes (Fig. 4.2):

(1) Those with a sharp decrease in growth at about 41 °C, which account for about 77 % of the mutants tested (Fig. 4.2*a*).

(2) Those with a gradual decrease in relative colony diameter with increase in temperature, which account for about 19 % of the mutants tested (Fig. 4.2*b*).

(3) Those which showed reduced growth at both lower and higher

Fig. 4.1. Examples of morphological effects of temperature-sensitive mutations on spore germination and growth in *A. nidulans* (Valentine, 1975). (*a*) Spores germinated for 48 h at 45 °C on minimal medium; (*b*) Spores germinated for 24 h at 25 °C followed by transfer to 45 °C for 24 h.

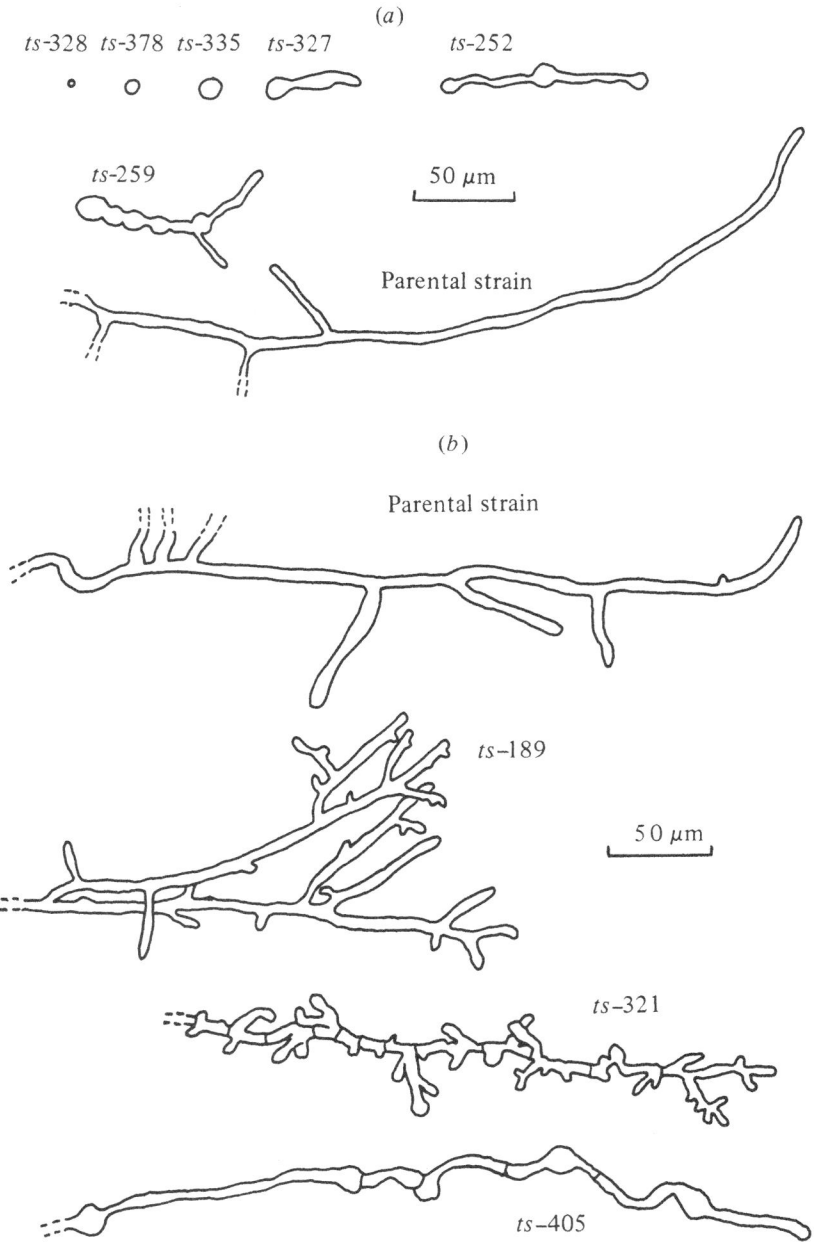

Fig. 4.2. Relative colony diameters of temperature-sensitive mutants of *A. nidulans* grown at different temperatures on minimal medium or minimal medium plus 6% sodium chloride. Conidial suspensions, 1×10^6 ml^{-1} were inoculated at three locations per plate. When the control plate had reached a diameter of 8 mm the mutant strain's diameter was measured using a Shadowmaster (Buck & Hickman). Two measurements were taken at right angles for each colony and the final value was calculated from at least three colonies. Values were expressed as a percentage of the control value (Valentine, 1975). (*a*) Mutant relieved by mannose, *mnrA455*; (*b*) mutant *ts-313*, total lysis after growth as for Fig. 4.1 (*b*); (*c*) mutant relieved by glucosamine, *glc-95*. Symbols: ○, minimal medium; ● minimal medium plus 6 % sodium chloride.

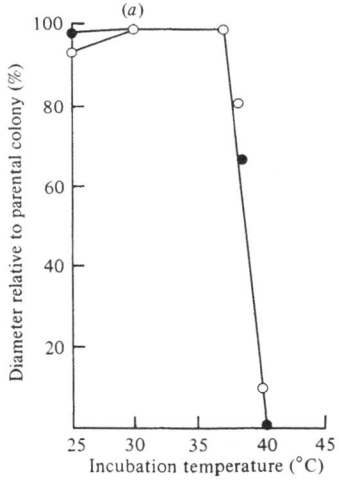

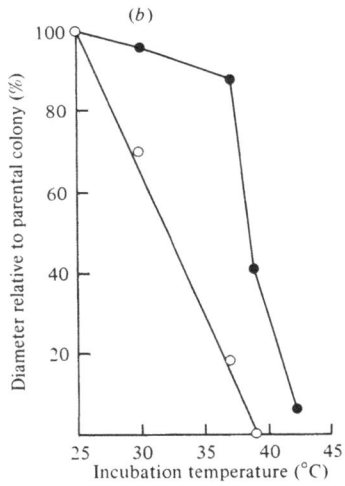

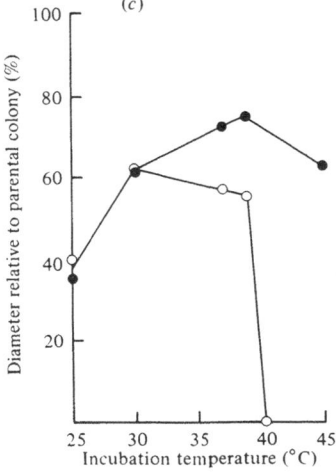

temperatures but showed relatively better growth at intermediate temperatures, accounting for 4 % of the mutants tested (Fig. 4.2*c*). Some mutants could be relieved by the addition of osmotic stabilizers such as sodium chloride (6 % w/v) and a mutant in this class could also be relieved by glucosamine, suggesting that a deficiency in chitin had caused a defect in the wall. This mutant, *glc-95*, resembles the mutant *ts6* reported by Cohen *et al.* (1969).

Analysis of mutants with 'balloons' or gross hyphal swelling

Fourteen mutants which produced 'balloons' and four with gross hyphal swelling were selected for detailed study because it was felt that they might represent mutants in which polarized growth of the wall had been eliminated. Complementation analysis in heterokaryons, forced by nutritional markers, showed that only four of the mutants failed to complement with each other, suggesting that there were at least 15 genetic loci involved in this phenotype (Valentine & Bainbridge, 1978). Mutant *mnrA455*, producing 'balloons', was relieved by mannose, *glc-95* with gross hyphal swelling was relieved by glucosamine (Valentine & Bainbridge, 1978) and *choC3* with balloons was relieved by choline (Fig. 4.3*b*; Markham & Bainbridge, 1978*a*). Biochemical and genetic tests (Markham & Bainbridge, 1978*b*) have indicated that *choC3* is a new locus, distinguishable from *choA* (Arst, 1968) and *choB* (Waldron & Roberts, 1974). It seems likely that alterations either in cell wall structure or membrane composition result in the abnormal phenotype seen in these mutants. Although the *choC3* strain was isolated as a temperature-sensitive mutant, it is now known that the original strain was a double mutant *choC3 ts-121*, and a strain *choC3* separated from *ts-121* showed no temperature sensitivity (P. Markham, unpublished data). The *choA* strain has also been shown to balloon (Markham & Bainbridge, 1978*a*; Fig. 4.3*a*).

Biochemical work has concentrated on the mutant relieved by mannose, *mnrA455*. This mutant strain failed to grow normally on minimal agar unless 1 % mannose was added (Table 4.1). In the absence of mannose, ballooning occurred (Fig. 4.3*b*) and the 'balloons' had a high affinity for the fluorescent dye Photine HV (Fig. 4.4), suggesting that there were qualitative differences in wall structure. Although the hyphal swellings have been called 'balloons' there is evidence that the wall is thicker than the normal wall (B. W. Bainbridge and P. Markham, unpublished results). In addition, 'balloons' could still be detected even after disruption of cultures in the French pressure cell and extensive

Table 4.1. *Morphology and growth of a selection of temperature-sensitive mutants of* A. nidulans. *(Valentine, 1975; Valentine & Bainbridge, 1978; Markham & Bainbridge, 1978a & b)*

Gene symbol	Morphology on MM[a] after 24 h at 25°C and 24 h at 45°C	Growth response at 42°C on MM plus:				
		Glucose	Mannose	Mannose + glucose	Glucosamine	Choline
mnrA455	Balloons	–	+	–	–	–
manA1	Balloons	–	–	+	–	–
glc-95	Swollen hyphae	–	–	–	+	–
choC3[b]	Balloons	–	–	–	–	+

– no growth
+ normal growth
[a] MM = minimal medium.
[b] Not a temperature-sensitive mutant.

Fig. 4.3. Cultures of choline mutants of *A. nidulans* growing on minimal agar medium in the absence of choline, showing the extent and distribution of 'balloons'. Mutants: (*a*) *choA1*; (*b*) *choC3*.

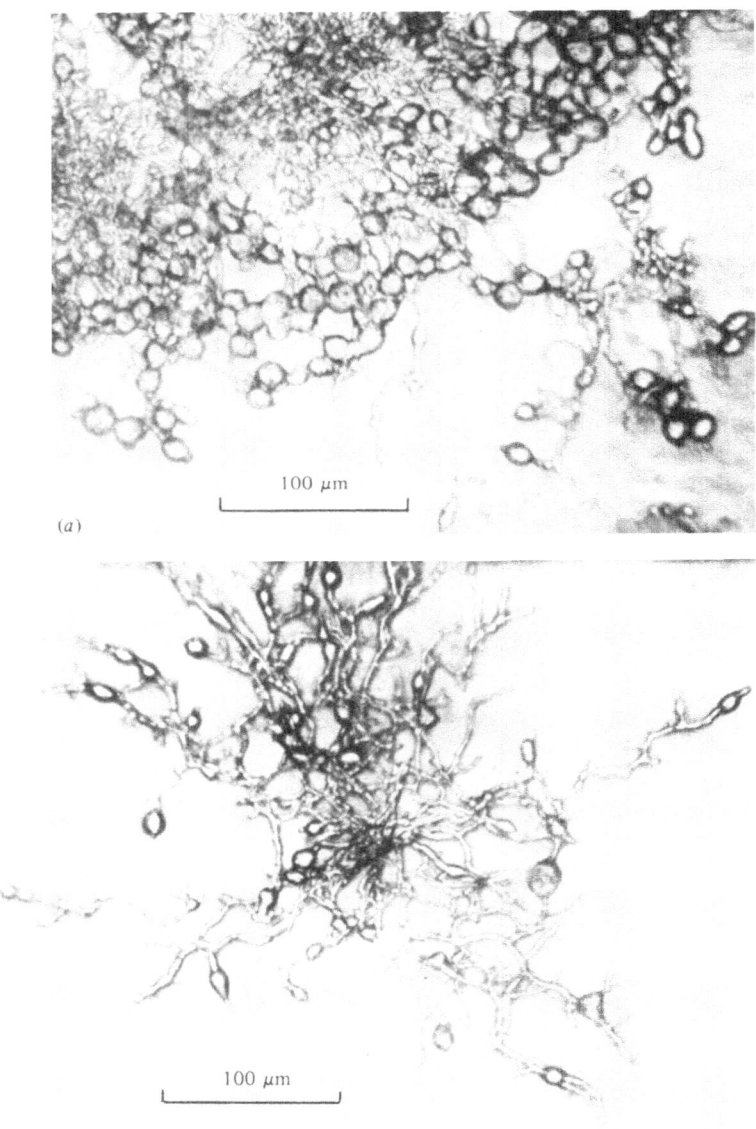

100 μm

(*a*)

100 μm

(*b*)

wall isolation and purification procedures. This suggests that 'balloons' are due to extensive, unpolarized, wall synthesis rather than to a simple stretching of the wall by internal pressure. This contrasts strongly with the glucosamine mutant reported by Cohen *et al.* (1969), which lyses in the absence of osmotic stabilizers, possibly due to autolytic enzyme activity in the absence of chitin synthesis (Katz & Rosenberger, 1971).

An analysis has been made of the composition of the wall of the

Fig. 4.4. Photomicrograph of hyphae of *A. nidulans* stained with 0.5 % (w/v) aqueous Photine HV and viewed under a Nikon Apophot microscope with fluorescent attachments. Photographs are negative prints taken from colour transparencies. (*a*) Parental strain; (*b*) mutant *mnr4A455*. (From Valentine & Bainbridge 1978.)

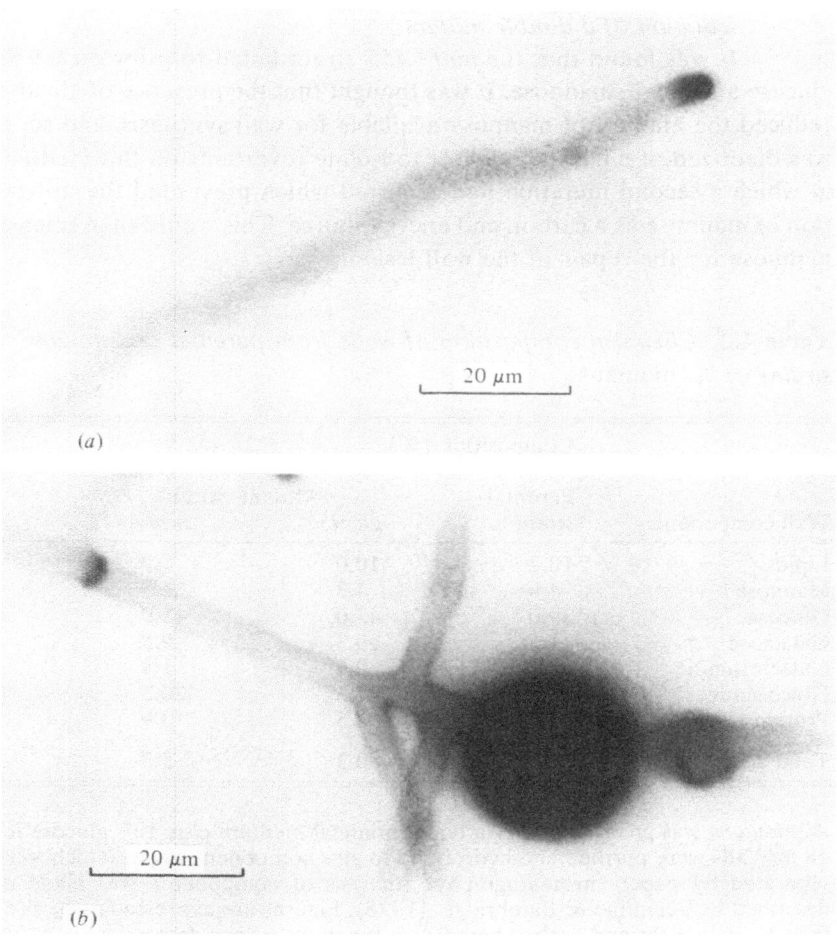

parental strain and the *glc-95* and *mnr A455* strains. They were grown in liquid minimal medium containing 1 % glucose for 48 h at 40 °C, which is a semi-permissive temperature, permitting sufficient growth to provide the wall material necessary for biochemical analysis. It should be realized that the choice of this temperature probably results in the presence of a mosaic of normal and mutant wall, thus partially masking any possible differences between the two types of wall. The *mnrA455* strain was found to contain about a third of the normal amount of mannose found in the normal strain, and the *glc-95* strain contained about half the amount of glucosamine (Table 4.2).

Influence of sugar mixtures on the growth of *mnrA*
Isolation of a double mutant
It was found that the *mnrA455* strain failed to grow on 0.9 % glucose and 0.1 % mannose. It was thought that the presence of glucose reduced the amount of mannose available for wall synthesis, and so, it was theorized, it might be possible to isolate revertants on this medium in which a second mutation had occurred which prevented the utilization of mannose as a carbon and energy source. This would then release mannose for the repair of the wall lesion.

Table 4.2. *Chemical composition of walls from parental and mutant strains of* A. nidulans

	Composition (%)		
	Parental	Mutant strains	
Wall components	strain	*glc-95*	*mnrA455*
Lipid	10.2	10.0	7.7
Mannose	4.0	4.3	1.5
Glucose	39.0	43.0	41.0
Galactose	9.5	10.3	9.5
Galactosamine	2.3	0.6	1.4
Glucosamine	13.5	7.6	20.3
Protein	3.5	4.5	3.9
Total accounted for	82.0	80.3	85.3

A. nidulans was grown at 40°C in liquid minimal medium plus 1 % glucose for 48 h. Walls were purified and hydrolysed to give component sugars which were separated by paper chromatography. Analysis of components was made as described in Valentine & Bainbridge. (1978). Figures are expressed as percentage of dry weight and each is based on a mean from four determinations.

Spores were treated with a mutagen and 50 revertants were obtained, one of which was the predicted double mutant *mnrA manA* (Valentine & Bainbridge, 1978). The rest of the revertants were wild-type, incidentally confirming that the *mnrA* mutation was a single gene change responsible for both the ballooning phenotype and the response to mannose. The *manA* mutation was epistatic to *mnrA*, suggesting that the second mutation was earlier in the pathway involving the utilization of mannose for wall synthesis. It was further found that the double mutant could only grow on a mixture of glucose and mannose and that it failed to grow on either sugar alone. This property will be discussed later.

The inhibition of sugar mutants by mixtures of sugars has also been shown for other systems. The glucosamine mutant *ts6* grows abnormally on mixtures of glucose and glucosamine (Katz & Rosenberger, 1970*a*), suggesting that perhaps there was imbalance between growth on glucose and wall synthesis using glucosamine. Inhibition of sugar mutants by mixtures of sugars has also been reported for glucose and fructose in *Neurospora crassa* (Murayama & Ishikawa, 1975) and glucose and mannose in *Saccharomyces cerevisiae* Hansen (Herrera, Pascual & Alvarez, 1976). Various explanations have been put forward for these phenomena. There may be competition for a common permease, resulting in the reduction in the uptake of the limiting sugar. Alternatively, there may be competition for phosphorylases in the cell, which reduces the phosphorylation of the limiting sugar, again lowering the availability of the sugar. A third possibility is that certain sugar phosphates accumulate in the mycelium and inhibit RNA synthesis (Maitra, 1971). A decision between these possibilities must await further experimentation.

The isolation of revertants has been used to analyse other systems, and Brody & Tatum (1966) obtained a semicolonial revertant at 24°C from a colonial mutant, *col-2*, of *Neurospora crassa*. This revertant was temperature-sensitive, as it was again colonial when the temperature was raised from 24 °C to 34 °C. The revertant had an altered glucose-6-phosphate dehydrogenase, thus providing further confirmation that this gene was the structural gene for the enzyme, and that a single change altered both the enzyme and the morphological lesions. Such an approach has also been used for the isolation of enzymes with altered properties in *Pseudomonas* (Clarke, 1974). It appears likely that similar isolation of revertants could be used to study a variety of enzymes involved in wall synthesis.

A pathway for mannose utilization in Aspergillus nidulans

The properties of the *mnrA manA* mutant strain were used to propose a pathway for the utilization of mannose for wall synthesis and as a carbon and energy source (Fig. 4.5). This explains the relief of the *mnrA* strain by mannose, as phosphorylation can occur in either the 1- or 6-position, thus by-passing the genetic block. The introduction of the second mutation, *manA*, could then prevent the utilisation of mannose as a carbon and energy source, which might release more of the mannose for wall synthesis. It could also prevent the accumulation of mannose-6-phosphate, which may be inhibitory as discussed above. As a consequence, the double mutant requires mannose, for wall synthesis, and glucose, as a carbon and energy source, since the two mutations would effectively prevent the interconversion of the sugars. Enzyme assays (Valentine & Bainbridge, 1978) have confirmed that the enzymes predicted in this scheme are in fact affected in the mutants. The *mnrA* strain has a thermolabile phosphomannose mutase (half-life, at 32 °C, 2.2 min compared to a wild-type value of 3.6 min). The *manA* mutant has been shown to have a forty-fold reduction in the specific activity of phosphomannose isomerase when grown at 43 °C, although the enzyme itself is not thermolabile, suggesting that the mutation is not

Fig. 4.5. A proposed pathway for the utilization of mannose for wall synthesis and as a carbon and energy source in *A. nidulans*. The broken lines indicate probable sites of blocks due to different mutants.

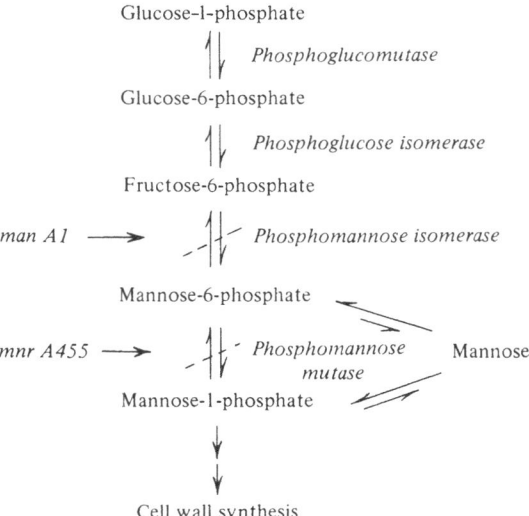

in the structural gene but that it affects the regulation of synthesis of the enzyme.

Radioactive labelling of fungal walls

Autoradiography

The double mutant *mnrA manA*, as argued above, requires mannose for wall synthesis, but cannot use mannose as a carbon source. Radioactively labelled mannose should therefore be specifically taken up into mannose in the fungal wall.

Spores of the double mutant were germinated in minimal medium containing 0.09 % glucose, 0.01 % mannose and [2-^{3}H]mannose. Autoradiography showed that radioactivity was taken up predominantly at the hyphal tip (Valentine & Bainbridge, 1978; Fig. 4.6). This suggests strongly that mannose is a wall precursor which is laid down in an area of active wall extension. The length of the labelled hyphal tip showed an exponential increase with time of labelling, as would be expected with hyphae which were growing exponentially.

Similar results have been obtained for acetylglucosamine labelling for various fungi (Gooday, 1971) and for galactose labelling of *gal-5* (Katz & Rosenberger, 1970*b*), suggesting that mannose, glucosamine and galactose are all laid down in polymers involved in tip extension. Trinci (1978) has suggested that radioactive precursors of primary wall polymers should be laid down at the tip while radioactive precursors of secondary wall polymers should be deposited in the subapical regions. However, predictions of this type must be treated with caution, as radioactive sugars can be converted into other sugars, reducing the specificity of the label. Thus Gooday (1971) showed that hyphae labelled with glucose also gave densely labelled tips but, subapically, had higher levels of label than was observed with radioactive acetyl-glucosamine. It is very likely that radioactive glucose was converted to mannose, glucosamine and galactose, giving dense label in the tip, and that subapical labelling was due to glucose incorporation into glucans. The walls should therefore be checked to see if the label is still in the labelled precursor added. A high degree of specificity of label can only be achieved if mutants are used to prevent the interconversion of sugars which would obscure the relationship between precursor and position in the hypha.

Labelling of wall components

The fate of radioactive wall precursors can be followed by purifying the wall, hydrolysing it, and then separating the individual sugars by chromatography. For example, Katz & Rosenberger (1970*b*) showed that wall labelled in the presence of [^{14}C]galactose contained labelled glucose and galactose. The data already presented for the *mnrA manA* strain suggested that it should be possible to label wall mannose

Fig. 4.6. Autoradiographs of hyphae of double mutant *mnrA manA* of *A. nidulans* grown at 43 °C for different times in the presence of [2-^{3}H]mannose. (*a*) 20 min; (*b*) 80 min. (From Valentine & Bainbridge, 1978.)

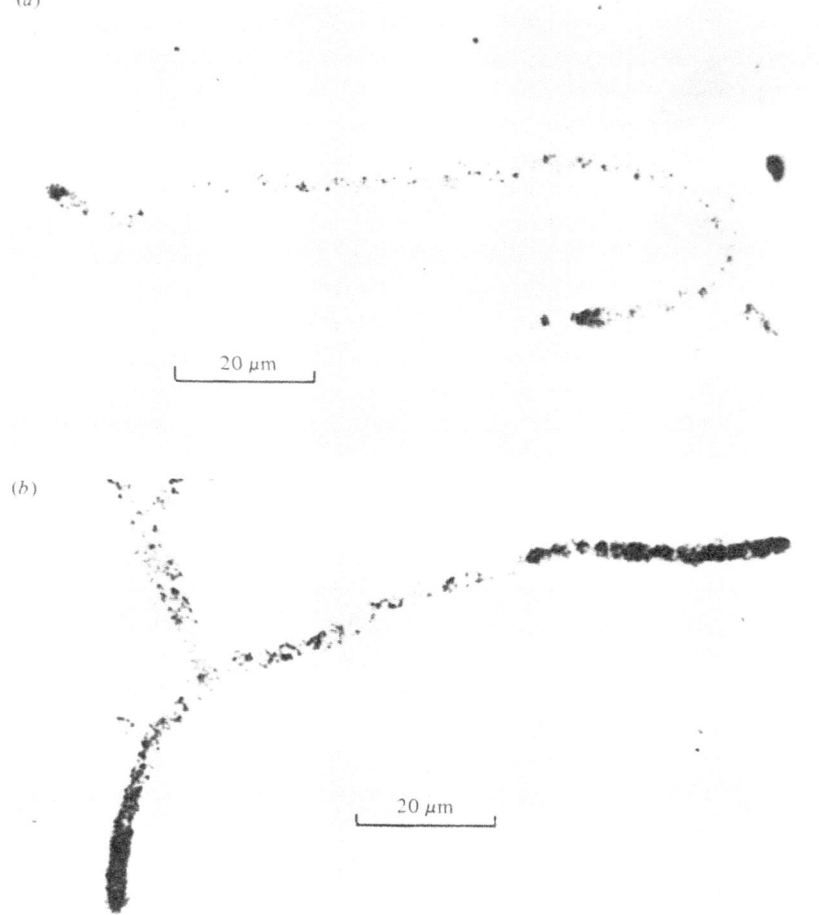

(*a*)

20 µm

(*b*)

20 µm

specifically at the non-permissive temperature. Growth of the wild-type and double mutant in the presence of radioactive mannose at 30 °C and 43 °C was followed by purification and hydrolysis of the walls (Valentine & Bainbridge, 1978). At 30 °C, the wild-type had 50 % of the total radioactivity in glucose and 15.4 % in mannose, whereas the double mutant at 43 °C had 9.4 % of the label in glucose and 83.9 % in mannose. This finding should allow the distribution of mannose to be followed during fractionation and separation of the wall polymers. It should also be possible to label wall preparations with mannose, glucosamine or choline in double labelling experiments.

A tentative model for ballooning

The phenomenon of ballooning has been observed to occur following mutation at twelve genetic loci. For three mutant strains, *mnrA, manA* and *choC3*, the balloning can be eliminated by a specific chemical added to the agar media. As discussed earlier, there is evidence for the *mnrA* strains that balloons result from unpolarized wall growth rather than by stretching of the wall by internal pressure. The wall of the balloon is thick and tough and does not lyse easily. There may, however, be many causes of ballooning which would result in a variety of wall thicknesses as well as variation in wall composition. Ballooning has also been reported in normal cultures in response to a variety of conditions such as nutrient starvation or elevated temperature. Giant spores in the microcycle of *Aspergillus niger* van Tiegham induced by incubation at 44 °C (Smith *et al.*, 1977) may be a related effect, but it should be pointed out that in *choC* strains ballooning occurs at 37 °C. There is no evidence that balloons formed by the mutants in *Aspergillus nidulans* differentiate into conidiophores as seen in the microcycle.

A preliminary time-lapse analysis of the development of 'balloons' has suggested the following tentative model (Fig. 4.7):

(1) Hypha growing at 25 °C with normal vesicle flow to the tip.
(2) Hypha transferred to 42 °C, vesicles are now depleted of an essential wall precursor such as mannose, so that linear extension of wall polymers at the tip ceases.
(3) Vesicles accumulate at the tip and, eventually, are diverted to a region behind the tip where wall rigidification is not yet complete. Vesicles fuse with the wall and unpolarized wall growth occurs.
(4) Wall rigidification is complete, and unpolarized wall growth ceases. Balloon has reached its maximum size.

Balloons can also occur at the tip and, on the theory proposed, this would be explained by assuming that these tips had stopped growth and that the extent of wall rigidification at the tip was equivalent to the subapical region in growing hyphae. Other models could be proposed but this model leads to testable predictions about wall thickness, distribution of vesicles and the structure of the wall. The requirements for mannose and choline for linear extension of wall polymers could be explained by the occurrence of mannose or glycolipid cross-links between the chitin microfibrils. In the absence of cross-links, coordinated intercalation of chitin microfibrils might not occur. Wessels (Chapter 2) has provided some evidence for the occurrence of such cross-linkages in the walls of *Schizophyllum commune* Fries and it is possible that cross-linkages may also occur in other fungal walls.

Temperature-sensitive mutations have been observed in *Bacillus*

Fig. 4.7. Tentative model for the origin of ballooning in mutants of *Aspergillus nidulans* (for details see text).

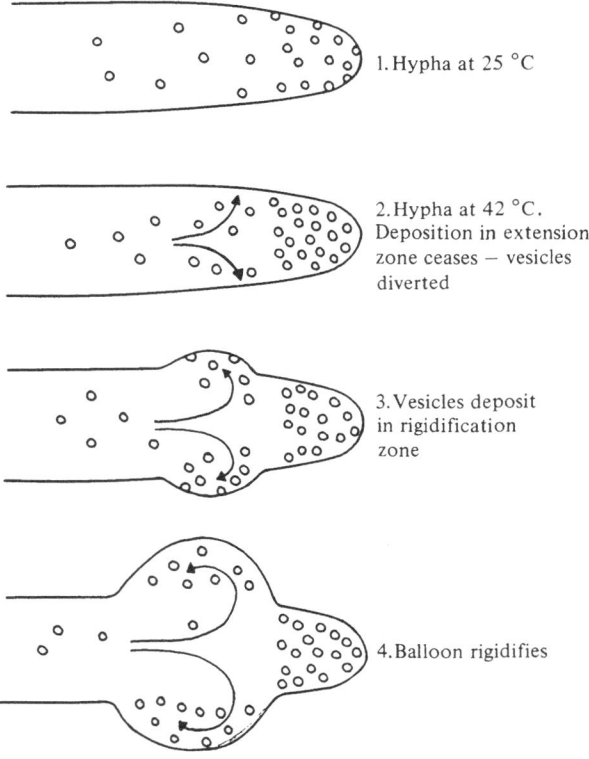

1. Hypha at 25 °C

2. Hypha at 42 °C. Deposition in extension zone ceases – vesicles diverted

3. Vesicles deposit in rigidification zone

4. Balloon rigidifies

subtilis (Ehranberg) Cohn which change rods into coccal-shaped cells (Rodgers, Ward & Burdett, 1978). On raising the temperature the normal pattern of wall synthesis by diffuse growth switches to growth at the septum only and the cells increase in diameter but not in length. It has been suggested that extension sites in the wall are inactivated at the non-permissive temperature and that wall synthesis is diverted to the septal region. A similar mechanism could be envisaged in the production of balloons, as there may be labile synthetic or lytic sites at the hyphal apex. Inactivation of these sites might result in the diversion of vesicles to other sites which may or may not be located at the septa.

Influence of osmotic stabilizers, sodium deoxycholate and actidione on temperature-sensitive mutants

It has been commonly observed that temperature-sensitive mutants can be repaired at the restrictive temperature by growth in the presence of osmotic stabilizers such as 15 % mannitol or 6 % sodium chloride. From a total of approximately 300 mutants, 34 % were found to have increased growth on a variety of osmotic stabilizers (Valentine, 1975), and Waldron & Roberts (1974) have reported that out of 75 cold-sensitive mutants 15 were repaired by molar sodium chloride. It must not be assumed from this, however, that all of these osmotic-remedial mutants have defective walls or membranes, as temperature-sensitive nutritional mutants have been shown to be repaired by osmotic stabilizers. In *Neurospora crassa*, a temperature-sensitive, adenine-requiring mutant produced higher levels of an altered adenylosuccinase when grown at a non-permissive temperature in the presence of molar sorbitol or mannitol (Martin & DeBusk, 1975). However, as mentioned previously, mutants defective in glucosamine synthesis are also relieved by osmotic stabilizers, although mannose mutants are not. Similar osmotic-remedial mutants affecting wall composition and membrane fragility have also been reported in *N. crassa* (Inoue & Ishikawa 1975). Katz & Rosenberger (1971) have proposed that during normal wall synthesis lytic enzymes degrade the wall at specific sites and that these weakened regions can be protected either by normal chitin synthesis or by osmotic stabilizers in the glucosamine-deficient strain.

Other temperature-sensitive strains have been found to be sensitive to sodium chloride, sodium deoxycholate or actidione. The choline mutant *choC3* showed sensitivity to all three compounds and sensitivity to deoxycholate and actidione has also been reported for cold-sensitive mutants (Waldron & Roberts, 1974). It seems likely that loss of osmotic relief in the presence of deoxycholate could be due to cell wall or membrane changes which increase permeability to deoxycholate, a mechanism which has been described in *Escherichia coli* (Migula) Castellani & Chalmers (Russell, 1972). These and related phenomena have received little attention in fungi and it is probable that a detailed analysis would be rewarding.

Attempts at isolating temperature-sensitive mutants affected in germination

In principle it should be possible to isolate temperature-sensitive mutants affected specifically in their spore germination. Mutants can be screened by pre-incubation at a permissive temperature, to allow germination, followed by transfer to a non-permissive temperature. Growth will continue if the temperature-sensitive lesion affects germination but not growth. In spite of an extensive search in *Aspergillus nidulans* (Waldron & Roberts, 1974; B. P. Valentine and B. W. Bainbridge, unpublished data) and in *Neurospora crassa* (Schmidt & Brody, 1976) no convincing mutations affected solely at germination have yet been identified. This failure can be explained in three ways. Firstly, there may be no functions specific to spore germination, so that mutants will never be isolated. Thus germ tube emergence would have exactly the same enzymes normally present in the growth of the hyphal apex. Secondly, there may be preformed mRNA in the spores which codes for the spore-specific functions. In the initial isolation of mutant spores, the mRNA would be of parental origin which would give normal germination so that the mutant would not be detected. Experiments designed to eliminate carry-over effects of this type have not yielded germination mutants. Thirdly, there may be very few germination-specific functions and the proteins may be very thermostable, making isolation of mutants difficult.

Conclusions

Models of wall synthesis in fungi are complex (Bartnicki-Garcia, 1973) and it seems likely that a combined approach using genetics, biochemistry, physiology and electron microscopy will be necessary to solve the problem of wall synthesis. A start has been made with the exploitation of temperature-sensitive mutants, but the full potential of these mutants has not yet been realized. It is hoped that future work will make full use of genetics as a tool for the study of wall synthesis.

Acknowledgement. We are grateful for financial support from the Science Research Council.

References

Arst, N. H. (1968). Genetic analysis of the first steps of sulphate metabolism in *Aspergillus nidulans. Nature (London),* **219,** 268–80.

Bainbridge, B. W. (1976) Estimation of the generation time and peripheral growth zone of *Aspergillus nidulans* and *Alternaria solani* hyphae from radial growth rates and ranges in apical cell length. *Journal of General Microbiology,* **97,** 125–7.

Bartnicki-Garcia, S. (1973). Fundamental aspects of hyphal morphogenesis. In *Microbial Differentiation* (23rd Symposium of the Society for General Microbiology, eds. J. M. Ashworth & J. E. Smith, pp. 245–67. Cambridge University Press.

Brody, S. & Tatum, E. L. (1966). The primary biochemical effects of a morphological mutation in *Neurospora crassa. Proceedings of the National Academy of Sciences, USA,* **56,** 1290–7.

Clarke, P. (1974). The evolution of enzymes for the utilization of novel substrates. In *Evolution in the Microbial World* (24th Symposium of the Society for General Microbiology), eds. M. J. Carlile & J. J. Skehel, pp. 183–217. Cambridge University Press.

Clutterbuck, A. J. (1970). Synchronous nuclear division and septation in *Aspergillus nidulans. Journal of General Microbiology,* **60,** 133–5.

Cohen, J., Katz, D. & Rosenberger, R. F. (1969). Temperature sensitive mutant of *Aspergillus nidulans* lacking amino-sugars in its cell wall. *Nature (London),* **224,** 713–15.

Fiddy, C. & Trinci, A. P. J. (1976). Mitosis, septation, branching and the duplication cycle in *Aspergillus nidulans. Journal of General Microbiology,* **97,** 169–84.

Gooday, G. W. (1971). An autoradiographic study of hyphal growth of some fungi. *Journal of General Microbiology,* **67,** 125–33.

Hartwell, L. H. (1967). Macromolecular synthesis in temperature-sensitive mutants of yeast. *Journal of Bacteriology,* **93,** 1662–70.

Hayes, W. (1968). *The genetics of Bacteria and Their Viruses,* 2nd edition. Oxford: Blackwell Scientific Publications.

Herrera, L. S., Pascual, C. & Alvarez, X. (1976). Genetic and biochemical studies of phosphomannose isomerase deficient mutants of *Saccharomyces cerevisiae. Molecular and General Genetics,* **144,** 223–30.

Inoue, H. & Ishikawa, T. (1975). Isolation of osmotic remedial temperature-sensitive mutants defective in cell wall or membrane in *Neurospora crassa. Journal of General and Applied Microbiology,* 21 389–91.

Katz, D. & Rosenberger, R. F. (1970*a*). A mutation in *Aspergillus nidulans* producing hyphal walls which lack chitin. *Biochimica et Biophysica Acta,* **208,** 452–60.

Katz, D. & Rosenberger, R. F. (1970*b*). The utilization of galactose by an *Aspergillus nidulans* mutant lacking galactose phosphate-UDP glucose transferase and its relation to cell wall synthesis. *Archiv für Mikrobiologie,* **74,** 41–51.

Katz, D. & Rosenberger, R. F. (1971). Lysis of an *Aspergillus nidulans* mutant blocked in chitin synthesis and its relation to wall assembly and wall metabolism. *Archiv für Mikrobiologie,* **80,** 284–92.

Maitra, P. K. (1971). Glucose and fructose metabolism in a phosphoglucose-isomeraseless mutant of *Saccharomyces cerevisiae. Journal of Bacteriology,* **107,** 759–69.

Markham, P. & Bainbridge, B. W. (1978*a*). A morphological lesion (ballooning related to a requirement for choline in mutants of *Aspergillus nidulans. Proceedings of the Society for General Microbiology,* **5,** 65.

Markham, P. & Bainbridge, B. W. (1978*b*). Characterization of a new choline locus in *Aspergillus nidulans* and its significance for choline metabolism. *Genetical Research, Cambridge,* **32,** 303–10.

Martin, C. E. & DeBusk, A. C. (1975). Temperature-sensitive, osmotic-remedial mutants of *Neurospora crassa*: osmotic pressure induced alterations of enzyme stability. *Molecular and General Genetics,* **136,** 31–40.

Mishra, N. C. (1977). Genetics and biochemistry of morphogenesis in *Neurospora. Advances in Genetics* **19,** 341–405.

Morris, N. R. (1976). Mitotic mutants of *Aspergillus nidulans. Genetical Research, Cambridge,* **26,** 237–54.

Murayama, T. & Ishikawa, T. (1975). Characterization of *Neurospora crassa* mutants deficient in glucose-phosphate isomerase. *Journal of Bacteriology,* **122,** 54–8.

Orr, E. & Rosenberger, R. F. (1976). Initial characterization of *Aspergillus nidulans* mutants blocked in the nuclear replication cycle. *Journal of Bacteriology,* **126,** 895–902.

Rodgers, H. J., Ward, B. J. & Burdett, I. D. J. (1978). Structure and growth of the walls of Gram-positive bacteria. In *Relations between Structure and Function in the Prokaryotic Cell* (28th Symposium of the Society for General Microbiology), eds. R. Y. Stanier, H. J. Rodgers & B. J. Ward, pp. 139–76. Cambridge University Press.

Russell, R. R. B. (1972). Temperature-sensitive osmotic-remedial mutants of *Escherichia coli. Journal of Bacteriology,* **112,** 661–5.

Schmidt, J. C. & Brody, S. (1976). Biochemical genetics of *Neurospora crassa* conidial germination. *Bacteriological Reviews,* **40,** 1–41.

Smith, J. E., Anderson, J. G., Deans, S. G. & Davis, B. (1977). Asexual development in *Aspergillus. Genetics and Physiology of Aspergillus* (British Mycological Society Symposium Series No. 1, 1977), eds. J. E. Smith & J. A. Pateman, pp. 23–58. New York, London: Academic Press.

Steele, G. C. & Trinci, A. P. J. (1977). Effect of temperature and temperature shifts on growth and branching of a wild type and a temperature sensitive colonial mutant (cot 1) of *Neurospora crassa. Archives of Microbiology,* **113,** 43–48.

Trinci, A. P. J. (1978). Wall and hyphal growth. *Science Progress, Oxford,* **65,** 75–99.

Unrau, P. & Holliday, R. (1970). A search for temperature-sensitive mutants of *Ustilago maydis* blocked in DNA synthesis. *Genetical Research, Cambridge,* **15,** 157–69.

Valentine, B. P. (1975). The isolation and characterization of temperature-sensitive mutants of *Aspergillus nidulans* with special reference to cell wall synthesis and mannose utilization. Ph.D. Thesis, University of London.

Valentine, B. P. & Bainbridge, B. W. (1975). Mutations affecting the incorporation of mannose into the cell wall of *Aspergillus nidulans. Proceedings of the Society for General Microbiology,* **2,** 90.

Valentine, B. P. & Bainbridge, B. W. (1978). The relevance of a study of a temperature-sensitive ballooning mutant of *Aspergillus nidulans* defective in mannose metabolism to our understanding of mannose as a wall component and carbon/energy source. *Journal of General Microbiology*, **109**, 155–168.

Waldron, C. & Roberts, C. F. (1974). Cold-sensitive mutants in *Aspergillus nidulans*. I. Isolation and general characterization. *Molecular and General Genetics*, **134**, 99–113.

5
Growth of rhizomorphs, mycelial strands, coremia and sclerotia

S.C.WATKINSON

Botany Department, University of Oxford, South Parks Road, Oxford, OX1 3RA, UK

Introduction

Mycelial strands, rhizomorphs, coremia and sclerotia are vegetative structures concerned with survival and spread. They are each formed by a wide variety of species of fungi (Ramsbottom, 1953). Butler (1966) reviewed vegetative aggregations in fungi, including a stimulating discussion of the work of earlier morphologists, and more recently Willetts (1972, 1978) and Chet & Henis (1975) reviewed sclerotium formation, the latter two from the point of view of metabolic control of morphogenesis. The development of all these structures represents a departure from the normal form of the primary vegetative colony. The individual hyphae composing an aggregate abandon, at some point, their original radial growth orientation, their fairly regular branching and even spacing over the substratum. Instead they cease to behave as individuals, merge into a larger structure and then differentiate further in a more or less regular pattern. Morphological change is accompanied by changes in the pattern of enzyme proteins (Wong & Willetts, 1974) and activities (Chet, Retig & Henis, 1972; Kritzman *et al.*, 1976). This process usually occurs at a distance from the primary hyphal apices of the growing colony, so that two kinds of hyphal growth are proceeding simultaneously in the same colony. A fundamental question in relation to the development of rhizomorphs, mycelial strands, coremia and sclerotia is: what morphological change initiates their formation, and how is this related in time and place to the stimulus that elicits it? It is a feature of these hyphal aggregations that they form in response to a wide variety of external changes. Table 5.1 shows the widespread effectiveness of alterations in nutrition for eliciting their development, although other particular stimuli may operate in indi-

Table 5.1. *Range of stimuli eliciting formation of vegetative aggregated structures*

Structure	Fungus	Stimulus	Reference
Mycelial strands	*Serpula lacrimans*	High C:N ratio sodium nitrate }	Watkinson (1975*b*)
	Serpula lacrimans *Helicobasidium purpureum* *Poria xantha*	Translocation in leading hyphae Translocation in leading hyphae Cu^{2+}	Watkinson (1971*a*) } Valder (1958) Hirt (1948)
Rhizomorphs	*Armillaria mellea*	Certain amino acids Sufficient nutrition } Critical C:N ratio } Ethanol Indoleacetic acid } Aminobenzoic acids }	Weinhold & Garraway (1966) Garrett (1953) Weinhold (1963) Garraway (1970)
Coremia	*Penicillium claviforme*	Light–dark transfer 3°C temperature } affect time of formation cycles Alcohols Detergents Glutamic and aspartic acids	Faraj-Salman (1970, 1971*a*, *b*)
	Penicillium isariiforme	Light (450–520nm) Carbon dioxide	Watkinson (1977) Carlile *et al.* (1962*a*) Piskorz (1967*a*, *b*); Bennink (1972) Piskorz (1972)
Sclerotia	*Sclerotium rolfsii*	Critical C:N ratio Threonine Lactose Light } affect time Glucose } of formation Ammonium Sclerin	Wheeler & Sharan (1965) Kritzman *et al.* (1976) Okon, Chet & Henis (1973) Humpherson-Jones & Cooke (1977*a*, *b*)
	Coprinus cinereus *Sclerotinia sclerotiorum*		Moore & Jirjis (1976) Marukawa *et al.* (1975)

vidual cases. Answers to questions about the mechanisms of morphogenesis can only come from studies of the time-sequence of processes that connect the eliciting stimulus and the resulting differentiation of a hypha or hyphae.

Mycelial strands
Morphogenesis

Mycelial strands are very common in Basidiomycetes, although those of *Serpula lacrimans* (Wulfen ex Fr.) Schroeter are the best-known. Their early stages and later development have been well described (Hartig, 1902; Falck, 1912; Butler, 1957, 1958). They will form in mycelium growing over glass slides and, since their development is acropetal, a mature strand visible in old mycelium displays early stages of formation near the mycelial margin, so the whole sequence from the earliest morphological stages can be seen. Three processes appear to be involved: the formation of three different types of hypha, a change in the orientation of growth of new hyphae, and the acquisition of adhesiveness between some or all hyphae. There are two types of hyphae in *S. lacrimans* mycelium even before stranding begins. Wide, empty-looking hyphae are visible in mycelium growing over glass; these differ from others in their staining properties both with methylene blue (Butler, 1958) and Nile blue (Helsby, unpublished). They can act as strand initials (in the sense of being distinguishable future sites of strands which may or may not develop further). Hyphae also differentiate into 'tendrils' that lose their original orientation and display thigmotropic growth around others. Much later the third type, fibre hyphae, appear; these run longitudinally through mature strands and with their thick walls and almost occluded lumina are similar to the skeletal hyphae of other polypores (Corner, 1932). They have only been described as components of strands. In the electron microscope, mature strands seem to be composed of hyphae of all ages – wide, thin-walled vessel hyphae and thick-walled fibre hyphae being interspersed with others that appear metabolically active. Between them is possibly adhesive, extrahyphal fibrillar material (Fig. 5.1). 'Vessels' and 'fibres' together make up the longitudinal system. It has so far not been possible to see by electron-microscopy the structures described by Hartig (1902) as beam and ring thickenings of the 'vessel' hyphae, analogous to the secondary thickenings of xylem elements.

Stimulus to strand formation

The first detectable morphological changes in strand formation
are the cohesion of younger hyphae round an older one, followed by
further localized growth over and around it. This cohesion results from
chance encounters, not from tropic responses. How are these steps
related to strand-eliciting stimuli? It is easy to see and count mature
strands and so describe some of the environmental changes that lead to
their formation. Prominent among such changes are those that result in
the hyphae themselves becoming the main nutrient sources for new
mycelium. Strands develop in mycelium growing from a food-base
through non-nutrient medium (Butler, 1958). Contact of the hyphal
margin with a fresh food-base stimulates stranding in the intervening
mycelium in *Serpula lacrimans* (Butler, 1958; Watkinson, 1971*a*) and

Fig. 5.1. Portion of the central region of a strand of *Serpula lacrimans*
showing hyphae embedded in extrahyphal fibrillar material (arrowed).
(By permission of L. Helsby.)

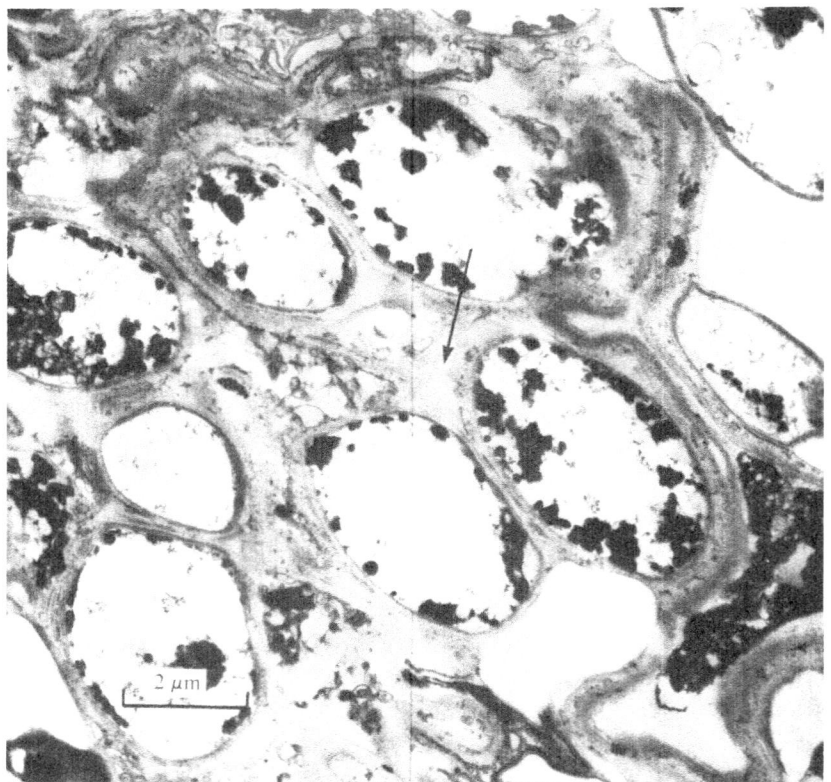

Helicobasidium purpureum Pat. (Valder, 1958). In initially unstranded *Serpula* mycelium, the bridging of two nutrient agar discs 1 cm apart by mycelium is followed within one day by increased branching near the point of contact with the fresh medium and within a week by formation of strands in the connecting mycelium, although not elsewhere. Translocation of ^{14}C can be demonstrated during the period of strand formation in the stranding mycelium (Watkinson, 1971*a*). Loss of some of the translocated material from the mycelium is evident both from experiments with radioisotopes and from observations of crystalline deposits on the outside of hyphae in dry conditions. If this exudate is utilizable by younger hyphae, further growth would be made possible around older translocating hyphae and stranding would result.

Stranding is also elicited in conditions where the hyphae become the chief repositories of substrate in an ageing mycelium, for example after prolonged growth (6–8 weeks) on nutrient agar medium. Strand formation is favoured by a higher carbon:nitrogen ratio than that which is optimal for growth, and also by the use as nitrogen source of sodium nitrate, which supports only moderate growth compared with amino acids and induces relatively early autolysis. On nitrate-containing media, stranding was observed to increase even while total dry weight decreased. Stranding could be prevented by perfusion of mycelial mats with fresh medium (Watkinson, 1971*a*). The explanation suggested for these observations (Watkinson, 1975*b*) is that stranding results from the limitation of new growth to the regions around existing hyphae as they become the main source of nutrients, possibly nitrogenous nutrients.

Strand formation might thereby be a means of conserving and recycling nitrogen in the nitrogen-poor medium of wood. The need that wood-rotting fungi have for some such mechanism has been demonstrated by Levi, Merill & Cowling (1968, 1969). The morphological changes occurring during strand development, described above, are consistent with this picture of new hyphae deriving nutrients from older, senescent and more permeable ones by lateral movement of materials within the strand. In summary, while the capacity of the fungus to form strands rests on its genetic ability to develop the appropriate differentiated hyphae, the sequence of development that results in stranding is elicited trophically. The mechanism that decides the developmental fate of individual hyphae – whether into 'vessel', 'tendril' or 'fibre' – remains unexplained. It may be significant that while 'vessel' hyphae develop in nutrient-poor conditions, 'fibre' hyphae only appear in the nutrient-rich environment of the strand.

Function of mycelial strands

Some discussion of function is relevant to developmental questions about any structure with obvious adaptive significance. This is because it is likely that its evolutionary adaptation will have included an ability to form in relation to appropriate stimuli. If we can assume this appropriateness of eliciting stimuli we can also make deductions from them about function.

Many workers have considered the special function of strands as channels for translocation of water or nutrients to be self-evident. This is partly because, in *Serpula lacrimans* in particular, strands often subtend growing fruit bodies and are the only surviving connection between a wood food-base and the growing mycelial margin (Cartwright & Findlay, 1958). It is also because of the presence of a longitudinal system of 'vessel' hyphae in strands (even though these are also present in unstranded mycelium). There is incontrovertible evidence that strands translocate water and nutrients (Butler 1958; Weigl & Ziegler, 1960; Watkinson 1975*a*, *b*) but it is less clear that they function as specialized channels for rapid transport of aqueous solutions (Butler, 1966). Jennings *et al.* (1974) have inferred an ability of *Serpula* to generate a hydrostatic pressure inside the hyphae which enables it to pump water and dissolved solutes along the hyphae. The structure of strands as seen by electron- and light microscopy, shows as possible channels for mass flow the 'vessel' hyphae which are infrequently septate and relatively wide (10μm), but which would not appear to be pressure-resistant as they are thin-walled and often appear to have collapsed (Helsby, unpublished). The 'fibre' hyphae have not been suggested as channels, probably because they sometimes appear completely occluded by wall thickening. Translocation in *Serpula* may be independent of gradients in water potential; translocation of ^{14}C was demonstrated both towards and away from the mycelial margin under conditions of uniform water potential (Day, 1968). Besides nutrient translocation, the growth habit of *Serpula* requires the conservation of nitrogen and resistance to desiccation.

These purposes are probably better served by a plectenchymatous mycelium than by dispersed, separate hyphae; the adaptive advantage of strands may be simply as the most efficient means of supporting mycelial advance from a distant food-base. This would be consistent with the nutrient regimes that cause stranding.

Rhizomorphs

Morphogenesis

Rhizomorphs and coremia are structures remarkable for the coordinated apical growth of the hyphal aggregate. In coremia and some rhizomorphs the individual hyphal tips lie parallel and distinguishable at the apex. However, in the well-known rhizomorph of *Armillaria mellea* (=*Armillariella mellea*) (Vahl. ex Fr.) Quélet normal apical growth of hyphae is replaced by growth from a meristem-like zone (de Bary, 1887). Motta (1969, 1971) showed that growth of *Armillaria* rhizomorphs occurred from a point 25 μm behind the extreme tip. He described an apical centre giving rise to three groups of derivatives analogous in behaviour to the histogens of the apical meristems of higher plants: anterior derivatives gave rise to a root-cap-like structure, lateral ones to an inner cortex and medulla, and posterior ones to an internal medulla of loosely-woven cells which near the apex react positively with periodic acid–Schiff stain but later collapse to leave a central lumen. In Fig. 5.2 this organization is contrasted with that of the *Sphaerostilbe repens* Berk & Br. rhizomorph investigated by Botton & Dexheimer (1977). This rhizomorph also grows submerged in the medium; it shows some apical organization but individual hyphae are discernible at the apex. In both rhizomorphs there is a central lacuna and both have mucilage between the hyphae, probably consisting in part of α1–4 and β1–4 linked glucans (Botton & Dexheimer, 1977), which is

Fig. 5.2 Rhizomorph structure. (*a*) Apex of *Sphaerostilbe repens* rhizomorph, redrawn from Botton & Dexheimer (1977); (*b*) apex of *Armillaria mellea* rhizomorph redrawn from Motta (1969). Shaded area represents 'meristematic' tissue; arrows show direction of production of derivatives.

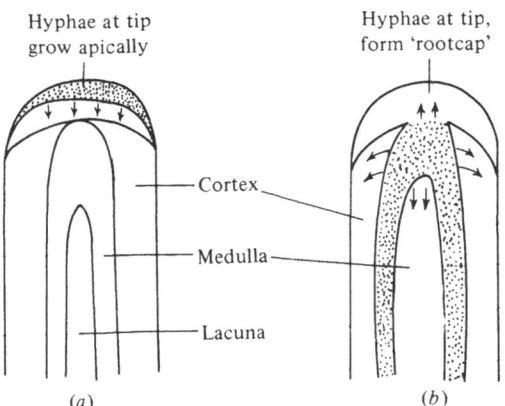

Hyphae at tip grow apically

Hyphae at tip, form 'rootcap'

Cortex

Medulla

Lacuna

(*a*)   (*b*)

produced also on the surface of the rhizomorph in *Armillaria,* where it darkens on exposure to oxygen (Smith & Griffin, 1971). The mature structure of fungal rhizomorphs has been described by Garrett (1970), and recent electron microscope studies of *Armillaria* have been made by Schmid & Liese (1970) and Wolkinger, Plank & Brunegger (1975). The central medulla contains large thin-walled cells interspersed with fibres, similar to those in strands of *Serpula lacrimans,* surrounded by an inner medulla of cells with large nuclei and numerous mitochondria. The cells of the outer medulla become concentrically thickened and surrounded by electron-dense mucilage, and the outermost rind consists of hyphae which are conspicuously laterally-connected, to the extent of having dolipores in adjacent lateral walls. The longitudinal system in mature *Armillaria* rhizomorphs is limited to the central lacuna and the fibres.

Rhizomorphs represent an extreme of fungal development in which there appears to be stable mitotic inheritance of differentiation. The apically organized rhizomorph, once established, is able to survive subculture and originate new apices by branching (Botton & Ly, 1976). How this stable pattern of division and differentiation is established and maintained is unknown. One factor that might give rise to a morphogenetic gradient is oxygen diffusing from the lacuna. Smith & Griffin (1971) have convincing evidence that this is an air channel, as originally assumed by de Bary (1887), and Botton & Dexheimer (1977) and Schmid & Liese (1970) found mitochondria concentrated in the inner medulla, surrounding the lacuna. The degree of aerobiosis has been implicated in other fungal systems as a morphogenetic factor (Turian, 1975).

An important question which awaits an answer is how this self-perpetuating apex is originally constituted from undifferentiated mycelium.

Stimuli to rhizomorph formation

The importance of nutritional control of rhizomorph initiation was demonstrated by Garrett (1953) using *Armillaria mellea.* Later, Weinhold (1963) and Weinhold & Garraway (1966) found that nutrient medium containing 500 ppm of ethanol stimulated rhizomorph formation, and demonstrated an interaction between ethanol and amino acids as rhizomorph-eliciting factors. Of a range of nitrogen sources, only casein hydrolysate supported good rhizomorph development in the absence of ethanol, while a few amino acids did so when the ethanol concentration was kept at 500 ppm. Above a certain concentration, nitrogen may become supraoptimal for rhizomorph formation (Garrett, 1953). Rhizomorphs have also been induced to form by the presence in the culture of other organisms (Pentland, 1965; Botton & El-Khouri, 1978).

Garraway (1969, 1970) found that compounds related to the shikimic acid pathway, and particularly indoleacetic acid, stimulated *Armillaria* rhizomorph production at concentrations too low to be nutritionally significant.

Vance & Garraway (1973) found that ethanol altered the phenol composition of the mycelium, reducing the amounts of phenols formed, and also that it inhibited glucose uptake and metabolism. They suggest that ethanol promotes rhizomorph and mycelial growth by redirecting metabolism away from the production of inhibitory phenolic compounds.

Rhizomorph growth and the experimental conditions that elicit their formation have now been so well-described that study of the intervening processes may well prove rewarding.

Function of rhizomorphs

The function of rhizomorphs as organs of aggressive colonization by root-infecting fungal parasites has been fully described and discussed by Garrett (1970). Their formation in response to ethanol and indoleacetic acid would seem to have obvious adaptive significance. So also would their capacity for very rapid growth compared with undifferentiated mycelium. Rishbeth (1968) reports a rate of growth of *Armillaria mellea* rhizomorphs of 9.8 mm d^{-1} compared with 0.75 mm d^{-1} for normal mycelium. This rate was for growth from a food-base through non-nutrient medium. What special features of rhizomorph organization make this possible? Improved efficiency of translocation and concentration of growth in a parallel longitudinal direction have both been suggested (Garrett, 1956, 1970). Growth through soil in itself demonstrates effective translocation mechanisms, although not necessarily faster ones than in undifferentiated mycelium. Rishbeth (1968) interprets his observation of differential temperature optima for rhizomorph growth through nutrient and non-nutrient media as a consequence of different optima for uptake and translocation. In a recent (1978) paper he has further discussed the ecological consequences of the fact that rhizomorph growth in *A. mellea* is limited to the temperature range 10–26 °C. The translocating mechanism, as in mycelial strands, remains obscure. The ultrastructural studies of Wolkinger *et al.* (1975) and Schmid & Liese (1970) do not reveal obvious channels for symplastic transport of organic materials, such as would be expected by analogy with the higher plant roots that rhizomorphs resemble.

Perhaps the speed of growth of rhizomorphs is a result not of their success at nutrient transport but of their special mode of growth, whereby multiple nuclear divisions in the generative zone are followed by formation of numerous septa to produce isodiametric cells which then enlarge, probably at the expense of intracellular reserves of glucan, so causing intercalary extension of the whole organ (Motta, 1971). While translocation of solutions in rhizomorphs is poorly understood, Smith & Griffin (1971) have demonstrated that transport of gases certainly occurs in them. Oxygen tension at the base of a rhizomorph of *Armillariella elegans* (Heim) J. B. Taylor, Hawkins & McLaren can affect its growth rate, which increases with the oxygen tension supplied. However, O_2 applied at the exterior surface of a rhizomorph tends to slow its growth rate and to cause deposition of brown pigments in the intercellular mucilage. This is prevented in damp conditions by a film of water over the rhizomorph apex, and a minimum O_2 tension in the soil atmosphere is required for rhizomorph growth (Morrison, 1976; Rishbeth, 1978). Rishbeth suggested that this was because a high rate of aerobic metabolism was needed to sustain it. The characteristic apical organization breaks down once the fungus has gained entry to the host root (Smith & Griffin, 1971).

Coremia
Morphogenesis

Coremia (synnemata) are aerial, multihyphal structures in which the apices of the component hyphae advance together. There is some differentiation of the hyphae, and the term is usually applied to structures which bear spores.

Coremia of the Penicillia have been most studied, although coremia are found in other groups of Deuteromycetes and Ascomycetes. In a series of papers Carlile and his coworkers described the morphogenesis of these coremia (Carlile, *et al.*, 1961; Carlile *et al.*, 1962*a*; Carlile, Dickens, & Schipper, 1962*b*). The most elaborate coremia are formed by *Penicillium isariiforme* Stolk & Meyer and *Penicillium claviforme* Bainier. In both cases, they arise as bundles of new branches behind the colony margin, which become aligned in unidirectional light owing to their positive phototropism. This is soon lost in the case of *P. claviforme*, which sporulates at the apex of a coremium, but in *P. isariiforme* unidirectional blue light remains essential to the maintenance of the alignment of the hyphal tips (Benninck, 1972; Carlile *et al*., 1962*a*), and sporulation occurs on side branches behind the coremium apex. The entire structure remains strongly phototropic throughout its indeterminate development.

The earliest stages of coremium development are hard to see when they form in a felt of established mycelium, but can be observed in young mycelium of *Penicillium claviforme* produced by spreading spores in a layer over a nutrient medium. About 4 h after germination, the potential sites of future coremia become visible as centres of radiating hyphae (Fig. 5.3). The proportion of these that develop further is influenced by nutrition (Watkinson, 1977). Coremia of *P. claviforme* develop in three stages. First, primordia form at the central points, as shown in Fig. 5.3. These consist of bundles of new, aligned hyphal branches about 1 mm high, which have acquired sufficient endogenous substrate to be capable of completing development in the absence of further supplies of substrate (Taber, 1960; Carlile *et al*., 1961). Then the coremium primordia elongate by apical growth. This is terminated by sporulation, initiated at the expense of endogenous substrates but continuing at the expense of exogenous substrate if this is supplied at the base of the coremium (Watkinson, 1975*a*). This sequence of events is a particularly attractive model system for the study of development in fungi because the effect of eliciting stimuli on the earliest stages of development can be observed; because the time and place of development is predictable; and because it can be made sufficiently synchronous for biochemical investigation of the different stages.

Stimuli to coremium formation

Several factors combine to decide the time and place of origin of a coremium in a colony of *P. claviforme*. First is the age of the supporting mycelium. Faraj-Salman (1970) showed that coremia were never formed more than 5 mm behind the margin of the advancing mycelium. When grown from mycelium of uniform age, coremia develop in a single synchronous flush (Watkinson, 1977). Second is an endogenous periodicity of coremium initiation which results in the formation of circles of coremia; this is to some extent genetically determined, and mutants with enhanced or absent rhythm exist (Faraj-Salman, 1966). This periodicity can also, however, be entrained by cyclic environmental changes, e.g. temperature cycles of 3 °C amplitude, between 22 °C and 25 °C, and light–dark cycles of 24 h duration, in which a circle of coremia is initiated at the end of a light period (Faraj-Salman, 1970). The periodicity can be accentuated by the presence in the medium of alcohols, detergents (Faraj-Salman 1971*a*, *b*), 2,4-dinitrophenol and high concentrations of Sørensen buffers (Watkinson, unpublished observation). However, the periodicity breaks down in the presence of very concentrated nutrient medium, when coremium initiation becomes continuous. Nutrient supply affects the development of initials into primordia. This is demonstrable in 30 h

Fig. 5.3. Potential site of coremium primordium approximately 4 h after germination of spores (*Penicillium claviforme*).

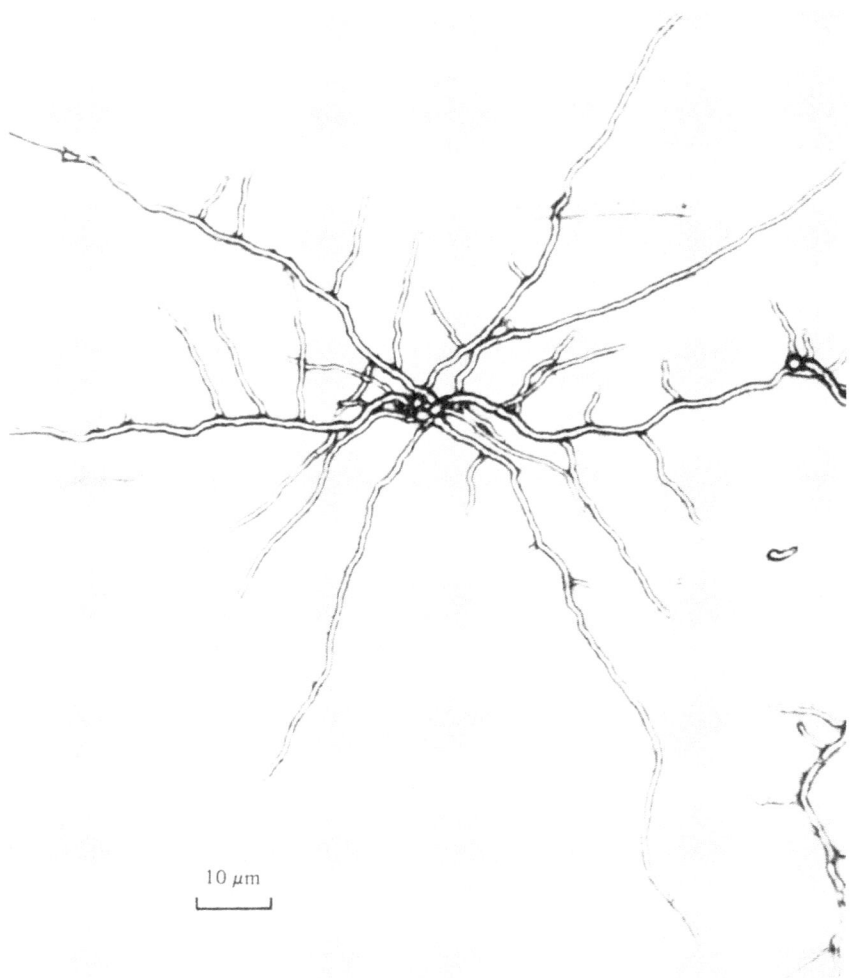

10 μm

synchronous mycelium of *P. claviforme* in which transfer to an appropriate medium is followed within 48 h by the production of a number of primordia. The nitrogen component of the medium determines this number (Fig. 5.4), which varies directly with the concentration of some amino acids (e.g. glutamate and asparagine) but remains low and constant at all concentrations of others (e.g. serine and glycine). With all four amino acids there is a similar increase in dry weight and a stimulation of activity of glutamate dehydrogenases as their concentrations in the medium are raised.

The apparent need for a critical intrahyphal level of certain amino acids or amino-acid-derived substances suggests the simple model for periodic coremium development present in Fig. 5.5. This supposes that uptake from fresh medium by the hyphal tips results in accumulation of the effective substances to a critical level, when localized upward branching is initiated. The ensuing elongation uses up the accumulated

Fig. 5.4. The influence of the nature and concentration of the nitrogen source on the numbers of coremia and dry weight of mycelium formed by *Penicillium claviforme*. (*a*) Nitrogen source monosodium L-glutamate; (*b*) L-serine; line, dry weight (3 cm² samples); histogram, number of primordia (1 cm² samples).

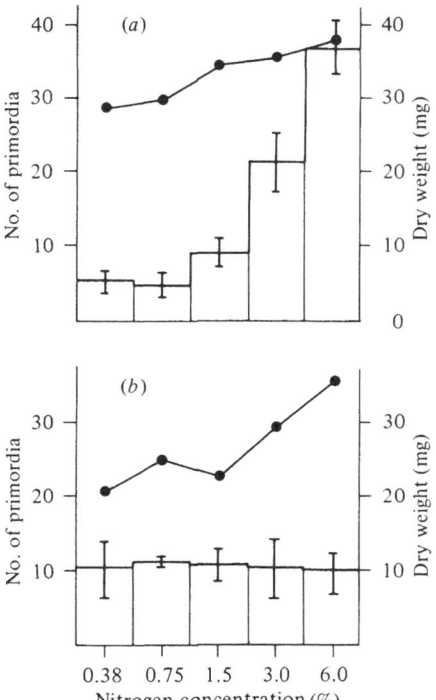

substance, and new primordia do not form until the hyphal tips have again taken up enough amino acid from the newly-exploited medium. This model is consistent with Fig. 5.6 which shows that the distance *a* (length of undifferentiated hypha behind the apex) increases as the nutrient medium is diluted, although the rate of extension growth remains constant. Accumulation of free amino acids has been suggested as the initiating step in morphogenesis (Watkinson, 1977). A similar mechanism was proposed for the glucose-sensitive periodicity of sclerotium initiation by *Sclerotium rolfsii* Sacc. (Humpherson-Jones & Cooke, 1977*a, b*).

A relationship between illumination, metabolism and morphogenesis has been demonstrated in *Penicillium isariiforme* (Piskorz, 1967*a*; Graafmans, 1973, 1974) and in *P. clavigerum* (Piskorz, 1972). When

Fig. 5.5. Model for periodic initiation of coremia in a colony depending on accumulation of a medium-derived precursor to a critical level. Dots indicate relative amounts of the hypothetical essential precursor and morphogenetic substance; *a* is the length of undifferentiated hypha behind the apex.

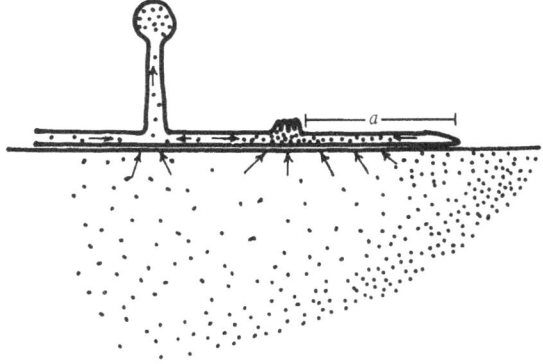

Fig. 5.6. Dependence of *a* on medium concentration (linear growth rate is constant). (i)–(v), serial dilutions of the medium by a factor of 2; *a* is the mean of 4 measurements.

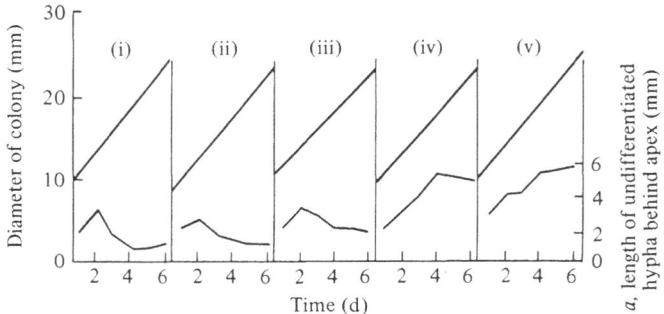

cultures of *P. isariiforme* are exposed to blue light, coremia form, and there is also a shift of metabolism from the fixation of CO_2 and excretion of citric acid into the medium, to the intrahyphal accumulation of lipids, nucleic acid and protein. A number of other coremium-forming fungi have been found to respond morphogenetically to particular nutrient substrates (Taber, 1960, 1961; Harris & Taber, 1970).

How the resulting changes in intermediary metabolism affect gene expression in morphogenesis remains an unanswered question.

Sclerotia

The morphogenesis of sclerotia has been well reviewed elsewhere (Townsend & Willetts, 1954; Butler, 1966; Willetts, 1972, 1978; Chet & Henis, 1975). Their earliest stages and the stimuli that elicit their formation are in several cases very similar to those of strands and coremia. Nutritional effects are paramount, nitrogen nutrition in particular, in initiating sclerotia (Rudolph, 1962; Wheeler & Sharan, 1965; Agnihotri & Vaartaga, 1968; Chet & Henis, 1968; Kritzman *et al.*, 1976). As in coremia, there appears to be an early definition of a site of formation (Townsend & Willetts, 1954) and sclerotium development proceeds by accumulation of materials at this point (Cooke, 1969; Waters, Moore & Butler, 1975). The initials in aerial sclerotia of *Coprinus* resemble early coremium primordia in that a radiating system of hyphae is established, followed by a local accumulation of glycogen at its centre. Waters, Moore & Butler (1975) showed that in *Coprinus* sclerotia glycogen is converted into wall material as the sclerotium matures, and they suggest that the thick walls act as a long-term carbohydrate store. Okon, Chet & Henis (1973) showed that in *Sclerotium rolfsii* local accumulation of translocate also preceded visible differentiation. Goujon (1970) found that, as in *Serpula lacrimans* strands, conditions leading to a local concentration of substrates inside the hyphae led to sclerotium formation at the probable site of accumulation, and also demonstrated an induction of sclerotium formation in young mycelium by a factor transmissible from older, sclerotium-containing mycelium. In all these fungi, accumulation of endogenous substrates, in some cases possibly a protein (Goujon, 1970; Okon, Chet & Henis, 1973) in others a carbohydrate (Waters, Butler & Moore, 1975), preceded sclerotium formation.

Illumination is also important in the initiation of sclerotia of many fungi (see Humpherson-Jones & Cooke, 1977*a, b*). Alterations of metabolism of phenolic substances have been observed during sclerotium metabolism and Marukawa *et al.*, (1975) isolated a phenolic metabolite which accumulated during sclerotium formation and which induced sclerotial initials when added to mycelium in *Sclerotinia sclerotiorum* (Lit.) De Bary.

Discussion

There are several striking similarities in the early stages of all types of vegetative aggregation reviewed here. One is, that morphogenesis is often, at least partly, at the expense of endogenous substrate. Secondly, it has been repeatedly observed that accumulation of reserves at the future site of development accompanies, *or precedes* visible morphogenesis. Apparently, morphogenetic metabolism cannot be sustained by direct uptake of nutrients from the medium. Morphogenesis of these structures can accordingly be seen as a two-stage process consisting of first, a fuelling phase without change in form, followed by a 'programmed starvation', as described by Wright (1973), during which morphogenesis proceeds and is regulated by the availability of precursors, activators and inhibitors to alternative metabolic pathways. Wright has postulated the need for a closed system within which a gradual decrease in the amount of endogenous metabolites controls the relative rates of different reactions without interference by exogenous substrate. In fact the conditions in which sclerotia, strands, coremia and rhizomorphs form *in vivo* are normally those in which the supply of exogenous material is reduced. In zoospore morphogenesis it has been shown how morphogenesis can be set in motion by exogenous metabolites (Cantino 1966) and how it can also become self-regulating (Selitrennikoff & Sonneborn, 1976).

The factors that initiate differentiation are likely to be those that localize the accumulation of endogenous substrates, possibly by directing translocation. Zalokar (1970) suggested that the polarity and branching of hyphae were based on the segregation of morphogenetic substances within them. Recent work on vegetative aggregations supports this hypothesis and it appears that normal intermediary metabolites may play morphogenetic roles (Moore & Jirjis, 1976). Morphogenesis is fundamentally the result of a physical organization and localization of metabolism; a better understanding of the localizing factors is now needed.

References

Agnihotri, V. P. & Vaartaga, O. (1968). Effects of nitrogenous compounds on sclerotium formation in *Aspergillus niger*. *Canadian Journal of Microbiology*, **14**, 1253–8.

Bary, A. de (1887). *Comparative Morphology and Biology of the Fungi, Mycetozoa and Bacteria*, pp. 22–44. Oxford: Clarendon Press.

Bennink, G. J. H. (1972). Photomorphogenesis in *Penicillium isariiforme*. II. The action spectrum for light-induced formation of coremia. *Acta Botanica Neerlandica*, **21**, 535–8.

Botton, B. & Dexheimer, J. (1977). Ultrastructure des rhizomorphes du *Sphaerostilbe repens* B. et Br. *Zeitschrift für Pflanzenphysiologie*, **85**, (5), 429–43.

Botton, B. & El-Khouri, M. (1978). Synnema and rhizomorph production in *Sphaerostilbe repens* under the influence of other fungi. *Transactions of the British Mycological Society*, **70**, 131–6.

Botton, B. & Ly, P. R. (1976). Régeneration des boutures d'Apex de rhizomorphes chez le *Sphaerostilbe repens* B. et Br. *Biologia Plantarum*, **18** (6), 436–41.

Butler, G. M. (1957). The development and behaviour of mycelial strands in *Merulius lacrymans* (Wulf.) Fr. I. Strand development during growth from a food-base through a non-nutrient medium. *Annals of Botany* (N.S.) **21**, 523–7.

Butler, G. M. (1958). The development and behaviour of mycelial strands in *Merulius lacrymans* (Wulf.) Fr. II. Hyphal behaviour during strand formation. *Annals of Botany* (N.S.), **22**, 219–36.

Butler, G. M. (1966). Vegetative structures. In *Fungi: An Advanced Treatise*, vol. II, *The Fungal Organism*, eds. G. C. Ainsworth and A. S. Sussman, pp. 83–112. London, New York: Academic Press.

Cantino, E. C. (1966). Morphogenesis in aquatic fungi. In *Fungi: An Advanced Treatise*, vol. II, *The Fungal Organism*, eds. G. C. Ainsworth and A. S. Sussman, pp. 283–337. London, New York: Academic Press.

Carlile, M. J., Lewis, B. G., Mordue, E. M. & Northover, J. (1961). The development of coremia. I. *Penicillium claviforme*. *Transactions of the British Mycological Society*, **44**, 129–33.

Carlile, M. J., Dickens, S. W., Mordue, E. M. & Schipper, M. A. A. (1962a). The development of coremia. II. *Penicillium isariaeforme*. *Transactions of the British Mycological Society*, **45**, 457–61.

Carlile, M. J., Dickens, S. W. & Schipper, M. A. A. (1962b). The development of coremia. III. *Penicillium clavigerum* with observations on *P. expansum* and *P. italicum*. *Transactions of the British Mycological Society*, **45**, 462–4.

Chet, I. & Henis, Y. (1968). The control mechanism of sclerotial formation in *Sclerotium rolfsii* Sacc. *Journal of General Microbiology*, **54**, 231–6.

Chet, I. & Henis, Y. (1975). Sclerotial morphogenesis in fungi. *Annual Review of Phytopathology*, **13**, 169–92.

Chet, I., Retig, N. & Henis, Y. (1972). Changes in total soluble proteins and in some enzymes during morphogenesis of *Sclerotium rolfsii*. *Journal of General Microbiology*, **72**, (3), 451–6.

Cooke, R. C. (1969). Changes in soluble carbohydrates during sclerotium formation by *Sclerotinia sclerotiorum* and *S. trifoliorum*. *Transactions of the British Mycological Society*, **53** (1), 77–86.

Cooke, R. C. (1971). Uptake of carbon-14 glucose and loss of water by sclerotia of *Sclerotinia sclerotiorum* during development. *Transactions of the British Mycological Society*, **57** (3), 379–84.

Corner, E. J. H. (1932). The fruit body of *Polystictus xanthopus*. *Annals of Botany*, **46**, 71–111.

Day, S. C. (1968). The morphogenesis of mycelial strands in the timber dry rot fungus *Merulius lacrymans* (Wulf.) Fr. Ph.D. thesis, University of Cambridge.

Falck, R. (1912). Die Meruliusfaule des Bauholzes. *Hausschwammforschung*, **6**, 1–405.

Faraj-Salman, A.-G. (1966). Induction einer endogenen Rhythmik der Koremienbildung durch Alkohol bei einer Mutante von *Penicillium claviforme* Bainier und einer varietät davon. *Biochimie und Physiologie der Pflanzen*, **161**, 42–9.

Faraj-Salman, A.-G. (1970). Einfluss von Licht auf die Koremienbildung und ihre Kreisformige Anordnung. *Kulturpflanze*, **18**, 89–97.

Faraj-Salman, A.-G. (1971a). Zur Induktion einer endogenen Rhythmik bei Mutanten des Pilzes *Penicillium claviforme* Bainier. I. Wirkungsweise von Alkoholen. *Archiv für Protistenkunde*, **113**, 306–13.

Faraj-Salman, A.-G. (1971b). Zur Induktion einer endogenen Rhythmik bei Mutanten des Pilzes *Penicillium claviforme* Bainier. II. Wirkungsweise von Detergenten. *Biochemie und Physiologie der Pflanzen*, **162**, 470–3.

Garraway, M. O. (1969). The influence of compounds related to the shikimic acid pathway on rhizomorph initiation and growth in *Armillaria mellea*. *Phytopathology*, **59**, 1027.

Garraway, M. O. (1970). Rhizomorph initiation and growth in *Armillaria mellea* promoted by *o*-amino benzoic and *p*-amino benzoic acids. *Phytopathology*, **60** (5), 861–5.

Garrett, S. D. (1953). Rhizomorph behaviour in *Armillaria mellea* (Vahl) Quel. I. Factors controlling rhizomorph initiation by *A. mellea* in pure culture. *Annals of Botany* (N.S.), **17**, 63–79.

Garrett, S. D. (1956). Rhizomorph behaviour in *Armillaria mellea* (Vahl) Quel. II. Logistics of infection. *Annals of Botany* (N.S.), **20**, 193–209.

Garrett, S. D. (1960). Rhizomorph behaviour in *Armillaria mellea* (Vahl) Quel. III. Saprophytic colonization of woody substrates in soil. *Annals of Botany* (N.S.), **24**, 275–85.

Garrett, S. D. (1970). *Pathogenic Root-Infecting Fungi.* Cambridge University Press.

Goujon, M. (1970). Méchanismes physiologiques de la formation des sclerotes chez le *Corticium rolfsii* (Sacc.) Curzi. *Physiologie Végétale*, **8** (3), 349–60.

Graffmans, W. D. J. (1973). The influence of carbon dioxide on morphogenesis in *Penicillium isariiforme*. *Archiv für Mikrobiologie*, **91**, 67–76.

Graafmans, W. D. J. (1974). Metabolism of *Penicillium isariiforme* on exposure to light, with special reference to citric acid synthesis. *Journal of General Microbiology*, **82** (2), 247–52.

Hartig, R. (1902). *Der echte Hausschwamm und andere das Bauholz zerstörende Pilze.* Berlin: Springer Verlag.

Harris, J. L. & Taber, W. A. (1970). Influence of certain nutrients and light on growth and morphogenesis of the synnema of *Ceratocystis ulmi*. *Mycologia*, **62** (1), 152–70.

Hirt, R. S. (1948). An isolate of *Poria xantha* on a medium containing copper. *Phytopathology*, **39**, 31–6.

Humpherson-Jones, F. M. & Cooke, R. C. (1977a). Morphogenesis in sclerotium-forming fungi. I. Effects of light on *Sclerotinia sclerotiorum*, *S. delphinii* and *S. rolfsii*. *New Phytologist*, **78** (1), 171–80.

Humpherson-Jones, F. M. & Cooke, R. C. (1977b). Morphogenesis in sclerotium-forming fungi. II. Rhythmic production of sclerotia by *Sclerotinia sclerotiorum*. *New Phytologist*, **78** (1), 181–7.

Jennings, D. H., Thornton, J. D., Galpin, M. F. & Coggins, C. R. (1974). Translocation in fungi. In *Transport at the Cellular Level* (28th Symposium of the Society for Experimental Biology), eds. W. A. Sleigh & D. H. Jennings, pp. 139–56. Cambridge University Press.

Jirjis, R. I. & Moore, D. (1976). Involvement of glycogen in morphogenesis of *Coprinus cinereus*. *Journal of General Microbiology*, **95** (2) 348–52.

Kritzman, G., Okon, Y., Chet, I., & Henis, Y. (1976). Metabolism of L-threonine and its relationship to sclerotium formation in *Sclerotium rolfsii*. *Journal of General Microbiology*, **95**, 78–86.

Levi, J. D. & Cowling, E. B. (1969). Role of nitrogen in wood deterioration. VII. Physiological adaptation of wood-destroying and other fungi to substrates deficient in nitrogen. *Phytopathology*, **59** (4), 460–8.

Levi, M. P., Merrill, W. & Cowling, E. B. (1968). Role of nitrogen in wood deterioration. VI. Mycelial fractions and model nitrogen compounds as substrates for growth of *Polyporus versicolor* and other wood-destroying and wood-inhabiting fungi. *Phytopathology* **58** (5), 626–34.

Marukawa, S., Funakawa, S., & Satomura, Y. (1975). Role of sclerin on morphogenesis in *Sclerotinia sclerotiorum* including *Sclerotinia libertiana*. *Agricultural and Biological Chemistry Bulletin*, **39**, 645–50.

Moore, D. & Jirjis, R. I. (1976). Regulation of sclerotium production by primary metabolites in *Coprinus cinereus*. *Transactions of the British Mycological Society*, **66** (3), 377–82.

Morrison, D. J. (1976). Vertical distribution of *Armillaria mellea* rhizomorphs in soil. *Transactions of the British Mycological Society*, **66**, 393–9.

Motta, J. J. (1969). Cytology and morphogenesis in the rhizomorph of *Armillaria mellea*. *American Journal of Botany*, **56**, (6), 610–19.

Motta, J. J. (1971). Histochemistry of the rhizomorph meristem of *Armillaria mellea*. *American Journal of Botany* **58**, 80–87.

Okon, Y., Chet, I., & Henis, Y. (1972). Lactose induced synchronous sclerotium formation in *Sclerotium rolfsii* and its inhibition by ethanol. *Journal of General Microbiology*, **71** (3), 465–70.

Okon, Y., Chet, I., & Henis, Y. (1973). Effects of lactose, ethanol & cycloheximide on the translocation pattern of radioactive compounds and on sclerotium formation in *Sclerotium rolfsii*. *Journal of General Microbiology*, **74**, 251–8.

Pentland, G. D. (1965). Stimulation of rhizomorph development of *Armillaria mellea* by *Aureobasidium pullulans* in artificial culture. *Canadian Journal of Microbiology*, **11**, 345–50.

Piskorz, B. (1967*a*). Investigations on the formation of coremia. I. Action of light on the formation of coremia in *Penicillium isariiforme*. *Acta Societatis Botanicae Poloniensis*, **36**, 123–31.

Piskorz, B. (1967*b*). Investigation on the action of light on the growth and development of *Penicillium claviforme* Bainier. *Acta Societatis Botanicorum Poloniae*, **36**, 677–98.

Piskorz, B. (1972). Comparative investigations on the development of fungi from the section of *Penicillium clavigerum*. III. Action of carbon dioxide on the morphology of coremia and dry weight of mycelium. *Acta Societatis Botanicorum Poloniae*, **41**, 341–55.

Ramsbottom, J. (1953). *Mushrooms and Toadstools*. London: Collins.

Rishbeth, J. (1968). The growth rate of *Armillaria mellea*. *Transactions of the British Mycological Society*, **51**, 575–86.

Rishbeth, J. (1978). Effects of soil temperature and atmosphere on growth of *Armillaria mellea* rhizomorphs. *Transactions of the British Mycological Society*, **70** (2) 213–20.

Rudolph, E. D. (1962). The effect of some physiological and environmental factors on sclerotial Aspergilli. *American Journal of Botany*, **49**, 71–8.

Schmid, R. & Liese, W. (1970). Feinstruktur der Rhizomorphen von *Armillaria mellea*. *Phytopathologisches Zeitschrift*, **68**, 221–31.

Selitrennikoff, C. P. & Sonneborn, D. R. (1976). Post-translational control of *de novo* cell wall formation during *Blastocladiella emersonii* zoospore germination. *Developmental Biology*, **54**, 37–51.

Smith, A. M. & Griffin, D. M. (1971). Oxygen and the ecology of *Armillariella elegans* Heim. *Australian Journal of Biological Sciences*, **24**, 231–62.

Taber, W. A. (1960). Studies on *Isaria cretacea*. Morphogenesis of the synnema and endogenous nutrition. *Canadian Journal of Microbiology*, **6**, 53–63.

Taber, W. A. (1961). Nutritional factors affecting the morphogenesis of the synnema. *Recent Advances in Botany*, **4**, 289–93.

Taber, W. A. & Vining, L. C. (1959). Studies on *Isaria cretacea*. Nutritional and morphological characteristics of two strains and morphogenesis of the synnema. *Canadian Journal of Microbiology*, **5**, 513–19.

Townsend, B. B. & Willetts, H. J. (1954). The development of the sclerotia of certain fungi. *Transactions of the British Mycological Society*, **37**, 213–21.

Turian, G. (1975). Differentiation in *Allomyces* and *Neurospora*. *Transactions of the British Mycological Society*, **64** (3), 367–80.

Valder, P. (1958). The biology of *Helicobasidium purpureum*. *Transactions of the British Mycological Society*, **41**, 283–308.

Vance, C. P. & Garraway, M. O. (1973). Growth stimulation of *Armillaria mellea* by ethanol and other alcohols in relation to phenol concentration. *Phytopathology*, **63** (6), 743–8.

Vance, C. P. & Garraway, M. O. (1974). The effect of ethanol on phenol oxidizing enzymes in *Armillaria mellea* in relation to rhizomorph development. *Proceedings of the American Phytopathological Society*, **1**, 109.

Waters, H., Butler, R. D. & Moore, D. (1975). Structure of aerial and submerged sclerotia of *Coprinus lagopus*. *New Phytologist*, **74**, 199–206.

Waters, H., Moore, D., & Butler, R. D. (1975). Morphogenesis of aerial sclerotia of *Coprinus lagopus*. *New Phytologist*, **74**, 207–13.

Watkinson, S. C. (1971*a*). The mechanism of mycelial strand induction in *Serpula lacrimans*: a possible effect of nutrient distribution. *New Phytologist*, **70** 1079–88.

Watkinson, S. C. (1971*b*). Phosphorus translocation in stranded and unstranded mycelium of *Serpula lacrimans*. *Transactions of the British Mycological Society*, **57**, 535–9.

Watkinson, S. C. (1975*a*). Regulation of coremium morphogenesis in *Penicillium claviforme*. *Journal of General Microbiology*, **87**, 292–300.

Watkinson, S. C. (1975*b*). The relation between nitrogen nutrition and the formation of mycelial strands in *Serpula lacrimans*. *Transactions of the British Mycological Society*, **64** (2), 195–200.

Watkinson, S. C. (1977). Effect of amino acids on coremium development in *Penicillium claviforme*. *Journal of General Microbiology*, **101**, 269–75.

Weigl, J. & Ziegler, H. (1960). Wasserhaushalt und Stoffleitung bei *Merulius lacrymans*. *Archiv für Mikrobiologie*, **37**, 124–33.

Weinhold, A. R. (1963). Rhizomorph production by *Armillaria mellea* induced by ethanol and related compounds. *Science*, **142**, 1065–6.

Weinhold, A. R. & Garraway, M. O. (1966). Nitrogen and carbon nutrition of *Armillaria mellea* in relation to the growth-promoting effects of ethanol. *Phytopathology*, **56**, 108–12.

Wheeler, B. E. J. & Sharan, N. (1965). The production of sclerotia by *Sclerotium rolfsii*. I. Effects of varying the supply of nutrients in an agar medium. *Transactions of the British Mycological Society* **48** (2), 291–301.

Willetts, H. J. (1972). The morphogenesis and possible evolutionary origins of fungal sclerotia. *Biological Reviews*, **47**, 515–36.

Willetts, H. J. (1978). Sclerotium formation. In *The Filamentous Fungi*, vol. III, eds. J. E. Smith & D. R. Berry, pp. 197–211. London: Arnold.

Wolkinger, F., Plank, S. & Brunegger, A. (1975). Rasterelektronenmikroskopische Untersuchungen von *Armillaria mellea*. *Phytopathologisches Zeitschrift*, **84**, 352–9.

Wong, A.-L. & Willetts, H. J. (1974). Polyacrylamide gel electrophoresis of enzymes during morphogenesis of sclerotia of *Sclerotinia sclerotiorum*. *Journal of General Microbiology*, **81** (1), 101–9.

Wright, B. E. (1973). *Critical Variables in Differentiation*. London: Prentice-Hall.

Zalokar, M. (1970). Some problems of differentiation in fungi. *Physiologie Végétale*, **8** (3), 449–59.

6
Wall growth during spore differentiation and germination

R.MARCHANT

School of Biological and Environmental Studies, New University of Ulster, Coleraine, County Londonderry, Northern Ireland, UK

Introduction

In discussing wall growth in relation to spore differentiation and germination it is all too easy to become preoccupied with the minor details and to lose sight of generalities and unifying principles. Gregory (1966) proposed that all fungal spores could be described as either memnospores – those which remained at the place of origin to tide over unfavourable periods, or xenospores – those which were dispersed to new geographic locations. Each of these spore types have specific adaptations which suit them to their function. The memnospores are relatively thick-walled, often require a particular set of conditions to stimulate germination and may not become detached from the parent mycelium. In contrast the xenospores are usually thin-walled, germinate easily, and are always completely separated from the parent thallus. It will be shown later how the functions of these two spore categories are reflected in the structure and growth of the spore and germ-tube walls.

The critical feature of all fungal spores is that they represent a discontinuous stage in the growth of the organism. This discontinuity lies between their production and germination to produce a new thallus. The period of the interruption in growth may be short or long depending on the type of spore and the particular set of conditions prevailing. Another important feature of many spores is that they become, at some stage, completely physically separated from the structures which produced them. They remain as discrete dormant entities separated from the thallus yet capable of reinitiating growth under suitable circumstances. Changes in biochemical processes accompanying spore production or germination have been fruitful areas of mycological research for several decades, and the wall structure must reflect these processes. It is clear that the wall has an important morphological role in the production and germination of fungal spores, and, even in those spores which are initially produced without walls, one of the major preliminary steps in encystment is wall synthesis. Because continued hyphal growth necessarily involves the steady synthesis of wall material, the sporulation discontinuity results in a disturbance of balanced wall synthesis. Similarly the germination process, with its formation of a germ tube, requires the initiation of wall synthesis in a structure where wall synthesis is inactive.

It is important to retain the concept of the wall as a dynamic structure and not as a fixed immutable part of the organism. While there may be little turnover of wall material, in the sense that is normally accepted for other cell constituents such as protein and RNA, even in actively growing cells there are undoubtedly extensive localized changes which occur in conjunction with morphogenetic events. Much of the analysis of wall participation in sporulation and germination has been hampered, at the ultrastructural level, by the restricted means of interpretation and analysis which have been available. There are still only a few examples of reasonably complete analyses of walls, and even these do not really extend to the different morphogenetic stages in the life cycles of fungi. There is, however, one situation where information exists on the possible effect of changing wall composition on growth form and pattern, namely in dimorphic organisms such as *Mucor rouxii* (Calmette) Wehm., which can exist in both yeast-like and mycelial forms (Bartnicki-Garcia, 1963). In *M. rouxii*, and other dimorphic fungi, changes in the proportion of various wall components are correlated with the changed morphology, and may point the way to a further understanding of the processes of sporulation and germination.

Wall growth during sporogenesis

The principal function of the processes of sporulation is to produce a discrete structure – a spore. In most instances this implies that at the completion of the sporulation sequence some growth functions cease. The synthesis of wall material is one of the processes which is halted for an indeterminate period when the spore is mature, normally until germination commences. One of the major difficulties associated with the analysis of wall synthesis in these small regions of differentiation is to interpret electron microscope images in terms of wall composition and structure. The majority of workers have resorted to identifying wall layers and trying to determine their continuity through the differentiating structures. This approach requires that the staining reactions of wall regions accurately reflect their structure and composition, but, until cytochemical methods are more widely used, this type of information must suffice. Examples of a single wall type treated in different fixing and staining regimes are shown in Fig. 6. 6*a, b* and *c*. In any case we have little information on the inter-relationships between wall composition in sporulating and vegetative regions of fungi.

A large proportion of fungal spores fall into one of two categories: those in which the cytoplasmic content of the spore is delimited at approximately its final volume before the wall is synthesized as a complete envelope, e.g. ascospores; and those in which wall synthesis takes place continuously as the spore increases in size by an extrusion process, e.g. basidiospores and phialospores. Exceptions to this classification are zoospores, in which wall synthesis is delayed until encystment, and zygospores and arthrospores, where delimitation of the spores occurs by a process similar to septation.

Endogenously produced spores

In these spore types the wall is synthesized around a delimited volume of cytoplasm. The mode of delimitation of the spore cytoplasm varies from species to species, and leads to the production of either xenospores or memnospores.

Ascospore formation. In the ascomycetes a large, peripheral, double-membrane sac is initially formed, enclosing most of the ascus contents. This sac then invaginates on itself to enclose each of the ascospore initials. The origin of the double-membrane sac is not entirely clear; Carroll (1967) and Oso (1969) claimed that membrane blebbed from nuclei migrated to the periphery of the ascus and there fused to produce the sac. Bandoni, Bisalputra & Bisalputra (1967) and Syrop & Beckett (1972) suggested that, in *Hansenula anomala* (Hansen) H. & P. Sydow and *Taphrina deformans* (Berk.) Tul. respectively, the ascospore-delimiting membranes were produced by invagination of the plasma membrane, while other workers (Wilsenach & Kessel, 1965; Lynn & Magee, 1970; Black & Gorman, 1971) have proposed that the endoplasmic reticulum gives rise to the so-called ascus vesicle. Greenhalgh & Griffiths (1970) even suggested that it is possible, although unlikely, that there exist variants of the process of ascospore delimitation. As more studies have been carried out, it seems increasingly likely that there are subtle differences in the process, and indeed in *T. deformans* the formation of an ascus vesicle prior to the segregation of separate ascospores does not occur (Syrop & Beckett, 1972).

Whatever the origin of the double-membrane system surrounding the ascospores, the later processes of wall formation appear to be similar. The double membranes become separated by the insertion of primary wall layer material between them. The deposition of this wall material is not always completely regular, and Lynn & Magee (1970) noted that in developing ascospores of *Saccharomyces cerevisiae* the greatest wall deposition occurred in regions where the endoplasmic reticulum lay close to the spore plasma membrane. Black & Gorman (1971) also noted, by the use of silver methenamine cytochemical staining, an irregular deposition of ascospore wall material in *Hansenula wingei* Wickerham. In many ascospores the primary wall layer becomes overlaid by secondary and tertiary wall layers, which may be pigmented. In *Saccobolus kerverni* (Cr.) Boud. (Carroll, 1969) secondary deposition takes the form of material between the ascospores producing a spore ball, which is then discharged intact from the ascus. In other species, e.g. *Hansenula anomala* (Bandoni *et al.*, 1967) and *H. wingei* (Black & Gorman, 1971) the secondary wall layers form an extension at one end of the ascospore to produce the typical hat-shaped structure. Wall development does not always proceed sequentially from the inside outwards; in *Sordaria fimicola* (Rob.) Ces. & de Not. Mainwaring (1972) reported the formation of an electron-opaque tertiary wall layer within the developing ascospore wall at the interface of the primary and secondary layers. In *Podospora anserina* (Rabh.) Niessl a tertiary wall structure is developed within the primary ascospore wall layer after an outer secondary layer has been formed (Beckett, Barton & Wilson, 1968). This tertiary material is pigmented and is in the form of blocks so that it does not form a complete layer around the spore. A similar development of an electron-opaque wall layer within the primary wall also occurs in *Poronia punctata* (L.) Fries (Stiers, 1974). The maturation of some ascospores involves the deposition of a final outer layer on the spore; in *S. cerevisiae* Hansen this has been shown to be a lipid layer (Lynn & Magee, 1970; Beckett, Illingworth & Rose, 1973), while in *Hypoxylon fragiforme* (Pers. ex. Fr.) Kickx (Greenhalgh & Evans, 1968), *S. fimicola* (Furtado & Olive, 1970; Mainwaring, 1972) and *Lophodermella sulcigena* (Rostr.) v. Höhn (Campbell, 1973) the ascospores have an outer layer, which has a vesiculated appearance, prior to discharge.

In summary, therefore, it is evident that ascospore wall synthesis involves the deposition of various wall layers between a pair of membranes during the maturation of the spore (Fig. 6.1). The greatest unresolved problem in these studies concerns the origin of the wall material. In most instances the amount of cytoplasm excluded by the delimitation of the spores is small, and it seems unlikely that it would contribute a large proportion of the spore wall material. However, it has been demonstrated that the outer ascospore lipid layer comes from this source in *S. cerevisiae* (Lynn & Magee, 1970; Beckett *et al.*, 1973). The major impediments to a satisfactory resolution of the problem of wall origin have been the reliance on permanganate fixation and the lack of cytochemical evidence. Even where freeze-etching techniques have been employed (Guth, Hashimoto & Conti, 1972) little useful information has been gained on wall synthesis and the cytological structures involved in the process.

Sporangiospore formation. Sporangiospore formation is a system which appears superficially similar to that of ascospore formation and also produces a xenospore. An enclosed mass of cytoplasm in the sporangium becomes cleaved into spore initials around which wall material is synthesized. The most complete investigation of this process has been carried out by Bracker (1966, 1968) in *Gilbertella persicaria* (Eddy) Hesseltine. He demonstrated that the cytoplasm of the sporangium is cleaved by the fusion and subsequent expansion of a system of vesicles. The wall material of the spores is then deposited outside the spore plasma membrane, which derives from the vesicle membrane. The wall deposition is thus at the outer surface of the membrane within the lumen of the sporangium and differs from the ascospore system of deposition between two opposed membranes.

In the development of the multispored merosporangia of *Syncephalastrum racemosum* Cohn ex Schroeter (Fletcher, 1972) the cleavage of the cytoplasm occurs by the enlargement of cleavage furrows, which finally fuse with the merosporangium plasmalemma. The cleavage furrow membranes are lined with particles as in *Gilbertella persicaria*. Wall material here is also deposited at the outer surface of the sporangiospore plasmalemma within the space created by the cleavage furrow.

Oospore formation. The oospore, normally a memnospore, represents another example of a spore which is produced within an enveloping structure, the oogonium. An oogonium may contain a single oospore, as

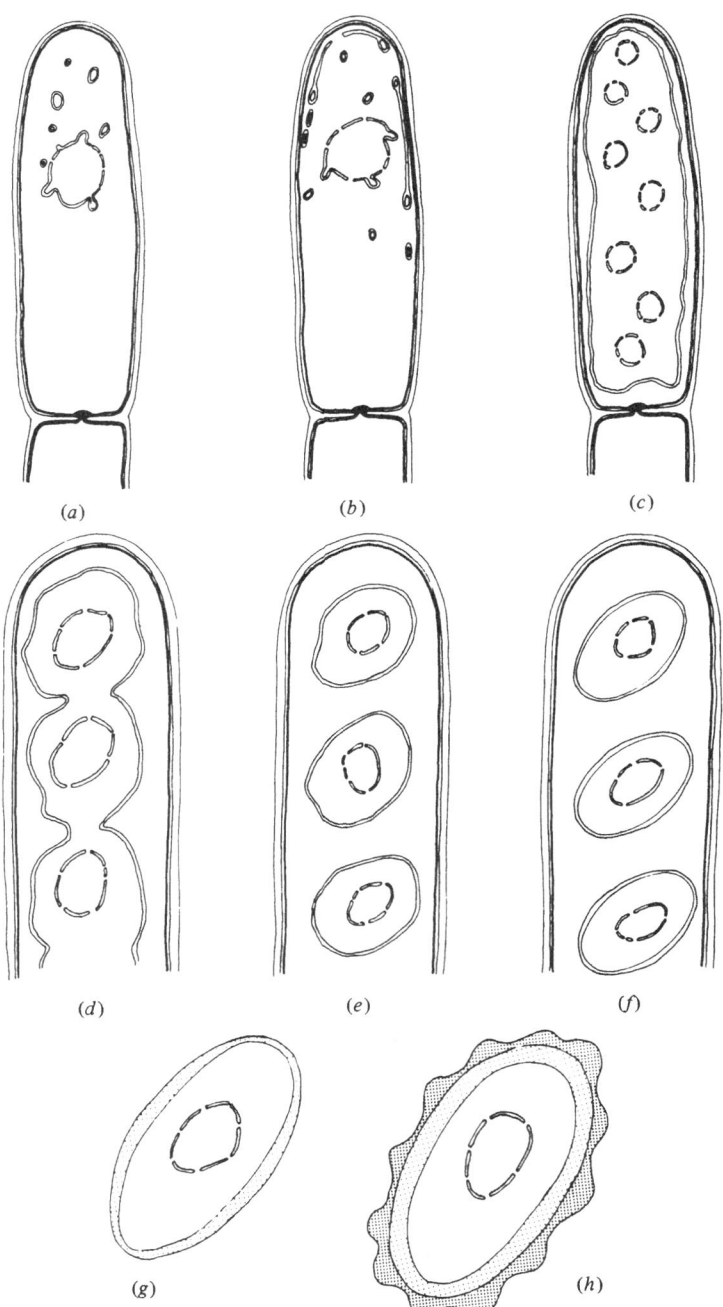

in *Pythium ultimum* Trow (Marchant, 1968) and *Phytophthora capsici* Leonian (Hemmes & Bartnicki-Garcia, 1975), or several as in *Saprolegnia terrestris* Cookson ex Seymour (Howard & Moore, 1970). A complete study of oospore development is not available for any species. Gay, Greenwood & Heath (1971) have shown that in *Saprolegnia furcata* Maurizio the cleavage of the oogonium contents, to produce a small number of oospheres, is achieved by fusion of cleavage vesicles – a mechanism similar to that for the production of zoospores in the same species. In the studies by Marchant (1968) and Hemmes & Bartnicki-Garcia (1975), where only one oosphere is produced, the mechanism by which the periplasm was isolated could not be determined. The result of the cleavage process is, however, to produce a situation similar to that in the sporangiospore where wall material is accreted at the outer surface of the spore plasma membrane.

The origin of wall material precursors is easier to determine in these oospore systems than in many others. The Oomycetes have readily recognizable dictyosomes and it has been indicated, from the studies of Heath, Gay & Greenwood (1971) and Vujičić (1971), that vesicles generated from dictyosomes within the developing oospore migrate to the spore plasma membrane and discharge their contents. Heath *et al.* (1971) have demonstrated that the dictyosome vesicles in these fungi do contain wall precursor material. The wall structure of mature oospores varies from species to species. Hemmes & Bartnicki-Garcia (1975) found that as the oospore of *Phytophthora capsici* matured, a thick, electron-transparent inner wall was deposited under the 'exospore' wall already formed. They postulated that the outer wall layer might be derived from the periplasm. In *Albugo candida* (Pers. ex Hooker) O. Kuntze the oospore has a five-layered wall, and additional wall layers are added to both the inner and outer surfaces of the primary oospore wall layer (Tewari & Skoropad, 1977). The periplasm is persistent in this

---

Fig. 6.1. Diagram showing ascospore development: (*a*) Vesicles blebbed from the nuclear envelope migrate towards the periphery of the ascus. (*b*) Fusion of the vesicles commences formation of the ascus vesicle. (*c*) Within the completed ascus vesicle nuclear division has produced eight nuclei. (*d*) Invagination of the ascus vesicles occurs between each nucleus. (*e*) Each nucleus is surrounded by a double membrane envelope derived from the ascus vesicle. (*f*) Deposition of ascospore wall material between the membrane commences. (*g*) Uneven deposition of the primary wall continues around each ascospore. (*h*) Secondary wall material added as the final stage in the maturation of the ascospore.

species, and it is suggested by Tewari & Skoropad (1977) that it plays an important role in the deposition of the oospore wall. In *Pythium ultimum* (Marchant, 1968) (Fig. 6.6*f*), *Pythium* sp. (McKeen, 1975) and *Saprolegnia terrestris* (Howard & Moore, 1970) the mature oospore wall does not show any major layering, although a thin, electron-opaque layer is evident on the outer surface of the *Pythium* oospores. As with many studies of fungal wall structure in the electron microscope, one of the biggest problems is to relate the observed images to any chemical structure. Lippman, Erwin & Bartnicki-Garcia (1974) have analysed the oogonium and oospore walls of *Phytophthora megasperma* Drechler var. *sojae* Hildebrand to show that approximately 80 % of the wall comprises insoluble β-glucan, with only a small proportion (<10 %) cellulose and about 12 % protein. The observed layering in the oospore wall therefore reflects the uneven distribution of the minor components throughout the wall in a manner as yet undetermined.

Summary. A review of the formation of these endogenous spore types reveals the paucity of knowledge concerning the synthesis of their walls. Basic knowledge of their wall composition and the structural arrangement of their components is lacking. It is clear that such walls are not static impervious structures, for there are examples of additional wall layers being added both at the outer and at the inner surfaces of existing wall layers and, indeed, examples also of new wall layers being generated within existing ones. A further point of contention, which remains unresolved in almost all instances, is the source of wall material precursors. In spores which develop within an enveloping structure there is always the possibility of wall material originating from within the developing spore or from the excluded cytoplasm. Until more knowledge of the spore wall structure and more refined cytochemical techniques are available the resolution of these problems will be difficult.

Exogenously produced spores

Spores of this type include basidiospores and most of the asexual spores of the imperfect fungi, e.g. phialospores, porospores etc. In contrast to the endogenous spores, the developing wall in these species is continuously synthesized as the spore expands. Also, because exospores are produced outside any cytoplasmic enclosing structure, the material for wall synthesis can only be derived from within the developing spore. It is immediately apparent, after only a cursory examination of this type of spore production, that their mode of production varies greatly. There is, however, a common basic function of all the systems, namely the production of a self-contained dispersive entity enclosed within a wall. Spores produced exogenously are often borne on specialized structures, e.g. basidium or conidiophore, and either the spores are produced successively from a single site or each site produces but a single spore. I do not intend to delve deeply into the wide range of different modes of spore production, but simply refer the reader to the Proceedings of the Kananaskis Symposium (Kendrick, 1971). Here, I shall try to highlight the important features of this type of spore generation.

If we consider the production of an exogenous spore as a localized extension of a pre-existing structure then we must examine how wall extension can occur to surround the developing spore. Examination of small spore initials indicates that they are enclosed within a wall layer, albeit very thin, from their inception. The formation of the spore wall can, therefore, only occur in one of three ways:

(1) By the developing spore being enclosed by a simple extension of the entire wall of the structure bearing it.
(2) By the spore wall comprising an extension of only part of the wall of the subtending structure, i.e. one or more of the underlying wall layers.
(3) By the spore wall being formed from a new wall layer, produced *de novo* beneath the wall of the structure bearing the spore, which pushes its way through the old wall.

If we analyse each system in this way we can see the basic similarities between the spore-production systems and, indeed, observe parallels with spore germination, as we shall see later.

Basidiospore formation. It is surprising that the development of basidiospores has not received more attention than it has. The best early studies were made by Wells (1964*a*, 1964*b*, 1965) on *Exidia nucleata* (Schw.) Rea and *Schizophyllum commune* Fries; however, the amount of information on wall development which can be gained from this work is strictly limited, despite the fact that Wells (1965) proposed a model for the explosive release of basidiospores. Several studies do show the early stages of sterigma outgrowth, e.g. *Lycoperdon perlatum* Pers. (Marchant, 1969) and *Coprinus cinereus* (Schaeff.) Fr. (McLaughlin, 1973). Both indicate that the sterigma wall is a continuation of the entire basidium wall (Fig. 6.8*d*). A more complete study of basidiospore development in *C. cinereus* has now been made by McLaughlin (1977) in which the asymmetric development of the spore on the sterigma is figured (Fig. 6.9*a*), particularly in relation to wall layers. The asymmetrical placement of the spore on the sterigma is brought about by a localized wall deposition in the region of the hilar appendix (Fig. 6.8*c*). Associated with this region is an electron-opaque structure in the cytoplasm of the spore, termed by McLaughlin the hilar appendix body. Similar structures have been observed in other *Coprinus* species (Hugueney, 1975*b*) and seem to disappear when the spore reaches full size. The role that the hilar appendix body plays, if any, in the localized wall thickening at the hilum is not known. McLaughlin (1977) suggests that peripheral endoplasmic reticulum and dictyosomes may function in the deposition of the spore wall during its rapid enlargement. The structures indicated by McLaughlin to be dictyosomes are not sufficiently defined to be so considered, although they may function in a similar manner. Most workers now seem to accept that terrestrial fungi do not have dictyosomes as such. The separation of the basidiospore from the sterigma was not illustrated by McLaughlin, but the development of the apiculus in *Coprinus congregatus* (Bull.) Fr. has been shown by Hugueney (1975*a*). The separation of the spore is achieved not by the synthesis of an abscission septum, as in many other spore systems, but by the formation of a less well-defined plug of material in the small opening through the thick wall at the base of the spore. A similar deposition of material at the base of the basidiospore in *Lentinus edodes* (Berk.) Suig. (Nakai & Ushiyama, 1974*a*; Nakai, 1975) allows the separation of the spore from the sterigma, although in this case the spore wall appears unlayered and the material deposited at the point of separation cannot be distinguished from the wall material of the spore.

Basidiospores are frequently characterized by a germination pore, which is a thin, or otherwise differentiated, region of the spore wall, through which the germ tube emerges. This pore has been shown by a number of workers (e.g. McLaughlin, 1977; Oláh & Reisinger, 1974) to be formed at the apex of the developing basidiospore, i.e. farthest from the point of attachment to the sterigma (Figs. 6.8*b* and 6.9*b*). The role of the germ pore will be discussed in a later section.

Wall layering seems to vary considerably between species and, indeed, in one species as observed by several authors, or by one author on several occasions. This has not, however, deterred workers from erecting elaborate schemes of nomenclature or numbering for these wall layers (Besson-Antoine & Kühner, 1972; Hugueney, 1972, 1975*a*;

Fig. 6.2. Diagrammatic representation of basidiospore formation in *Coprinus cinereus*: (*a*) The basidiospore initial arises as an asymmetric swelling at the tip of the sterigma. The hilar appendix body lies to one side of the initial. (*b*) Further growth of the initial occurs with particular wall synthesis at the hilum region. (*c*) Additional wall layers are added to the developing basidiospore, with further development at the hilum and with the initial formation of the germ pore cap. (*d*) In the nearly mature basidiospore the melanized wall layer is formed, the germ pore structure has developed and the hilum is further differentiated.

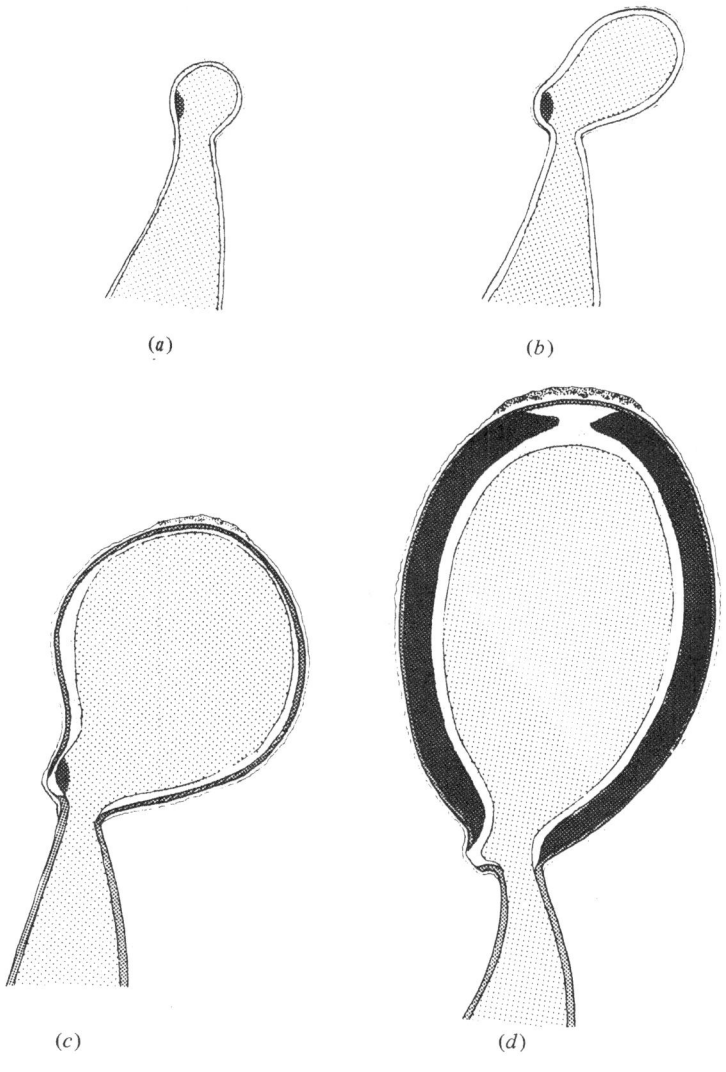

(*a*)   (*b*)

(*c*)   (*d*)

McLaughlin, 1977) and then attempting to make the systems compatible. Only Oláh & Reisinger (1974) appear to have attempted any cytochemical analysis of this type of wall system in a number of melanospored agarics. There seems little point in attempting a complex nomenclature of wall layers until there is structural and biochemical information to allow comparisons to be made and a standard system of nomenclature to be devised. The formation of basidiospores in *Coprinus cinereus* is summarized in Fig. 6.2.

The formation of conidia in Deuteromycetes. The diversity of asexual spore types is large and there have been many studies of their formation and development. Hughes (1953) produced a classification scheme for the different spore types based on their ontogeny. This system has since been discussed and enlarged on at the Kananaskis Symposium (Kendrick, 1971). Despite the complexity and diversity of asexual sporulation mechanisms it is possible to divide them broadly into two types: the blastic type, in which the conidium initial enlarges prior to its delimitation by a septum; and the thallic type, which may result from the enlargement of a conidial initial, but only after the initial has been delimited by a septum or septa. The thallic conidium thus differentiates from a whole cell. Furthermore, within these two types it is possible to distinguish at least two mechanisms by which the conidium wall is produced. It may arise either as an extension of (or incorporate all) the wall layers of the structure bearing or forming it, i.e., holoblastic or holothallic types; or the conidium wall may comprise initially only some, or one, of the wall layers of the subtending structure, i.e. enteroblastic or enterothallic types. It is thus possible to assign most conidium types easily to one of these four categories over and above the more complex system of Hughes (1953) (Fig. 6.3).

Blastic conidiogenesis. The blastic type of conidium has been widely studied. Some of the earliest studies examined the so-called porospore type. In these conidia there is a conspicuous pore between the conidiophore and the mature conidium. In *Alternaria brassicicola* (Schw.) Wiets (Campbell, 1968), *Dendryphiella vinosa* (Berk. & Curt.) Reisinger (Reisinger & Mangenot, 1969) and *Drechslera sorokiniana* (Sacc.) Sutram & Jain (Cole, 1973*a*) the porospores are formed enteroblastically by the extrusion of the inner wall layer of the conidiophore through a preformed channel in the thickened outer wall. Subsequently a holoblastic mechanism for porospore formation was reported in

Fig. 6.3. Diagram illustrating the four basic types of conidial formation: holothallic, enterothallic, holoblastic and enteroblastic.

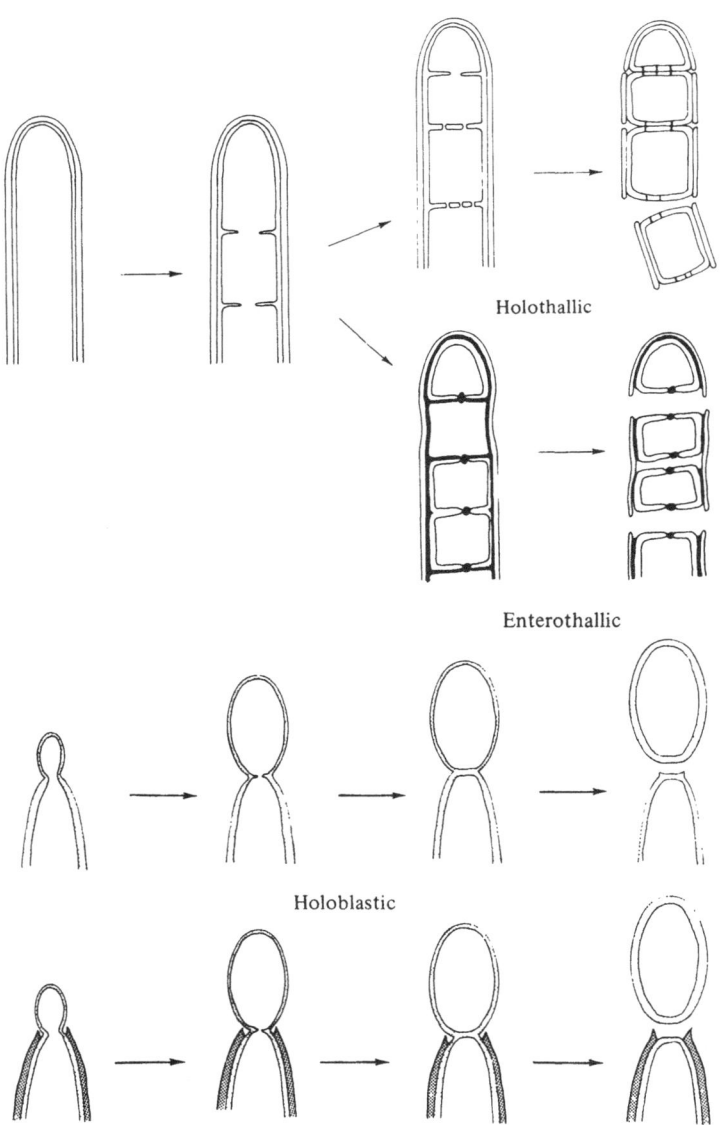

Holothallic

Enterothallic

Holoblastic

Enteroblastic

Stemphylium botryosum Wallr. (Carroll, 1972) and *Ulocladium atrum* (Pr.) Sacc. (F. E. Carroll & G. C. Carroll, 1974). In all these studies of enlarging conidia it is not easy to determine the source of material for the rapidly expanding walls, although in many cases there is obviously a large amount of activity at the plasma membrane of the spore, involving fusion of endoplasmic reticulum-derived vesicles (Campbell, 1969). Carroll (1972) found paramural bodies (Marchant & Robards, 1968) in the developing conidia of *S. botryosum* and suggested that, in support of the theory proposed by Marchant, Peat & Banbury (1967) and Marchant & Robards (1968), they were involved in secondary wall deposition in the conidium. Because the connection between the porospore and its conidiophore is narrow, the separation of the spore does not include the formation of a highly organized septum. In some systems, however, either a bilayered septum is produced which splits to allow conidium secession, e.g. *Fusarium culmorum* (W. G. Smith) Sacc. (Marchant, 1975), or two septa are laid down delimiting a region (disjunctor) either between the conidium and its condiophore, e.g. *Gonatobotryum apiculatum* (Peck.) Hughes (Cole, 1973*b*), or between successive conidia, e.g. *Phialomyces macrosporus* Misra & Talbot (Moore, 1969) (Figs. 6.8*a* and 6.9*d*).

A further common type of conidium is the phialospore. A succession of spores are produced from a fixed condiogenous locus (Subramanian, 1971); the first-formed phialospore is produced by a slightly different mechanism to the later ones, as it involves the rupture of the outer wall at the phialide apex. The walls of later-produced phialospores are considered to arise *de novo* within the phialide as figured by G. C. Carroll & F. E. Carroll (1974) (Fig. 6.8*a*). The development of the phialospore normally involves the formation of a secessional septum at the base of the spore and leads to the production of a collarette around the junction between the spore and the phialide (e.g. Hammill, 1972*a*, 1974; Cole, 1976). A number of modifications of these types of blastic conidium formation occur, with conidia produced on a sympodially proliferating conidiogenous cell (Hammill 1972*d*; Cole, 1976) or producing a succession of separation scars (annellophores) resulting from the involvement of the secession septum (Hammill, 1972*b*, 1972*c*).

In these blastic conidia it is reasonably easy to determine which types are holoblastic, but in many of the others it is difficult to determine whether the extending wall of the conidium is produced from an extension of a thin inner wall layer of the conidiogenous cell or from a completely new wall layer. A similar difficulty occurs with yeast budding systems; bud formation in *Saccharomyces cerevisiae* could clearly be called holoblastic (Marchant & Smith, 1968*a*), while in the basidiomycetous yeasts, such as *Rhodotorula glutinis* (Fres.) Harrison, there is dispute as to whether the bud wall is derived from a new wall layer within the mother cell (Marchant & Smith, 1967) or from an extension of the inner wall layer of the mother cell (Kreger-van Rij & Veenhuis, 1971). As discussed later, a similar difficulty exists in the interpretation of wall layers in spore germination studies.

Thallic conidiogenesis. The thallic mode of conidium formation has been examined in the holothallic species *Geotrichum candidum* Link ex Pers. (Fig. 6.9*c*) and in the enterothallic species *Sporendonema purpurascens* (Bonord) Mason & Hughes and *Briosia cubispora* (Bark & Curt.) von Arx (Cole, 1975). In the conversion of fertile hyphae to holothallic arthrospores in *G. candidum*, the inner wall becomes thickened and produces secessional cross-walls with micropores; when the arthrospores separate, all the wall layers of the original hypha are incorporated in the spore wall. In contrast to this mechanism the enterothallic system, as exemplified by *B. cubispora,* displays a different sequence of events. The fertile hypha becomes subdivided by cross-walls produced by a centripetal ingrowth of the secondary wall layer. When the subdivision is complete some cells develop an inner, thick, tertiary wall layer and become arthrospores, while adjacent cells which do not have thickened walls degenerate and effectively become disjunctors. During the deposition of the tertiary spore wall layer the original secondary wall layer of the fertile hypha is autolysed so that the arthrospores come to lie free within the envelope of the primary hyphal wall (Fig. 6.3).

Studies of *Geotrichum candidum* arthrospores by freeze-etching have revealed abundant plasmalemmal grooves (Cole, 1975) and these seem to be extremely variable in size, quantity and distribution (Figs. 6.7*c* and *d*). Cole (1975) has suggested that the grooves are formed in readiness for the activity of germination, as they are more abundant in older arthrospores. In addition to the information on cell membranes provided by freeze-etch studies, the technique has also revealed that the surface of many spores is covered by a complex pattern of rodlets (e.g. Sassen, Remsen & Hess, 1967; Hess, Sassen & Remsen, 1968). Similar rodlet patterns have been observed on the surface of hyphae of *Schizophyllum commune* (Wessels *et al.*, 1972) where they have been shown to be the so-called *S*-glucan component ($a(1-3)$ glucan), although McKeen *et al.* (1966) suggested that the rodlets on the conidial wall of *Penicillium levitum* Raper & Fennell consisted of lipids, and Hashimoto, Wu-Yuan & Blumenthal (1976) identified similar structures on the micronidial walls of *Trichophyton mentagrophytes* (Robin) Blanchard as glycoprotein. Cole (1973c) has now shown that the rodlet orientation on the surface of conidia may change as they increase in size during development. During the rapid expansion of the blastic conidium of *Gonatobotryum apiculatum* the rodlets become temporarily reorientated in the apical half, in a fashion reminiscent of the changes of orientation of microfibrils in plant walls during growth. Even in the thallic conidia, which Cole has examined, some rodlet reorientation occurs in particular areas.

Secondary differentiation of spore walls

Up to this point I have been mainly concerned with the initial formation of spores, although I have, in passing, discussed some spore-wall differentiation. Even a cursory examination of fungal spores shows that many have elaborate surface ornamentation or appendages, or are pigmented. These elaborations of the spore wall structure generally occur towards the end of the spore's differentiation phase. I shall now review what is known of the mechanisms underlying some of these differentiation processes. Returning to the opening remarks, it can be seen that these elaborations of the spore wall are a feature often associated with memnospores. The incorporation of large amounts of secondary wall material may provide the spore with the resistance which it requires to fulfil the functions of a memnospore.

Perhaps one of the best known memnospores which develops an enormously thickened wall is the zygospore. Zygospore development has been best studied in *Rhizopus sexualis* (Smith) Callen, where Hawker & Gooday (1968) first showed the verrucose nature of the zygospore surface by scanning electron microscopy. The verrucose wall is produced through the formation of 'warts' at the inner surface of the primary wall (Hawker & Beckett, 1971). The primary wall of the zygospore is initially two-layered; the warts first appear as discs of electron-opaque material at the inner surface of the primary wall and by growth at their margins become cup-shaped. They become progressively pushed through the primary wall layers and further wall layers become added at the inner surface. Finally the rough, black, warty surface becomes largely exposed through the degeneration of the primary wall layers. This mode of formation of a discontinuous pigmented series of structures should be contrasted with that mentioned earlier for the ascospores of *Podospora anserina* (Beckett, Barton & Wilson, 1968), where the pigmented blocks of material are formed between existing wall layers. Other ascospores, however, may have elaborate surface ornamentations (Hawker, 1968).

The teleutospores of *Tilletia* spp. have a reticulate pattern visible on the surface of the spore, which is developed initially by the deposition of interconnecting ridges of material at the inner surface of the primary wall. When the ridges are developed they become joined by a continuous inner layer of the same material and then finally covered on the inner surface by a further different wall layer (Hess & Weber, 1976). This system is thus somewhat similar to that of the zygospore, except that loss of the primary wall does not reveal the ridged wall layer. The spines on the uredospores of *Melampsora lini* (Ehrenb.) Desm. have also been shown to develop internally and to emerge through the primary wall layers (Littlefield, 1971; Littlefield & Bracker, 1971), while in the conidia of *Wallemia sebi* (Fr.) von Arx the spines are more superficial (Hawker & Madelin, 1976). Basidiomycete spores also, in general, provide a rich source of spore-surface ornamentation for study (Grand & Moore, 1970, 1971, 1972; Pegler & Young, 1972a, 1972b); however, the mechanism by which these shapes and ornamentations are formed has not been extensively examined.

In contrast to the wall elaborations produced internally, certain spore types develop electron-opaque or pigmented wall layers at the outer surface. An example of this type are the ascospores of *Neurospora tetrasperma* Shear & Dodge (Sussman, 1966), while in the ascospores of *Daldinia concentrica* (Bolt. ex. Fr.) Ces & De Not. one of the inner wall layers is extensively thickened and may contain sporopollenin (Beckett 1976a). At a more superficial level there are materials deposited just at the outer surface of spores which may materially affect the properties of the spores in respect of dispersal, infection potential etc. The conidia of the powdery mildew fungus *Erysiphe graminis* (DC.) Mérat var. *tritici* Marchal have an electron-opaque surface layer which is significantly affected by lipid extraction solvents (Johnson, Weber & Hess, 1976), while the macroconidia of *Fusarium culmorum* have a mucilaginous layer (Fig. 6.6a), composed of polysaccharide, which seems to play an important role in germination and dispersal in the soil (Marchant, 1966a, 1966b; Marchant & White, 1966).

Internal subdivision of spores

The majority of fungal spores are composed of a single cell, but there are many common organisms which have multicellular spores. The subdivision of these spores is achieved by wall growth during the development and differentiation of the spore. The formation of parallel septa, as in *Fusarium culmorum* (Marchant, 1975), produces one type of system in which the septal pores (Fig. 6.7*b*) are plugged, and each cell appears to act independently during germination (Figs. 6.6*d* and 6.7*a*). Other more complex spore subdivisions occur in species such as *Alternaria brassicicola* (Campbell, 1970), where once again the septal pores are plugged, but in this species the primary wall layer of both the external walls and the internal septa become heavily melanized.

Mangenot & Reisinger (1976) have classified the septation of conidia into three categories: euseptation, distoseptation, and a double process. In the euseptation process, the cross-wall is formed essentially by an extension of the existing wall, while in the distoseptation process the cross-wall is produced from a new wall layer which later extends around the entire inner surface of each cell of the conidium. The double process is one in which the cross-wall is initially formed by an extension of the existing wall and then becomes overlaid by an additional inner wall layer.

Spore germination

A quite extensive review of the fine structure of the spore wall and of changes during germination has been published by Akai *et al.* (1976). In this review the authors treated spores according to their taxonomic origins. There seems little point in repeating this approach and I shall, therefore, try to adopt a more general treatment of spore germination.

Sussman (1966) defined germination as 'The first irreversible stage which is recognizably different from the dormant organism, as judged by morphological, cytological, physiological or biochemical criteria.' The terms of this article confined us to looking at these criteria as they apply to wall growth and, because of limitations of present techniques, we are largely restricted to morphological changes. One of the great problems associated with ultrastructural studies of spores about to undergo germination is the difficulty of fixation. Often permanganate fixation provides an aesthetically acceptable image, but one in which information, particularly concerning wall deposition, is lacking. Aldehyde fixatives, by contrast, often prove unsatisfactory because an electron-opaque cytoplasm results. Freeze-etching provides a potential solution, yet in practical terms the information provided, particularly in respect of wall structure, is not always entirely satisfactory. Thus studies of fungal spore germination for many years have been hampered by technical difficulties and will doubtless remain so for some time.

In discussing asexual spore formation it was possible to simplify the system, in terms of wall growth, to a consideration of which wall layers of the supporting cell contributed to the spore. A similar system can be adopted with spore germination in describing the origin of the germ tube wall. Again, there are only three possible sources of the germ tube wall:

(1) As a continuation of all the spore wall layers.
(2) As a continuation of only one or some of the spore wall layers.
(3) Formed *de novo*.

In many instances the exact resolution of the origin of the germ tube wall is difficult to determine and has led to some confusion. In practice only types 2 and 3 of the above scheme seem to occur regularly, but the different types do not seem to be restricted to individual groups of fungi (Fig. 6.4).

In certain spores the initial stage in the germination process involves a considerable swelling of the spore, which in turn leads to a stretching of the wall (Ekundayo & Carlile, 1964; Marchant & White, 1966). Hess & Weber (1973) examined sporangiospores of *Rhizopus arrhizus* A. Fischer and observed that the longitudinally arranged ridges on ungerminated spores disappeared as the spores swelled. The germ tube, which is produced as the spore germinates, may be continuous with existing layers or a layer of the spore wall as in *Rhizopus* species (Hawker & Abbott, 1963); *Mucor rouxii* (Bartnicki-Garcia, Nelson & Cota-Robles, 1968*a*); *Aspergillus nidulans* (Eidam) Wint. (Border & Trinci, 1970); *Alternaria brassicicola* (Campbell, 1970); *Lentinus edodes* (Nakai & Ushiyama, 1974*b*); *Coprinus lagopus* Fr. (*cinereus*) (Heintz & Niederpruem, 1971); *Penicillium griseofulvum* Dierckx (Fletcher, 1971) and many others. Alternatively a new wall layer may be developed restricted to the area of germ tube emergence as in *Fusarium culmorum* (Marchant, 1966*a*, 1966*b*) (Fig. 6.6*e*); *Gilbertella persicaria* (Hawker, 1966); *Cunninghamella elegans* Lendner (Hawker, Thomas & Beckett, 1970): *Geotrichum candidum* (Steele & Fraser, 1973); *Daldinia concentrica* (Beckett, 1976*b*); *Tilletia* species (Hess & Weber, 1976) and others. Some support for the hypothesis that the germ tube wall is of a different composition to the spore walls can be gained from immunofluorescence studies of germinating macroconidia of *Fusarium culmorum* (Marchant & Smith, 1968*b*).

In the majority of fungi, the germ tube emerges through the spore wall at a point which is not previously differentiated from the remainder of the spore wall. In certain instances, however, the secondary differentiation of the wall which occurs would make it difficult for the germ tube to emerge satisfactorily. To overcome this problem some spores have predetermined points of germ tube emergence, so-called germ pores or

Fig. 6.4. Diagrams, prepared from published electron micrographs, of different spore germination types.

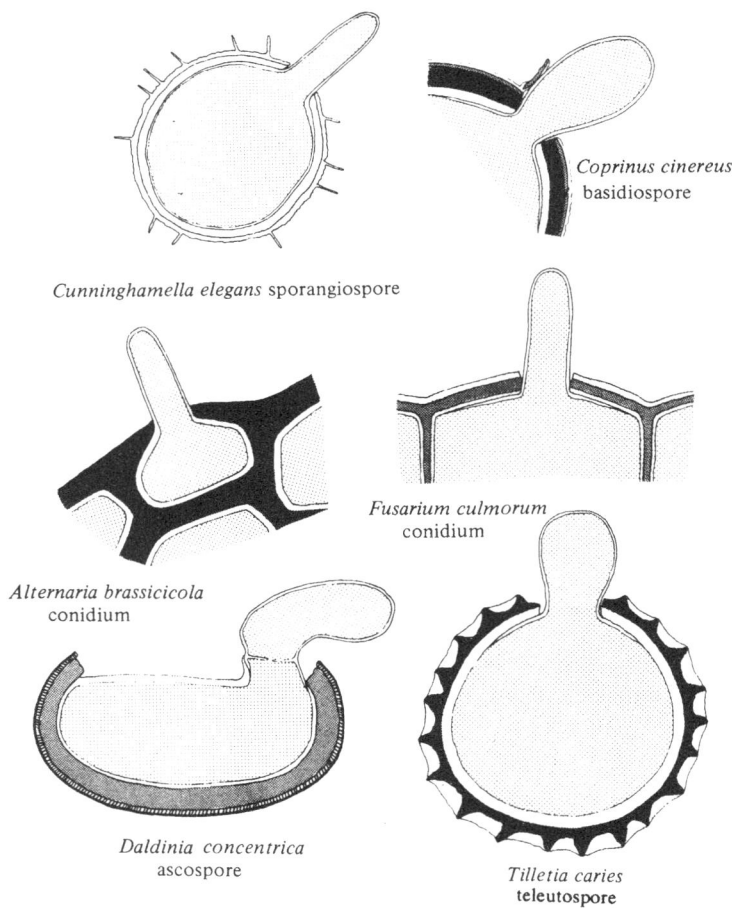

Coprinus cinereus
basidiospore

Cunninghamella elegans sporangiospore

Fusarium culmorum
conidium

Alternaria brassicicola
conidium

Daldinia concentrica
ascospore

Tilletia caries
teleutospore

germ slits (Fig. 6.5). The term 'germ pore' is somewhat misleading, for the spore wall does not have a complete aperture through it, but only a region of different composition. Germ pores occur commonly in rust spores (aecidiospores, uredospores and teleutospores), in the resistant spores of smuts, the basidiospores of many agarics and, less commonly, in ascomycetes, e.g. *Chaetomium* spp. (Millner, Motta & Lentz, 1977). Germ pore structure has been illustrated by Meléndez-Howell (1966, 1967*a*, 1967*b*, 1969) in both basidiospores and ascospores and by von Hofsten & Holm (1968) in aecidiospores (Fig. 6.5). The thickened and pigmented wall of the ascospore of *Daldinia concentrica* splits to reveal a germ slit through part of which the germ tube emerges (Beckett, 1976*b*) (Fig. 6.5). Heintz & Niederpruem (1971) have documented the emergence of the germ tube through the electron-transparent region of the *Coprinus lagopus* (*cinereus*) basidiospore (Figs. 6.8*b* and 6.9*b*). This germ pore is covered by a pore cap which is ruptured by the germ tube as it pushes through the non-pigmented channel in the wall.

Little work has been directed specifically towards determining how the wall of the growing germ tube is formed. Hess & Weber (1973) reported that there were few vesicles present at the tip of the germ tube, although vesicles were abundant in hyphal tips. In general, however, workers who have used suitable preparative techniques have reported

Fig. 6.5. Diagrams, prepared from published electron micrographs, of different spore germ pores, slits and furrows.

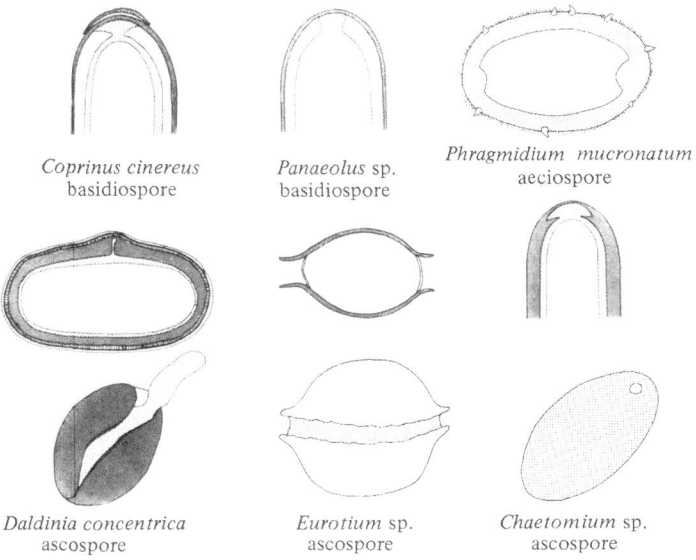

Coprinus cinereus
basidiospore

Panaeolus sp.
basidiospore

Phragmidium mucronatum
aeciospore

Daldinia concentrica
ascospore

Eurotium sp.
ascospore

Chaetomium sp.
ascospore

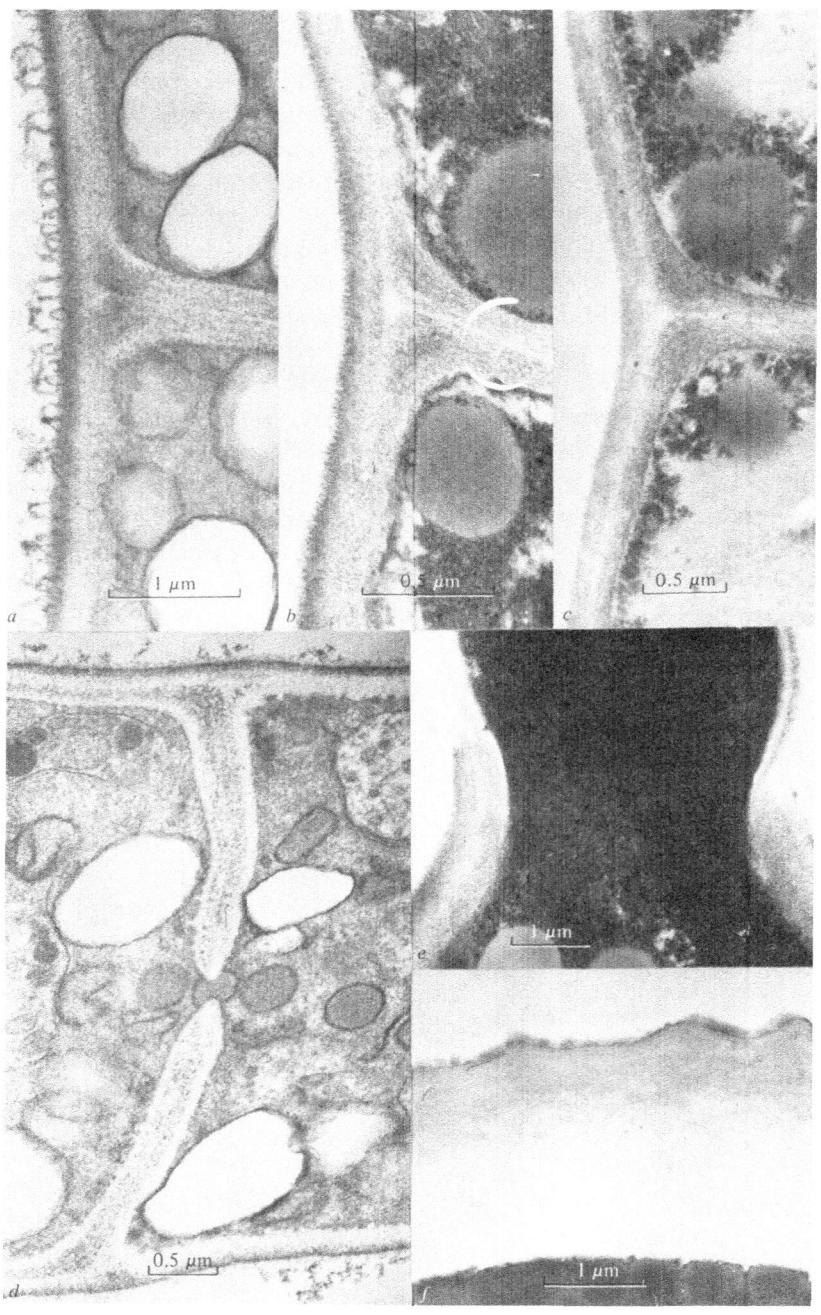

the presence of vesicles in the region of germ-tube emergence. Although the critical experiments have not been done to establish unequivocally the origin of these vesicles, it seems clear that they are derived from endoplasmic reticulum in all cases (except for some of the aquatic fungi). The functioning of these endoplasmic reticulum systems is clearly analogous to that of the dictyosomes in plants. There do not seem to be any specific cytological structures associated with germ-tube emergence and growth. The apical corpuscle reported by Bartnicki-Garcia, Nelson & Cota-Robles (1968*b*) in germ tubes of *Mucor rouxii* has not been regularly observed in other systems.

The budding of yeast cells

The budding process in yeast cells is not normally taken to be a sporulation process, but only a process of vegetative increase. It is clear, however, that even those species which are permanently yeast-like and have no mycelial phase are derived from both ascomycete and basidiomycete origins. There are also sufficient dimorphic organisms, which can exist in both yeast and mycelial forms, to suspect that the true, single-state, yeasts simply represent an extreme reduction along this pathway. It is, therefore, useful to consider how the budding process may have arisen and from what part of the normal growth cycle it may have been derived. If the sporulation nomenclature is adopted, then ascomycetous yeasts with unlayered walls, such as *Saccharomyces cerevisiae*, display a holoblastic mode of budding. In most ascomycetes, or imperfect conidial ascomycetes, this ontogeny prevails only during conidiation; spore germination being enteroblastic. In the basidiomycetous yeasts, e.g. *Rhodotorula glutinis*, budding can be described as enteroblastic. In the basidiomycetes proper, basidiospore formation is normally holoblastic, while germination appears to be enteroblastic. Thus it is possible to interpret the budding process in yeasts as having arisen as a modification either of sporulation or of the

Fig. 6.6. (*a–c*) Sections through the conidium wall of *Fusarium culmorum*, illustrating the different images produced by three preparative protocols. (*a*) 1.5 % potassium permanganate; (*b*) 3% acrolein/1 % osmium tetroxide, lead citrate; (*c*) Thiéry cytochemical staining for polysaccharide. (*d*) Longitudinal section through a conidium of *Fusarium culmorum* showing the plugged septal pore. (*e*) Section through an emerging germ tube of a conidium of *Fusarium culmorum*, showing the tapered ends of the germ tube wall entering the conidium. (*f*) Section through the largely unlayered wall of the oospore of *Pythium ultimum*.

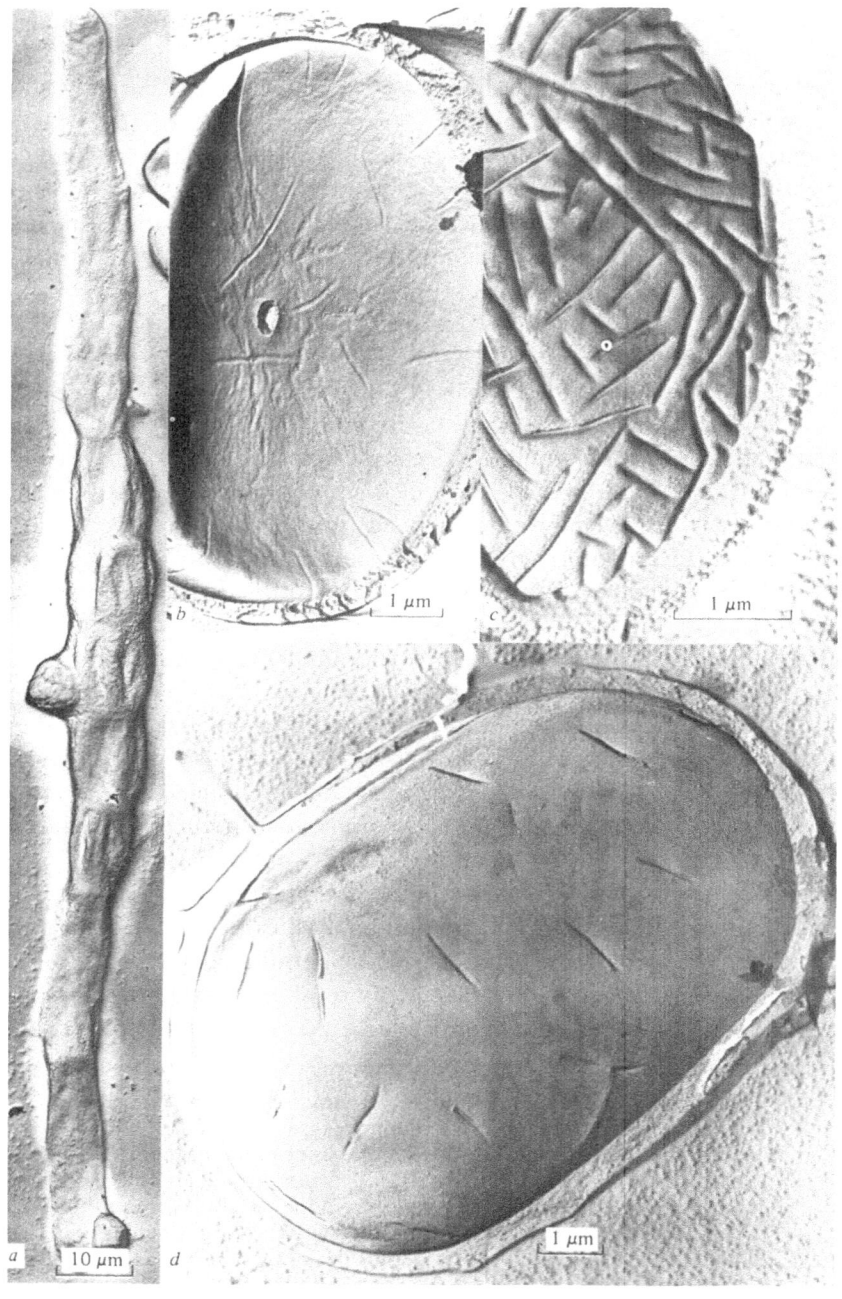

germination phases of the life-cycle. Yeast cells can, therefore, be divided between those that can be considered to undergo a continual asexual sporulation until the surface of the cell is completely exhausted by bud scars, and those that have a recurrent germination process in which the germ tube secedes before it attains any real length.

The above hypothesis is difficult to test in the true yeasts, but it is possible to gain support for the idea from other sources. In *Fusarium culmorum* it is possible to adjust the conditions in which spores germinate so that they produce extremely short germ tubes bearing a single conidium (Marchant, unpublished results). Therefore, although the macroconidia are multicellular they approach the situation of basidiomycetous yeasts of recurrent germination.

Conclusions

There can be no doubt that even if the wall does not determine the morphology of fungal spores it is the factor which maintains their shape. It is, therefore, part of the cell system which deserves investigation. It is not surprising that there are many pieces of work on fungal walls and, equally, not surprising that there are many gaps in our knowledge. In very few fungi is there anything like a complete knowledge of the chemical composition of the wall at any stage of the life cycle and yet it is quite clear that, in many instances, the spore wall composition differs markedly from that of the vegetative structures. Without some knowledge of the chemical composition of walls, attempts to define wall layers in developing structures, and to relate wall layers in different systems, have little real meaning.

The developing and germinating spore carries out some of the most rapid changes at any stage in fungal morphogenesis. Perhaps the most important point to be drawn from these studies is that the wall should not be considered as an inert envelope surrounding the fungal protoplast. New wall layers may be deposited at the inner or outer surface of an existing wall and even internal modifications may take place, to give

---

Fig. 6.7. (*a*) Replica of a germinating conidium of *Fusarium culmorum*, showing the septate conidium with three germ tubes emerging from different cells. (*b*) Freeze-etch micrograph of a conidial septum of *Fusarium culmorum*, showing the centrally placed pore. (*c*) Freeze-etch micrograph of the plasmalemma of an old arthrospore of *Geotrichum candidum*, showing abundant plasmalemmal grooves. (*d*) Freeze-etch micrograph of a young arthrospore of *Geotrichum candidum* with few plasmalemmal grooves.

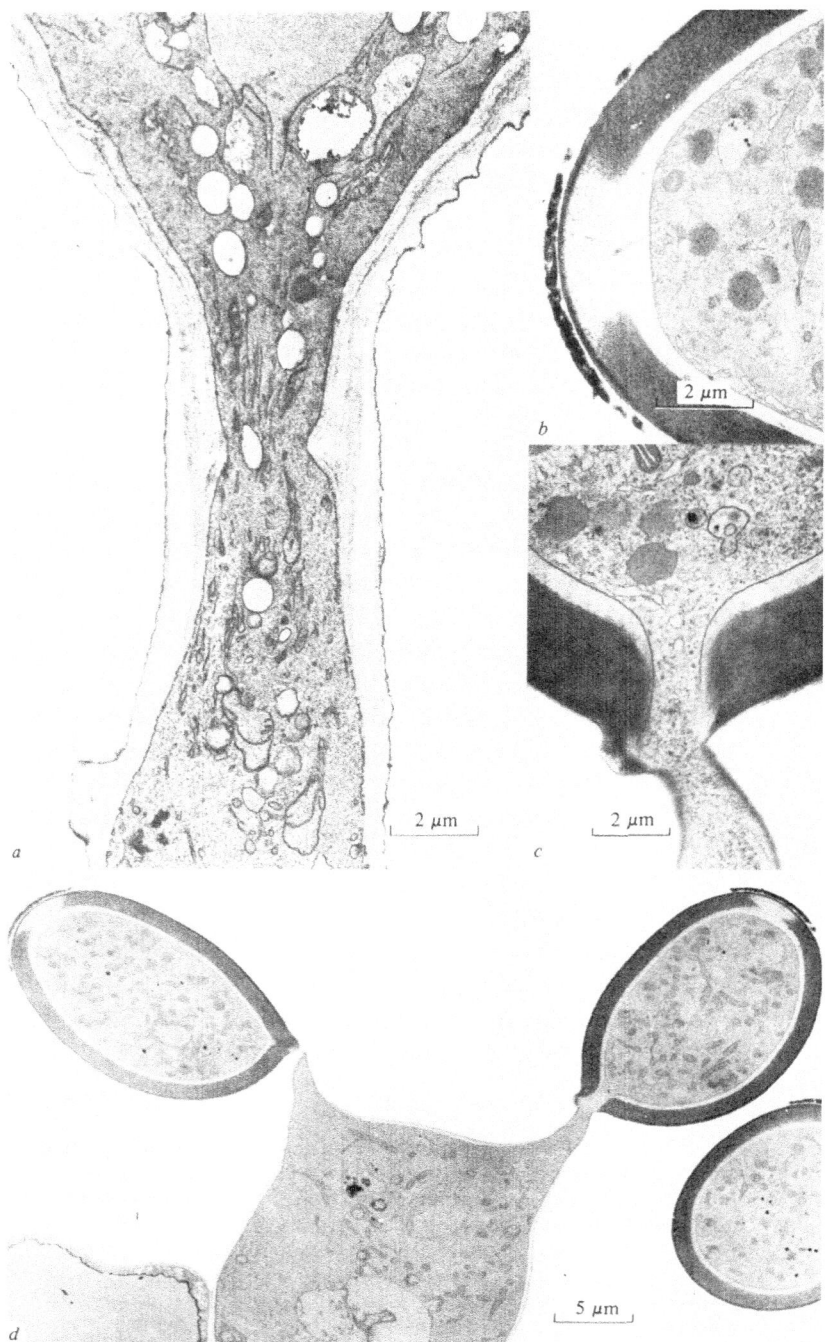

a   2 µm

b   2 µm

c   2 µm

d   5 µm

a new wall layer between existing ones. The information obtained from electron microscopy on the layering of walls relies heavily on differences in electron-opacity produced in sections by fixatives and stains. The bulk of wall components, the polysaccharides, appear electron-transparent in most protocols, and the electron-opacity is imparted by the minor components of the wall, protein, lipid, etc. In describing layering one may only be describing the distribution of these minor components of the wall.

Despite the fact that wall deposition is often extremely rapid in these sporulating and germinating situations not much information has been obtained about the mechanisms of wall synthesis from them. This is partly because of fixation difficulties caused by thick walls and condensed cytoplasm. However, in the majority of fungi the role of transport of wall material precursors appears to be assumed by endoplasmic reticulum cisternae which produce vesicles, while in the Oomycetes true golgi dictyosomes function in the same way. There is also some evidence that structures associated with the plasmalemma, so-called lomasomes, may operate in some instances in the secondary deposition of walls.

Knowledge of the role of the wall in spore systems is at present only fragmentary, but is continually being added to. The next major improvement in understanding of the functioning and synthesis of the wall will require the application of different techniques and approaches.

Acknowledgements—I wish to thank Mr S. F. Lowry, Experimental Officer in the University of Ulster Ultrastructure Unit, for preparing most of the micrographs specifically for this chapter. Thanks are also due to Dr R. T. Moore for providing the previously unpublished Figs. 6.8*a* and 6.9*d*.

Fig. 6.8. (*a*) Longitudinal section through the conidiophore and base of a developing conidium of *Phialospora macrospora*. The continuation of wall layers from the conidiophore to the spore can be seen, together with the start of cross-wall formation to produce the disjunctor cell. (*b*) Section through the apical germ pore of a nearly mature basidiospore of *Coprinus cinereus*, showing the pore cap. (*c*) Section through the point of attachment of a nearly mature basidiospore to the sterigma in *Coprinus cinereus*. (*d*) Section through a basidium of *Coprinus cinereus*, showing the attachment of two basidiospores.

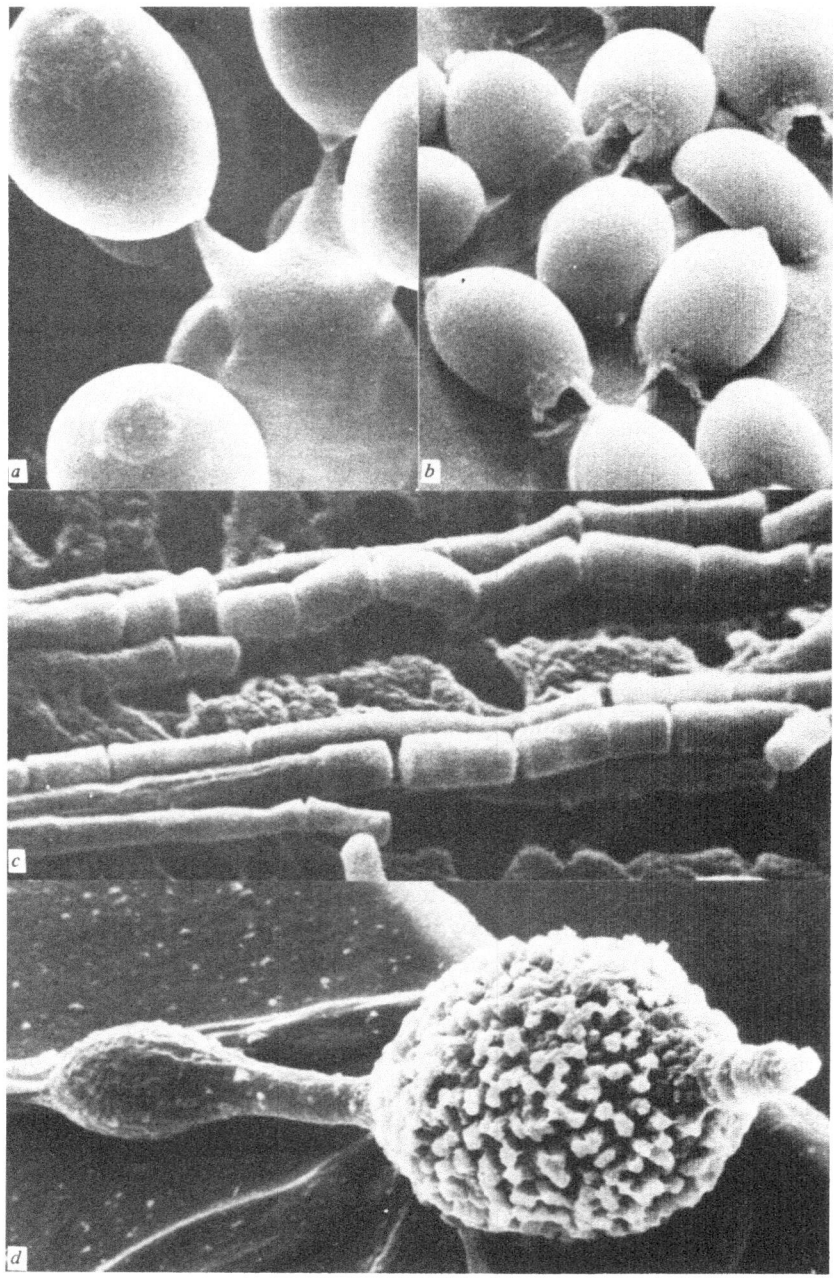

References

Akai, S., Fukutomi, M., Kunoh, H. & Shiraishi, M. (1976). Fine structure of the spore wall and germ tube change during germination. In *The Fungal Spore*, eds. D. J. Weber & W. M. Hess, pp. 355–410. New York: Wiley.

Bandoni, R. J., Bisalputra, A. A. & Bisalputra, T. (1967). Ascospore development in *Hansenula anomala*. *Canadian Journal of Botany*, **45**, 361–6.

Bartnicki-Garcia, S. (1963). Symposium on biochemical bases of morphogenesis in fungi. III. Mold–Yeast dimorphism of *Mucor*. *Bacteriological Reviews*, **27**, 293–304.

Bartnicki-Garcia, S., Nelson, N. & Cota-Robles, E. (1968a). Electron microscopy of spore germination and cell wall formation in *Mucor rouxii*. *Archiv für Mikrobiologie*, **63**, 242–55.

Bartnicki-Garcia, S., Nelson, N. & Cota-Robles, E. (1968b). A novel apical corpuscle in hyphae of *Mucor rouxii*. *Journal of Bacteriology*, **95**, 2399–402.

Beckett, A. (1976a). Ultrastructural studies on exogenously dormant ascospores of *Daldinia concentrica*. *Canadian Journal of Botany*, **54**, 689–97.

Beckett, A. (1976b). Ulstructural studies on germinating ascospores of *Daldinia concentrica*. *Canadian Journal of Botany*, **54**, 698–705.

Beckett, A., Barton, R. & Wilson, I. M. (1968). Fine structure of the wall and appendage formation in ascospores of *Podospora anserina*. *Journal of General Microbiology*, **53**, 89–94.

Beckett, A., Illingworth, R. F. & Rose, A. H. (1973). Ascospore wall development in *Saccharomyces cerevisiae*. *Journal of Bacteriology*, **113**, 1054–7.

Besson-Antoine, M. & Kühner, R. (1972). La paroi sporique ches les Coprinacées (Agaricales). *Comptes Rendus Hebdomadaires des Séances de l' Académie des Sciences, Paris*, **275**, 21–5.

Black, S. H. & Gorman, C. (1971). The cytology of *Hansenula*. III Nuclear segregation and envelopment during ascosporogenesis in *Hansenula wingei*. *Archiv für Mikrobiologie*, **79**, 231–48.

Border, D. J. & Trinci, A. P. J. (1970). Fine structure of the germination of *Aspergillus nidulans* conidia. *Transactions of the British Mycological Society*, **54**, 143–6.

Bracker, C. E. (1966). Ultrastructural aspects of sporangiospore formation in *Gilbertella persicaria*. In *The Fungus Spore*, ed. M. F. Madelin, 18th Symposium of the Colston Research Society, pp. 39–58. London: Butterworths.

Bracker, C. E. (1968). The ultrastructure and development of sporangia in *Gilbertella persicaria*. *Mycologia*, **60**, 1016–67.

Campbell, R. (1968). An electron microscope study of spore structure and development in *Alternaria brassicicola*. *Journal of General Microbiology*, **54**, 381–92.

Fig. 6.9. (*a*) Scanning electron micrograph of basidiospores of *Coprinus cinereus,* illustrating the asymmetric attachment of the spores to the sterigmata and the apical germ pore caps.
(*b*) Scanning electron micrograph of detached basidiospores of *Coprinus cinereus* in which the germ pore caps are ruptured and in some cases the point of attachment to the sterigma can be seen. (*c*) Scanning electron micrograph of fertile hyphae of *Geotrichum candidum* which are separating into holothallic arthrospores. (*d*) Scanning electron micrograph of a developing, rough-surfaced, conidium of *Phialospora macrospora*, showing the remains of the disjunctor cell at the apex and the early development of a second conidium.

Campbell, R. (1969). Further electron microscope studies of the conidium of *Alternaria brassicicola*. *Archiv für Mikrobiologie,* **69,** 60–8.

Campbell, R. (1970). An electron microscope study of exogenously dormant spores, spore germination, hyphae and conidiophores of *Alternaria brassicicola*. *New Phytologist,* **69,** 287–93.

Campbell, R. (1973). Ultrastructure of asci, ascospores, and spore release in *Lophodermella sulcigena* (Rostr.) V. Hohn. *Protoplasma* **78,** 69–80.

Carroll, F. E. (1972). A fine-structural study of conidium initiation in *Stemphylium botryosum* Wallroth. *Journal of Cell Science,* **11,** 33–47.

Carroll, F. E. & Carroll, G. C. (1974). The fine structure of conidium initiation in *Ulocladium atrum*. *Canadian Journal of Botany,* **52,** 443–6.

Carroll, G. C. (1967). The ultrastructure of ascospore delimitation in *Saccobolus kerverni*. *Journal of Cell Biology,* **33,** 218–24.

Carroll, G. C. (1969). A study of the fine structure of ascosporogenesis in *Saccobolus kerverni*. *Archiv für Mikrobiologie,* **66,** 321–39.

Carroll, G. C. & Carroll, F. E. (1974). The fine structure of conidium development in *Phialocephala dimorphospora*. *Canadian Journal of Botany,* **52,** 2119–28.

Cole, G. T. (1973*a*). Ultrastructure of conidiogenesis in *Drechslera sorokiniana*. *Canadian Journal of Botany,* **51,** 629–38.

Cole, G. T. (1973*b*). Ultrastructural aspects of conidiogenesis in *Gonatobotryum apiculatum*. *Canadian Journal of Botany,* **51,** 1677–84.

Cole, G. T. (1973*c*). A correlation between rodlet orientation and conidiogenesis in hyphomycetes. *Canadian Journal of Botany,* **51,** 2413–22.

Cole, G. T. (1975). The thallic mode of conidiogenesis in the Fungi Imperfecti. *Canadian Journal of Botany,* **53,** 2983–3001.

Cole, G. T. (1976). Condiogenesis in pathogenic hyphomycetes I. *Sporothrix, Exophiala, Geotrichum* and *Microsporum*. *Sabouraudia,* **14,** 81–98.

Ekundayo, J. A. & Carlile, M. J. (1964). The germination of sporangiospores of *Rhizopus arrhizus*; spore swelling and germ tube emergence. *Journal of General Microbiology,* **35,** 261–9.

Fletcher, J. (1971). Fine-structural changes during germination of conidia of *Penicillium griseofulvum* Dierckx. *Annals of Botany,* **35,** 441–9.

Fletcher, J. (1972). Fine structure of developing merosporangia and sporangiospores of *Syncephalastrum racemosum*. *Archiv für Mikrobiologie,* **87,** 269–84.

Furtado, J. S. & Olive, L. S. (1970). Ultrastructure of ascospore development in *Sordaria fimicola*. *Journal of the Elisha Mitchell Scientific Society,* **86,** 131–8.

Gay, J. L., Greenwood, A. D. & Heath, I. B. (1971). The formation and behaviour of vacuoles (vesicles) during oosphere development and zoospore germination in *Saprolegnia*. *Journal of General Microbiology,* **65,** 233–41.

Grand, L. F. & Moore, R. T. (1970). Ultracytotaxonomy of basidiomycetes. I. Scanning electron microscopy of spores. *Journal of the Elisha Mitchell Scientific Society,* **86,** 106–17.

Grand, L. F. & Moore, R. T. (1971). Scanning electron microscopy of basidiospores of species of Strobilomycetaceae. *Canadian Journal of Botany,* **49,** 1259–61.

Grand, L. F. & Moore, R. T. (1972). Scanning electron microscopy of *Cronartium* spores. *Canadian Journal of Botany,* **50,** 1741–2.

Greenhalgh, G. N. & Evans, L. V. (1968). The developing ascospore wall of *Hypoxylon fragiforme*. *Journal of the Royal Microscopical Society,* **88,** 545–56.

Greenhalgh, G. N. & Griffiths, H. B. (1970). The ascus vesicle. *Transactions of the British Mycological Society,* **54,** 489–92.

Gregory, P. H. (1966). The fungus spore: what it is and what it does. In *The Fungus Spore,* ed. M. F. Madelin, 18th Symposium of the Colston Research Society, pp. 1–13. London: Butterworths.

Guth, E., Hashimoto, T. & Conti, S. F. (1972). Morphogenesis of ascospores in *Saccharomyces cerevisiae*. *Journal of Bacteriology*, **109**, 869–80.

Hammill, T. M. (1972*a*). Electron microscopy of phialoconidiogenesis in *Metarrhizium anisopliae*. *American Journal of Botany*, **59**, 317–26.

Hammill, T. M. (1972*b*). Fine structure of annellophores. V. *Stegonosporium pyriforme*. *Mycologica*, **64**, 654–7.

Hammill, T. M. (1972*c*). Fine structure of annellophores II. *Doratomyces nanus*. *Transactions of the British Mycological Society*, **59**, 249–53.

Hammill, T. M. (1972*d*). Electron microscopy of conidiogenesis in *Chloridium chlamydosporis*. *Mycologia*, **64**, 1054–65.

Hammill, T. M. (1974). Electron microscopy of phialides and conidiogenesis in *Trichoderma saturnisporum*. *American Journal of Botany*, **61**, 15–24.

Hashimoto, T., Wu-Yuan & Blumenthal, H. J. (1976). Isolation and characterization of the rodlet layer of *Trichophyton mentagrophytes* microconidial wall. *Journal of Bacteriology*, **127**, 1543–9.

Hawker, L. E. (1966). Germination: morphological and anatomical changes. In *The Fungus Spore*, ed. M. F. Madelin, 18th Symposium of the Colston Research Society, pp. 151–161. London: Butterworths.

Hawker, L. E. (1968). Wall ornamentation of ascospores of species of *Elaphomyces* as shown by the scanning electron microscope. *Transactions of the British Mycological Society*, **51**, 493–8.

Hawker, L. E. & Abbott, P. McV. (1963). An electron microscope study of maturation and germination of sporangiospores of two species of *Rhizopus*. *Journal of General Microbiology*, **32**, 295–8.

Hawker, L. E. & Beckett, A. (1971). Fine structure and development of the zygospore of *Rhizopus sexualis* (Smith) Callen. *Philosophical Transactions of the Royal Society of London*, **B263**, 71–100.

Hawker, L. E. & Gooday, M. A. (1968). Development of the zygospore wall in *Rhizopus sexualis* (Smith) Callen. *Journal of General Microbiology*, **54**, 13–20.

Hawker, L. E. & Madelin, M. F. (1976). The dormant spore. In *The Fungal Spore*, eds. D. J. Weber & W. M. Hess, pp. 1–70. New York: Wiley.

Hawker, L. E., Thomas, B. & Beckett, A. (1970). An electron microscope study of structure and germination of conidia of *Cunninghamella elegans* Lendner. *Journal of General Microbiology*, **60**, 181–9.

Heath, I. B., Gay, J. L. & Greenwood, A. D. (1971). Cell wall formation in the Saprolegniales: cytoplasmic vesicles underlying developing walls. *Journal of General Microbiology*, **65**, 225–32.

Heintz, C. E. & Niederpruem, D. J. (1971). Ultrastructure of quiescent and germinated basidiospores and oidia of *Coprinus lagopus*. *Mycologia*, **63**, 745–66.

Hemmes, D. E. & Bartnicki-Garcia, S. (1975). Electron microscopy of gametangial interaction and oospore development in *Phytophthora capsici*. *Archives of Microbiology*, **103**, 91–112.

Hess, W. M. & Weber, D. J. (1973). Ultrastructure of dormant and germinated sporangiospores of *Rhizopus arrhizus*. *Protoplasma*, **77**, 15–33.

Hess, W. M. & Weber, D. J. (1976). Form and function of basidiomycete spores. In *The Fungal Spore*, eds. D. J. Weber & W. M. Hess, pp. 645–713. New York: Wiley.

Hess, W. W., Sassen, M. M. A. & Remsen, C. C. (1968). Surface characteristics of *Penicillium* conidia. *Mycologia*, **60**, 290–303.

von Hofsten, A. & Holm, L. (1968). Studies on the fine structure of aeciospores I. *Grana Palynologica*, **8**, 235–51.

Howard, K. L. & Moore, R. T. (1970). Ultrastructure of oogenesis in *Saprolegnia terrestris*. *Botanical Gazette*, **131**, 311–36.

Hughes, S. J. (1953). Conidiophores, conidia and classification. *Canadian Journal of Botany*, **31**, 577–659.

Hugueney, R. (1972). Ontogenèse des infrastructures de la paroi sporique de *Coprinus cineratus* Quel. var. *nudisporus* Kühner (Agaricales). *Comptes Rendus Hebdomadaires des Séances de l'Académie des Sciences, Paris*, **275**, 1495–8.

Hugueney, R. (1975a). Ultrastructure de la paroi sporique de *Coprinus congregatus* Bull. ex. Fr. et ses variations liées a la symetrie bilaterale de la spore. *Bulletin Mensuel de la Société Linnéenne de Lyon*, **7**, 196–202.

Hugueney, R. (1975b). Morphologie, ultrastructure et developpement de l'apicule des spores de quelques copriniacees: étude particuliere du punctum lacrymans. *Bulletin Mensuel de la Société Linnéenne de Lyon*, **8**, 249–56.

Johnson, D., Weber, D. J. & Hess, W. M. (1976). Lipids from conidia of *Erysiphe graminis tritici* (powdery mildew). *Transactions of the British Mycological Society*, **66**, 35–43.

Kendrick, B. (ed.) (1971). *Taxonomy of Fungi Imperfecti*. Toronto: University of Toronto Press.

Kreger-van Rij, N. J. W. & Veenhuis, M. (1971). A comparative study of the cell wall structure of basidiomycetous and related yeasts. *Journal of General Microbiology*, **68**, 87–95.

Lippman, E., Erwin, D. C. & Bartnicki-Garcia, S. (1974). Isolation and chemical composition of oospore–oogonium walls of *Phytophthora megasperma* var. *sojae*. *Journal of General Microbiology*, **80**, 131–41.

Littlefield, L. J. (1971). Scanning electron microscopy of uredospores of *Melampsora lini*. *Journal de Microscopie*, **10**, 225–8.

Littlefield, L. J. & Bracker, C. E. (1971). Ultrastructure and development of uredispore ornamentation in *Melampsora lini*. *Canadian Journal of Botany*, **49**, 2067–73.

Lynn, R. & Magee, P. T. (1970). Development of the spore wall during ascospore formation in *Saccharomyces cerevisiae*. *Journal of Cell Biology*, **44**, 688–92.

McKeen, W. E. (1975). Electron microscopy studies of a developing *Pythium* oogonium. *Canadian Journal of Botany*, **53**, 2354–60.

McKeen, W. E., Mitchell, N., Jarvie, W. & Smith, R. (1966). Electron microscopy studies of conidial walls of *Sphaerotheca macularis, Penicillium levitum*, and *Aspergillus niger*. *Canadian Journal of Microbiology*, **12**, 427–8.

McLaughlin, D. J. (1973). Ultrastructure of sterigma growth and basidiospore formation in *Coprinus* and *Boletus*. *Canadian Journal of Botany*, **51**, 145–50.

McLaughlin, D. J. (1977). Basidiospore initiation and early development in *Coprinus cinereus*. *American Journal of Botany*, **64**, 1–16.

Mainwaring, H. R. (1972). The fine structure of ascospore wall formation in *Sordaria fimicola*. *Archiv für Mikrobiologie*, **81**, 126–35.

Mangenot, F. & Reisinger, O. (1976). Form and function of conidia as related to their development. In *The Fungal Spore*, eds. D. J. Weber & W. M. Hess, pp. 789–846. New York: Wiley.

Marchant, R. (1966a). Fine structure and spore germination in *Fusarium culmorum*. *Annals of Botany*, **30**, 441–5.

Marchant, R. (1966b). Wall structure and spore germination in *Fusarium culmorum*. *Annals of Botany*, **30**, 821–30.

Marchant, R. (1968). An ultrastructural study of sexual reproduction in *Pythium ultimum*. *New Phytologist*, **67**, 167–71.

Marchant, R. (1969). The fine structure and development of the fructification of *Lycoperdon perlatum*. *Transactions of the British Mycological Society*, **53**, 63–8.

Marchant, R. (1975). An ultrastructural study of 'phialospore' formation in *Fusarium culmorum* grown in continuous culture. *Canadian Journal of Botany*, **53**, 1978–87.

Marchant, R. & Robards, A. W. (1968). Membrane systems associated with the plasmalemma of plant cells. *Annals of Botany*, **32**, 457–71.

Marchant, R. & Smith, D. G. (1967). Wall structure and bud formation in *Rhodotorula glutinis*. *Archiv für Mikrobiologie*, **58**, 248–56.

Marchant, R. & Smith, D. G. (1968*a*). Bud formation in *Saccharomyces cerevisiae* and a comparison with the mechanism of cell division in other yeasts. *Journal of General Microbiology*, **53**, 163–9.

Marchant, R. & Smith, D. G. (1968*b*). A serological investigation of hyphal growth in *Fusarium culmorum*. *Archiv für Mikrobiologie*, **63**, 85–94.

Marchant, R. & White, M. F. (1966). Spore swelling and germination in *Fusarium culmorum*. *Journal of General Microbiology*, **42**, 237–44.

Marchant, R., Peat, A. & Banbury, G. H. (1967). The ultrastructural basis of hyphal growth. *New Phytologist*, **66**, 623–9.

Meléndez-Howell, L. M. (1966). Ultrastructure du pore germinatif sporale dans le genre *Coprinus* Link. *Comptes Rendus Hebdomadaires de Séances de l'Académie des Sciences, Paris*, **263**, 717–20.

Meléndez-Howell, L. M. (1967*a*). Les rapports entre le pore germinatif sporal et la germination chez les basidiomycètes en microscopie electronique. *Comptes Rendus Hebdomadaires des Séances de l'Académie des Sciences, Paris*, **264**. 1266–9.

Meléndez-Howell, L. M. (1967*b*). Recherches sur le pore germinatif des basidiospores. *Annales des sciences naturelles (Botanique) Paris*, **8**, 487–638.

Meléndez-Howell, L. M. (1969). Rapports entre le pore germinatif sporal et les fentes de germination longitudinales ou circulaires chez les ascospores. *Comptes Rendus Hebdomadaires des Séances de l'Académie des Sciences, Paris*, **268**, 1273–4.

Millner, P. D., Motta, J. J. & Lentz, P. L. (1977). Ascospores, germ pores, ultrastructure, and thermophilism of *Chaetomium. Mycologia*, **69**, 720–33.

Moore, R. T. (1969). Conidiogenesis in *Phialomyces*. In *Proceedings XI International Botanical Congress, Seattle*, ed. R. C. Starr. Washington D.C.: International Botanical Congress.

Nakai, Y. (1975). Fine structure of shiitake, *Lentinus edodes* (Berk.) Sing IV. External and internal features of the hilum in relation to basidiospore discharge. *Report of the Tottori Mycological Institute, Japan*, **12**, 41–5.

Nakai, Y. & Ushiyama, R. (1974*a*). Fine structure of shiitake, *Lentinus edodes*. (Berk.) Sing. II. Development of basidia and basidiospores. *Report of the Tottori Mycological Institute, Japan*, **11**, 7–15.

Nakai, Y. & Ushiyama, R. (1974*b*). Fine structure of shiitake, *Lentinus edodes* (Berk.) Sing. III Germination of basidiospores. *Report of the Tottori Mycological Institute, Japan*, **11**, 16–22.

Oláh, G. M. & Reisinger, O. (1974). L'ontogénie des téguments de la paroi sporale en relation avec le stérigmate et la gouttelette hilaire chez quelques agarics mélanosporés. *Comptes Rendus Hebdomadaires des Séances de l'Académie des Sciences, Paris*, **278**, 2755–8.

Oso, B. A. (1969). Electron microscopy of ascus development in *Ascobolus*. *Annals of Botany*, **33**, 205–9.

Pegler, D. N. & Young, T. W. K. (1972*a*). Basidiospore form in the British species of *Inocybe. Kew Bulletin*, **26**, 499–537.

Pegler, D. N. & Young, T. W. K. (1972*b*). Basidiospore form in the British species of *Galerina* and *Kuehneromyces. Kew Bulletin*, **27**, 483–500.

Reisinger, O. & Mangenot, F. (1969). Analyses morphologiques au microscope électronique à balayage et étude de l'ontogénie sporale chez *Dendryphiella vinosa* (Berk. et Curt.) Reisinger. *Comptes Rendus Hebdomadaires des Séances de l'Académie des Sciences Paris*, **269**, 1843–1845.

Sassen, M. M. A., Remsen, C. C. & Hess, W. M. (1967). Fine structure of *Penicillium megasporum* conidiospores. *Protoplasma,* **64,** 75–88.

Steele, S. D. & Fraser, T. W. (1973). Ultrastructural changes during germination of *Geotrichum candidum* arthrospores. *Canadian Journal of Microbiology,* **19,** 1031–4.

Stiers, D. L. (1974). Fine structure of ascospore formation in *Poronia punctata. Canadian Journal of Botany,* **52,** 999–1003.

Subramanian, C. V. (1971). The phialide. In *Taxonomy of Fungi Imperfecti,* ed. B. Kendrick, pp. 92–114. Toronto: University of Toronto Press.

Sussman, A. S. (1966). Dormancy and spore germination. In *The Fungi,* vol. 2, *The Fungal Organism,* eds. G. C. Ainsworth & A. S. Sussman, pp. 733–64. New York: Academic Press.

Syrop, M. J. & Beckett, A. (1972). The origin of ascospore-delimiting membranes in *Taphrina deformans. Archiv für Mikrobiologie,* **86,** 185–91.

Tewari, J. P. & Skoropad, W. P. (1977). Ultrastructure of oospore development in *Albugo candida* on rapeseed. *Canadian Journal of Botany,* **55,** 2348–57.

Vujičić, R. (1971). An ultrastructural study of sexual reproduction in *Phytophthora palmivora. Transactions of the British Mycological Society,* **57,** 525–30.

Wells, K. (1964a). The basidia of *Exidia nucleata.* I. Ultrastructure. *Mycologia,* **56,** 327–41.

Wells, K. (1964b). The basidia of *Exidia nucleata.* II. Development. *American Journal of Botany,* **51,** 360–70.

Wells, K. (1965). Ultrastructural features of developing and mature basidia and basidiospores of *Schizophyllum commune. Mycologia,* **57,** 236–61.

Wessels, J. G. H., Kreger, D. R., Marchant, R., Regensburg, B. A. & de Vries, O. M. H. (1972). Chemical and morphological characterization of the hyphal wall surface of the basidiomycete *Schizophyllum commune. Biochimica et Biophysica Acta,* **273,** 346–58.

Wilsenach, R. & Kessel, M. (1965). The role of lomasomes in wall formation in *Penicillium vermiculatum. Journal of General Microbiology,* **40,** 401–4.

7
Chitosomes and chitin synthesis

S. BARTNICKI-GARCIA, J. RUIZ-HERRERA*,
AND
C. E. BRACKER†

Department of Plant Pathology, University of California, Riverside, California 92521, USA
* *Departamento de Genética y Biologia Molecular, Centro de Investigación y Estudios Avanzados del Instituto Politécnico Nacional, México D. F., México*
† *Department of Botany and Plant Pathology, Purdue University, West Lafayette, Indiana, 47907, USA*

Introduction

Microfibrils are the skeletal components of the cell walls of most fungi (Frey-Wyssling & Mühlethaler, 1950; Aronson, 1965; Bartnicki-Garcia, 1968; see also Chapter 1); knowledge about their synthesis and assembly is of paramount importance to understand the mechanisms of wall growth and morphogenesis in the fungi.

This article is about the synthesis of chitin microfibrils. Our eventual goal is to understand how these exceedingly long and highly crystalline structures are synthesized, assembled, and integrated harmoniously with other cell wall components. The assembly of chitin microfibrils is ostensibly a question of three-dimensional biochemistry: elucidation of the biosynthetic pathway is clearly not enough, we must also unravel the underlying cytological apparatus responsible for the spatial localization and orientation of microfibril synthesis.

The first evidence for chitin biosynthesis *in vitro* was obtained over two decades ago by Glaser & Brown (1957) who demonstrated that uridine diphosphate *N*-acetyl-D-glucosamine (UDP-GlcNAc) was the precursor for the synthesis of a polymer recognized as chitin by its insolubility and by its digestibility with chitinase. Subsequent researchers have repeatedly confirmed the ability of UDP-GlcNAc to serve as a precursor for chitin synthesis in cell-free extracts of other fungi (e.g., Porter & Jaworski, 1966; Camargo *et al.*, 1967; McMurrough, Flores-Carreón & Bartnicki-Garcia, 1971; Keller & Cabib, 1971; Jan, 1974; Gooday & DeRousset-Hall, 1975; Peberdy & Moore, 1975; Lopez-Romero & Ruiz-Herrera, 1976).

Before 1974, evidence on chitin biosynthesis in fungi was based exclusively on radiotracer studies showing the incorporation of radioactivity from UDP-GlcNAc (labelled in the glycosyl or acetyl moieties) into an insoluble product. The sources of chitin synthetase were various crude particulate fractions obtained by centrifugation of cell-free extracts. With these highly heterogeneous preparations, neither the physical structure of the product nor the subcellular localization of the enzyme could be ascertained. The product was only identified by its insolubility and by its digestibility with chitinase. In 1974, studies on the chitin synthetase from the yeast form of *Mucor rouxii* (Calmette) Wehm (Ruiz-Herrera & Bartnicki-Garcia, 1974, 1976) revealed that the chitin synthetase of crude membrane preparations (e.g. mixed membrane fraction) was highly zymogenic and usually stable. By exposing the mixed membrane fraction to the substrate of chitin synthetase (UDP-GlcNAc), in the cold, a procedure was discovered that separated the enzyme from other components (Ruiz-Herrera & Bartnicki-Garcia, 1974). A transparent 'solution' of chitin synthetase was thus obtained which, upon incubation at 20–25 °C, produced a copious precipitate of microfibrils. These microfibrils made *in vitro* were indistinguishable by electron microscopy (shadow-casting) or X-ray diffraction (powder camera) from native chitin microfibrils.

Subsequent studies (Ruiz-Herrera *et al.,* 1975) showed that the chitin synthetase 'solutions' were in fact diluted suspensions of exceedingly small particles (<100 nm). We named these particles *chitosomes*, and we showed that they undergo remarkable transformations during the formation of chitin microfibrils (Bracker, Ruiz-Herrera & Bartnicki-Garcia, 1976). The structure of chitosomes from yeast cells of *Mucor rouxii* and the process of fibril formation have been summarized elsewhere (Bartnicki-Garcia, Bracker & Ruiz-Herrera, 1978*a*). We believe that chitosomes are cytoplasmic microvesicles that serve as conveyors of latent chitin synthetase (zymogen) to its destination at the cell surface, where chitin microfibril assembly takes place.

In this article, we shall review our evidence for the occurrence of chitosomes in fungi other than *Mucor rouxii*, and present recent studies on the dissociation and reaggregation of chitosomes from *M. rouxii*, plus some thoughts on the origin and fate of chitosomes in the living cell.

Isolation of chitosomes from various fungi

For this study, we chose representatives of five major groups of chitinous fungi:
Chytridiomycetes, mycelium of *Allomyces macrogynus* (Emerson) Emerson & Wilson;

Zygomycetes, mycelium of *Mucor rouxii*;
Hemiascomycetes, yeast cells of *Saccharomyces cerevisiae* Hansen;
Euascomycetes, mycelium of *Neurospora crassa* Shear & Dodge;
Homobasidiomycetes, basidiocarp of *Agaricus bisporus* (J. Lange) Pilát.

The procedure used was essentially the same as that employed for isolating chitosomes from yeast cells of *Mucor rouxii* (Fig. 7.1). In all cases, the cells were broken with glass beads in an MSK homogenizer. After removal of cell walls (2000 *g*) and the major membrane fraction (54 000 *g*), a supernatant was obtained which contained a population of fine particulate matter (<100 nm) – the 'miniorganelles'. The miniorganelles are a mixture of macromolecular complexes and microvesicles suspended in the soluble portion of the cell-free extract. When the 54 000 *g* supernatant was applied to a Biogel A-5m column, the miniorganelles were separated (exclusion volume) from cytosol components. In this manner, a protein inhibitor of chitin synthetase was eliminated (Lopez-Romero, Ruiz-Herrera & Bartnicki-Garcia, 1978).

Ribosomes are the overwhelming component of the miniorganelle population and, hence, their removal is necessary to permit observation and characterization of other members of the population. Ribosomes were largely or entirely eliminated by treatment with high concentrations of ribonuclease. The extent of elimination depended on the

Fig. 7.1. Procedure for the isolation of chitosomes. Samples from the fractionated sucrose gradient were used for chitin synthetase assays and electron microscopic examinations by negative staining (Bracker *et al.*, 1976; Ruiz-Herrera *et al.*, 1977).

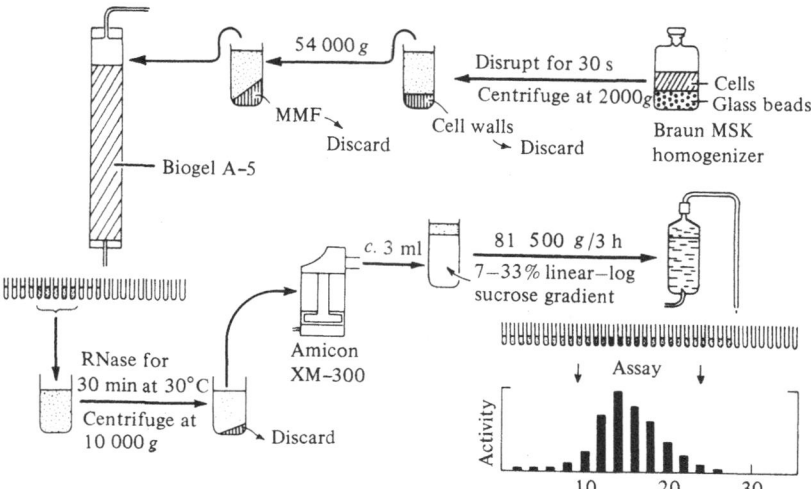

particular fungus. After removing precipitated ribonucleoproteins and concentrating the supernatant by ultrafiltration, the suspension was fractionated by density gradient centrifugation, and the fractions were assayed for chitin synthetase activity. Chitosome characterization was assessed by the following criteria: (1) sedimentation behaviour in a sucrose density gradient; (2) morphology in negative-stained preparations; (3) microfibril production; and (4) formation of fibroid coils.

(1) For all fungi examined, similar profiles of chitin synthetase were obtained in the sucrose density gradients (Fig. 7.2), an indication of the

Fig. 7.2. Sedimentation profiles of chitosomes from various fungi prepared as shown in Fig. 7.1. Data redrawn from Bartnicki-Garcia *et al.* (1978*b*). Values of chitin synthetase activity were normalized to a peak value of 1.0 for all preparations.

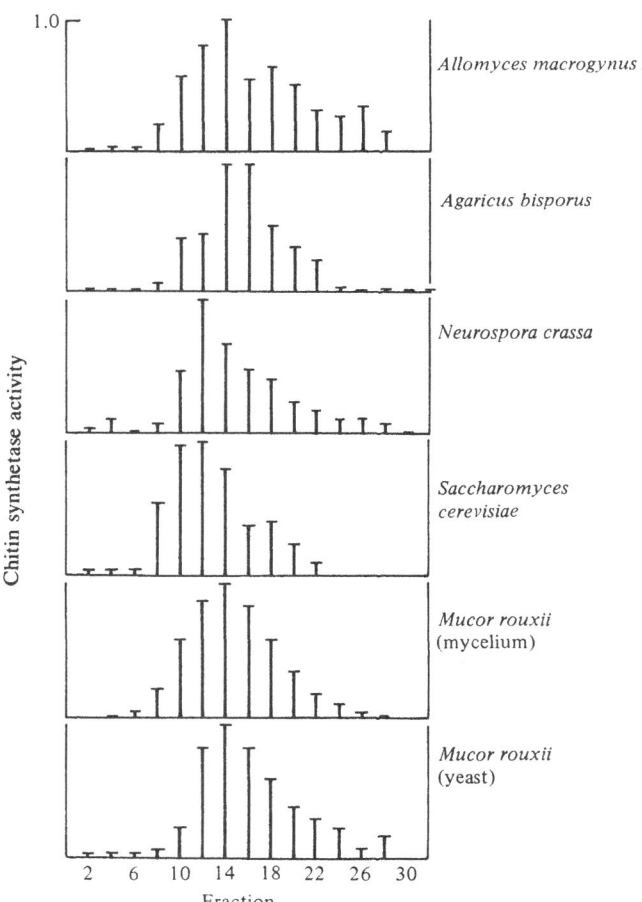

presence of chitin-synthesizing structures possessing the same sedimentation characteristics.

(2) Electron-microscopic examination of peak fractions of chitin synthetase revealed chitosome structures similar to those described for yeast cells of *Mucor rouxii*. Both the proctoid and cycloid appearances of chitosomes were evident. In all samples, the number of chitosomes per field was considerably lower than that routinely found in similar preparations from *M. rouxii* yeast cells. Fig. 7.3 shows examples of proctoid chitosomes from the various fungi and samples of the abundant chitosome population from yeast cells of *M. rouxii*.

Chitosomes treated with a high concentration of trypsin undergo a marked transformation. The membrane, or shell, of the chitosome revealed two distinct layers, a thin outer veil and a thicker, more robust inner layer. In most trypsin-treated chitosomes there was a conspicuous pouch formed by an invagination of the chitosome shell. The invaginated membrane was thinner than the rest of the chitosome shell. Evidently the strong trypsin treatment produced a partial disorganization of the chitosome structure with a total loss in chitin synthetase activity (Bartnicki-Garcia *et al.*, 1978*b*). Although the meaning of this transformation is not yet known, it may be construed as evidence that the chitosome is a complex structure with differentiated regions. We have used this treatment as an empirical test to compare the response of presumed chitosomes from other fungi and found that in all cases, a high concentration of trypsin caused a similar transformation. Examples are shown in Fig. 7.4.

(3) Upon incubation with UDP-GlcNAc, GlcNAc (as an allosteric activator (Porter & Jaworski, 1966; McMurrough & Bartnicki-Garcia, 1971)) and an activating protease (see below), chitosome preparations from all fungi produced microfibrils (Bartnicki-Garcia *et al.*, 1978*b*) (some examples in Fig. 7.5). There were differences in the appearance of the microfibrils (length, width, regularity). It would be premature to conclude that these differences reflect intrinsic differences in the morphology of microfibrils synthesized by each fungus, for we have seen that microfibril morphology in chitosome preparations from yeast cells of *Mucor rouxii* varies greatly with length of incubation (Bracker *et al.*, 1976).

(4) The most convincing evidence that chitosomes make microfibrils is in images which establish an unmistakable connection between the product and its synthesizing particle (Fig. 7.6). These are the 'fibroids' or coiled microfibrils found inside chitosomes. Fibroid coils are prob-

Fig. 7.3. Chitosomes from various fungi. Negatively stained and thin-sectioned preparations of peak samples from sucrose density gradients (see Fig. 7.2). (*a, b, c*) Fields of chitosomes from *Mucor rouxii*: (*a*) most with a proctoid appearance, negatively stained; (*b*) most with a cycloid appearance, negatively stained; (*c*) in thin section. Examples of proctoid chitosomes from: (*d*) mycelium of *M. rouxii*; (*e*) mycelium of *Allomyces macrogynus*; (*f*) budding cells of *Saccharomyces cerevisiae*; (*g*) mycelium of *Neurospora crassa*; (*h*) basidiocarp of *Agaricus bisporus*. *a, b, d–h* from Bartnicki-Garcia *et al.* (1978*b*).

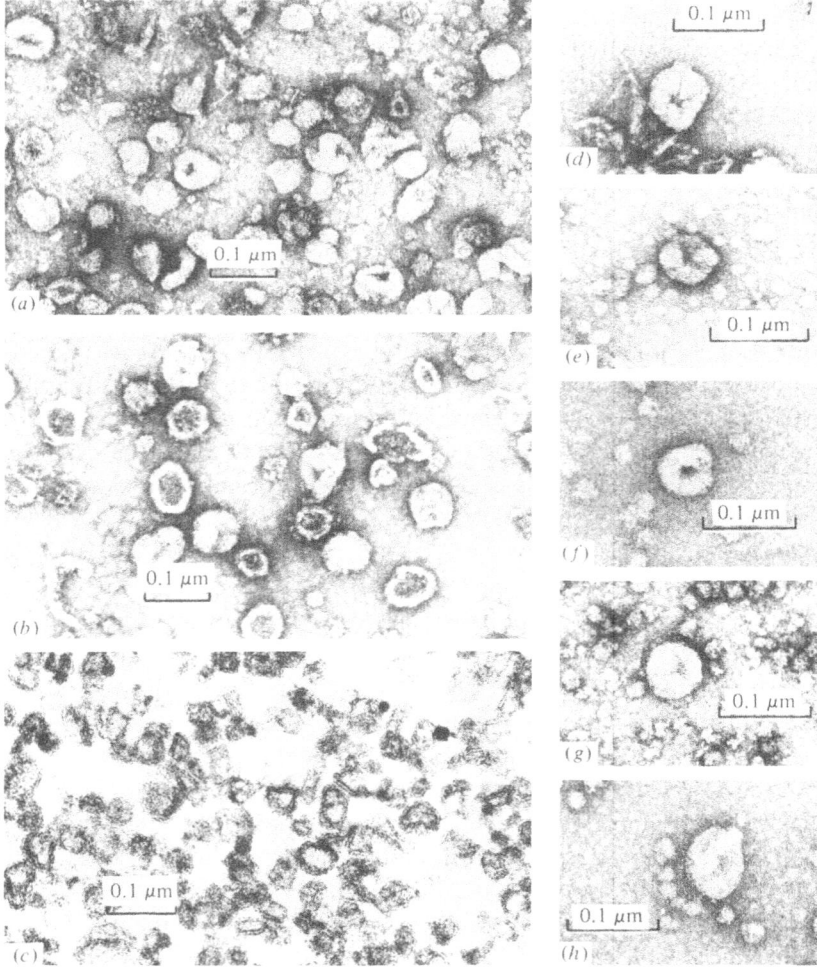

Fig. 7.4. Chitosomes after treatment with a high concentration of trypsin (0.8–1.0 mg ml⁻¹). Negatively stained samples. (*a*) mycelium of *Mucor rouxii*; (*b*) *Allomyces macrogynus*; (*c*) *Saccharomyces cerevisiae*; (*d*) *Neurospora crassa*; (*e*) *Agaricus bisporus*. From Bartnicki-Garcia *et al.* (1978*b*).

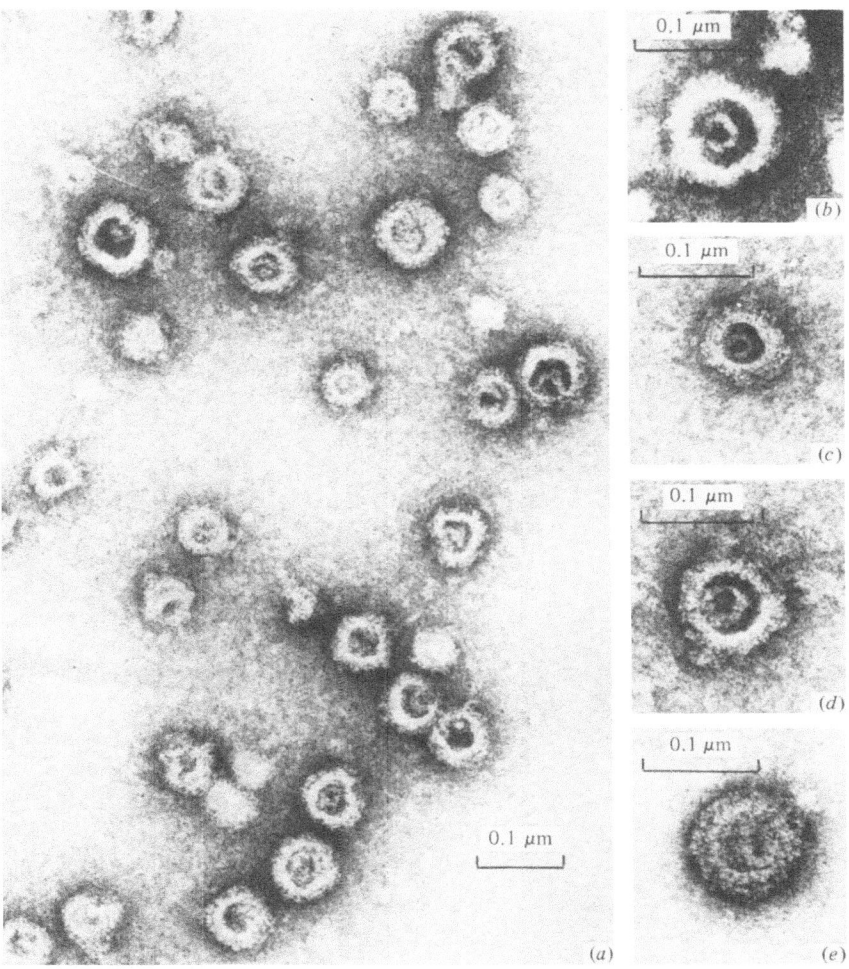

Fig. 7.5. Chitin microfibrils synthesized *in vitro* by chitosome preparations. Negatively stained samples from: (*a*) yeast cells of *Mucor rouxii*; (*b*) mycelium of *M. rouxii*; (*c*) budding cells of *Saccharomyces cerevisiae*; (*d*) mycelium of *Allomyces macrogynus*. From Bartnicki-Garcia *et al.* (1978*b*).

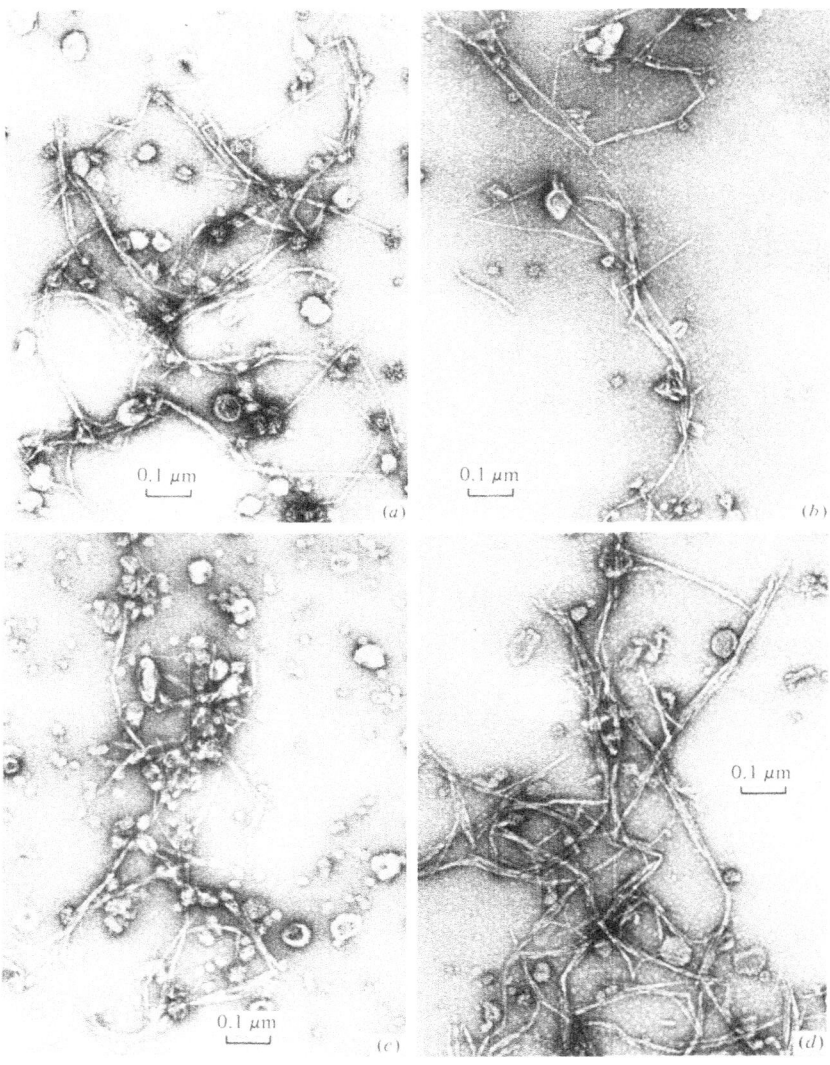

Fig. 7.6. Physical connection between chitosomes and microfibrils. (a, b) extended microfibrils arising from partly opened chitosomes. Note that the straight microfibrils are continuous with a microfibrillar coil – 'fibroid' – inside the chitosomes, (c) Naked 'fibroid' with its associated microfibril; the chitosome shell or membrane is usually lost during chitin synthesis. Figs. a–c are from chitosome samples from yeast cells of *M. rouxii* incubated with substrate and activators. Other examples of fibroids in: (d) *Saccharomyces cerevisiae*; (e) *Allomyces macrogynus*; (f) *Neurospora crassa*; (g) mycelium of *Mucor rouxii*; (h) *Agaricus bisporus*. From Bartnicki-Garcia *et al*. (1978b).

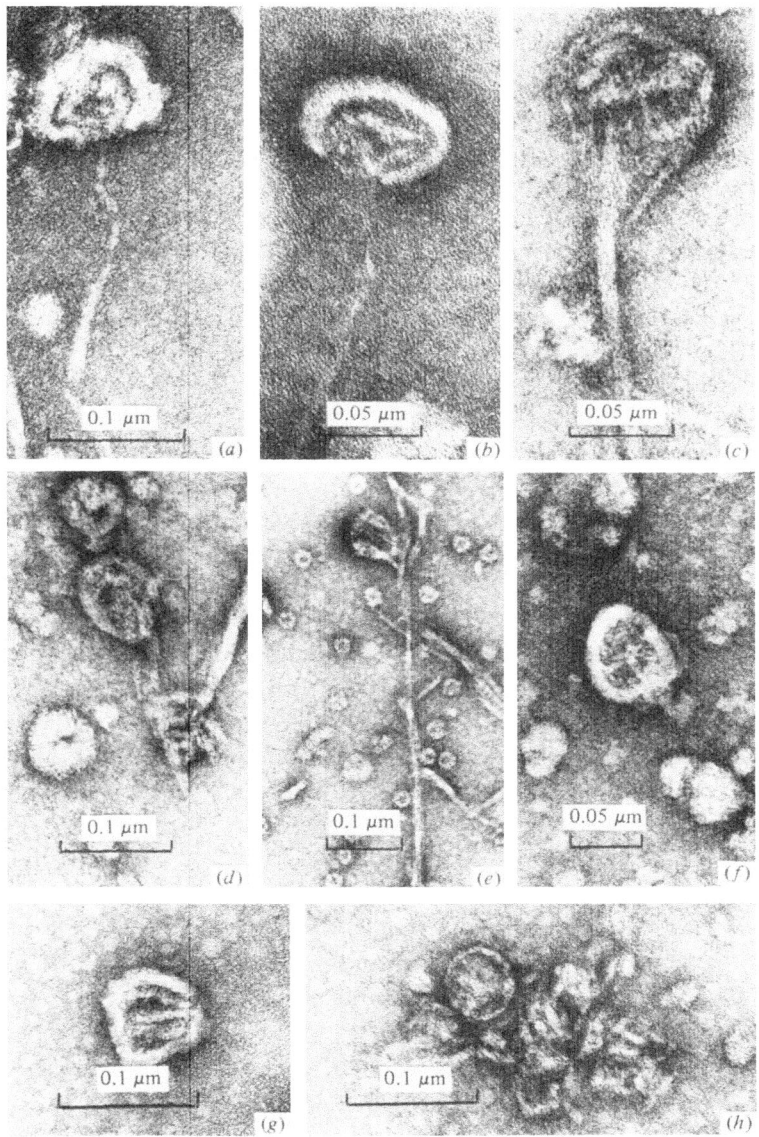

ably the result of confinement of chitin synthesis to the chitosome lumen. In most images, fibroid coils are continuous with a long extended microfibril. Most chitosome shells disintegrate during fibrillogenesis leaving behind naked fibroids (Fig. 7.6c). Most revealing, though more rarely found, are fibroid coils enveloped by a recognizable piece(s) of chitosome shell (Figs. 7.6a, b). Images of microfibrils connected to fibroid coils within or without pieces of the original chitosome shell were found in chitosome samples from all fungi (Fig. 7.6).

Zymogenicity

Cabib & Farkas (1971) discovered a latent, zymogenic form of chitin synthetase in crude preparations from *Saccharomyces*. The zymogen was activated by limited proteolysis. We have found evidence of zymogenicity in chitosomal preparations of all fungi examined in that treatment with protease(s) caused a marked rise in the activity of chitin synthetase. The chitin synthetase zymogens from the various fungi responded differently to neutral or acid proteases. The zymogen from *Mucor rouxii* was activated to higher levels by Rennilase (a crude preparation of acid protease from *Mucor miehei* Cooney & Emerson) than by trypsin; conversely, the zymogen of *Neurospora crassa* and *Agaricus bisporus* responded better to trypsin than rennilase. There was considerable variation in the ratio of zymogen to active enzyme in the purified chitosome pools, but these differences are probably the result of uncontrolled activation by endogenous proteases released during cell breakage. Cabib & Farkas (1971) found that in crude particulate preparations from *Saccharomyces* spp., chitin synthetase is in an essentially inactive form. Likewise, in preparations of chitosomes from yeast cells of *M. rouxii* and *S. cerevisiae*, more than 90 % of the chitin synthetase is in the zymogen from (Ruiz-Herrera *et al.*, 1977; Bartnicki-Garcia *et al.*, 1978b). Accordingly, we suggest that fungal chitin synthetase is produced as a zymogen* and packaged into a chitosome for delivery to the cell surface.

Chitosome dissociation

Chitosomes are the smallest entities (most of them measure 40–70 nm in diameter) with chitin synthetase activity that have been isolated from cell-free extracts of *Mucor rouxii*; so far, we have found no

---

* The term zymogen is used in its broadest sense to indicate a naturally occurring inactive enzyme precursor. The exact manner by which proteases activate the chitin synthetase zymogen has yet to be elucidated.

evidence for chitin synthetase in the soluble portion of cell-free extracts (Table 7.1).

Electron micrographs of a single microfibril 'arising' from an individual chitosome (Ruiz-Herrera, *et al.*, 1975; Bracker *et al.*, 1976) indicate that the chitosome is a large enzyme complex equipped with as many chitin synthetase units as are necessary to synthesize the many chains that constitute a microfibril. Presumably, each chitin chain is made by a separate chitin synthetase unit and the nascent chains collectively crystallize into a microfibril as they are synthesized. Accordingly, we thought it would be possible to dissociate a chitosome into its chitin-synthesizing units (chitosome subunits). Following the success of Gooday & De Rousett-Hall (1975) and Duran, Bowers & Cabib (1976) in using digitonin to 'solubilize' chitin synthetase from crude preparations from *Coprinus cinereus* (Schaeff.) Fr. and *Saccharomyces cerevisiae*, respectively, we were able to dissociate chitosomes from *Mucor rouxii* with 0.5 % digitonin (Ruiz-Herrera *et al.*, 1976; Bartnicki-Garcia, Bracker & Ruiz-Herrera, 1977). For this purpose, chitosomes were first concentrated by precipitation with polylysine and then treated with 0.5 % digitonin and centrifuged at $160\,000\,g$. Despite substantial losses in activity, we recovered 15.9 % of the initial activity in the $160\,000\,g$ supernatant. Electron microscopy of this supernatant showed that it contained particles *c.* 7–12 nm in diameter. Upon centrifugation in a calibrated sucrose density gradient, the particles with chitin synthetase activity – chitosome subunits – separated as a sharp band with a sedimentation coefficient of 16 S (equivalent to that of a spherical protein of M.W. *c.* 500 000).

After dissociation with digitonin, the stability of chitin synthetase was

Table 7.1. *Distribution of chitin synthetase activity in cell-free extracts of yeast cells of* Mucor rouxii

	Separated by		Particle size (nm)	Total chitin synthetase[a] (%)
Fraction	Sedimentation	Gel filtration		
Cell walls	2 000 g pellet	–	10^3–10^4	5–10
Mixed membranes	54 000 g pellet	–	10^2–10^3	40–60
Miniorganelles	54 000 g	Excluded	>10^1–10^3	30–50
Cytosol	supernatant	Included	<10^1	0

[a] Percentage range in a number of fractionations.

drastically reduced (in contrast, purified chitosomes from yeast cells of *Mucor rouxii* are remarkably stable; we now have samples kept for 3 years at $-80\,°C$ with no apparent loss of activity). Because of the instability of chitosomal chitin synthetase after dissociation with digitonin, a much larger and more readily accessible source of dissociated chitin synthetase was needed. We found that chitin synthetase could be easily prepared in a dissociated form by treating mycelial walls of *M. rouxii* with 0.5 % digitonin. The chitin synthetase thus extracted was applied to a 5–20 % sucrose density gradient containing 0.25 % digitonin throughout the gradient. A single large peak of chitin synthetase was observed (Fig. 7.7) with the same sedimentation coefficient (16 S) and electron microscopic morphology (Fig. 7.8*a*) as that of a comparable sample prepared from purified chitosomes.

Reassembly of chitosome-like structure

To investigate the effect of digitonin removal on particle aggregation, we centrifuged preparations of digitonin-dissociated chitin synthetase (from either purified chitosomes or from mycelial walls) through a sucrose density gradient (5–20 %) *without* digitonin. In such gradients, one or more additional bands of chitin synthetase were detected deeper into the gradient (Fig. 7.7). Electron microscopy of these bands showed that removal of digitonin had caused a reaggregation of the 16 S subunits into planar structures (Fig. 7.8*b*), many of which had a distinct curvature and resembled large chitosome fragments. Occasionally, a seemingly complete vesicle was formed by reaggregation.

Upon incubation with UDP-GlcNAc, GlcNAc, and an activating protease, chitin microfibrils were produced by the reaggregated chitosome-like structures. These microfibrils (Fig. 7.8*d*) were, in general, shorter than those obtained from isolated intact chitosomes.

From a pooled sample of dissociated and subsequently reaggregated chitin synthetase, isolated from mycelial walls, a tiny pellet was separated by centrifugation at $196\,000g$ (R_{avg}) for 1 h. This was fixed, embedded and sectioned. The reaggregated structures exhibited a distinct tripartite membranous character (Fig. 7.8*c*) similar to that of the chitosome shell in thin section. Frequently, the reaggregated structures showed a distinct vesiculoid profile.

Clearly, the 16 S chitin synthetase units, isolated from chitosomes or from mycelial walls, have the intrinsic capacity to associate with one another *in vitro* and reassemble into vesiculoid structures capable of synthesizing microfibrils.

Reality and role of chitosomes

Are chitosomes *bona fide* subcellular components or artifical structures produced by fragmentation and vesiculation of other cell components? We do not have decisive evidence to show, unequivocally, chitosomes in the cytoplasm of whole cells. However, our evidence supports the existence of chitosomes *in vivo*. A crucial piece of supportive evidence is the fact that thin sections of *Mucor rouxii* and other fungi show cytoplasmic microvesicles that are morphological correlates of isolated chitosomes (Bracker *et al.,* 1976).

The main reservation one can voice against chitosomes being cellular organelles is to regard them as fragments of other larger cellular

Fig. 7.7. Sedimentation pattern of digitonin-dissociated chitin synthetase in 5–20 % sucrose gradients with (solid line) and without (broken line) 0.25 % digitonin. Preparations of chitin synthetase isolated from mycelial walls of *Mucor rouxii* by treatment with 0.5 % digitonin were applied to the gradient (35 ml) and centrifuged at 95 000 *g* for 23 h. The gradient was fractionated into 1 ml fractions and assayed for chitin synthetase.

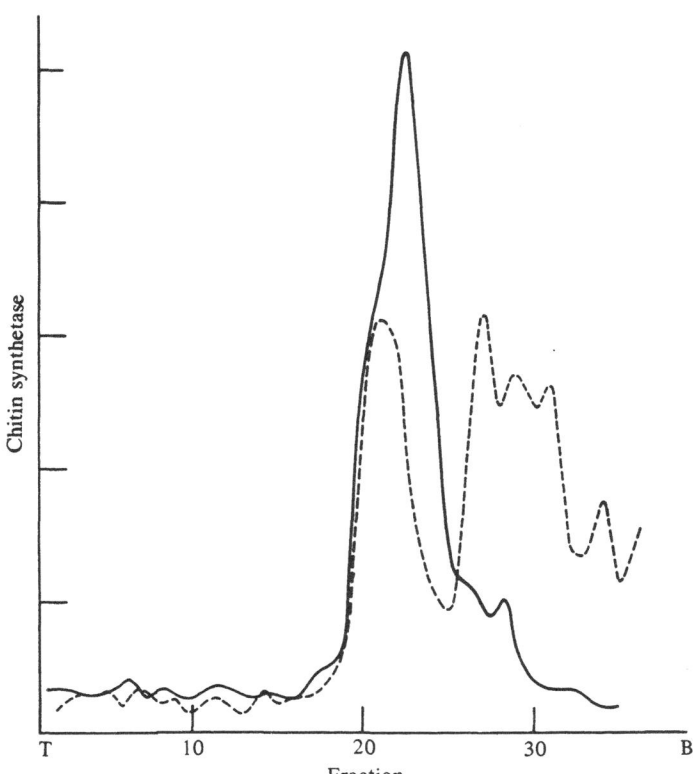

Fig. 7.8. Reassembly of chitosome-like structures from digitonin-dissociated chitin synthetase. (*a*) Appearance of peak sample of digitonin-dissociated chitin synthetase purified by sedimentation through a sucrose gradient containing digitonin similar to that in Fig. 7.7. These 16 S particles measure 7–12 nm. (*b*) Reaggregated structures produced by sedimenting digitonin-dissociated enzyme on a sucrose gradient without digitonin. This sample corresponds to fraction 29 in Fig. 7.7. Note that some of the aggregates have a vesiculoid appearance, reminiscent of chitosomes. (*c*) Pellet of reaggregated material (fractions 28–31, Fig. 7.7) fixed with glutaraldehyde + OsO$_4$, embedded and thin-sectioned. Note the tripartite profiles, some of which delimit vesiculoid structures. (*d*) Microfibrils produced by reaggregated structures from (*b*) incubated with UDP-GlcNAc and activators (see Bartnicki-Garcia *et al.*, 1977).

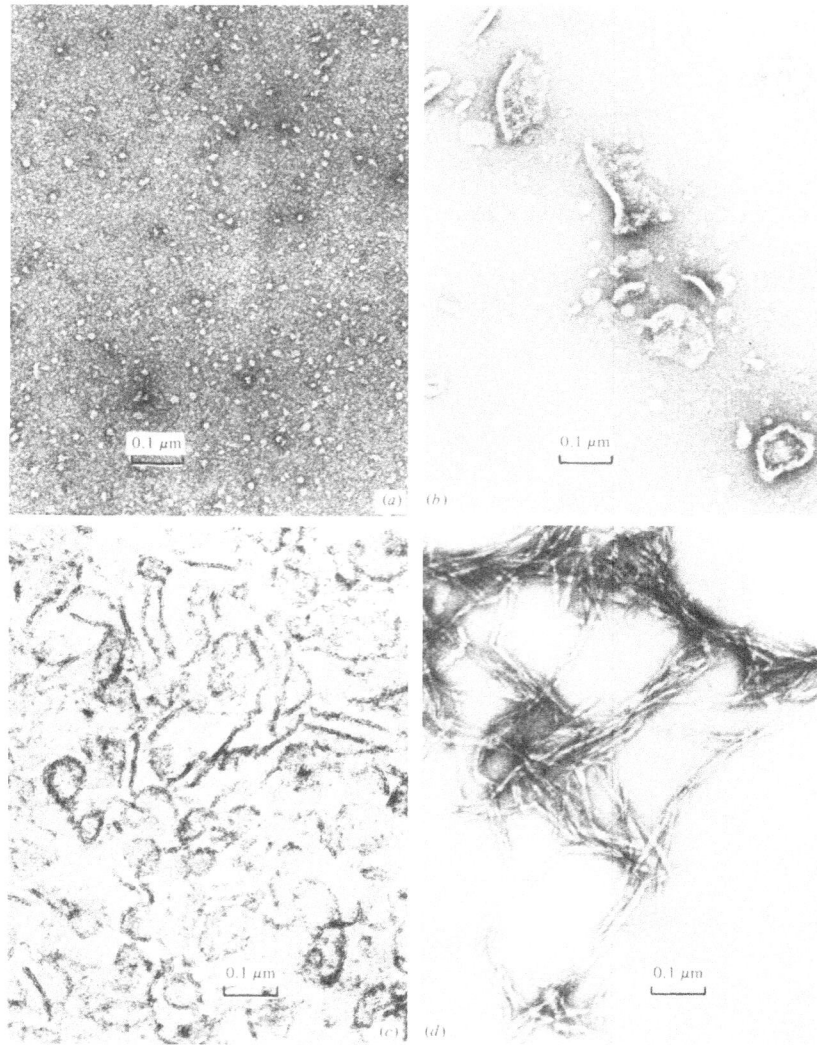

structures produced by the harsh ballistic disruption that has been routinely employed to prepare chitosomes (Bracker *et al.*, 1976; Ruiz-Herrera *et al.*, 1977). However, milder breakage procedures, e.g. osmotic lysis of either *Mucor rouxii* protoplasts or the slime mutant of *Neurospora crassa*, or low shear disruption of mycelium of *M. rouxii* with a razor-blade chopper in a hypertonic medium, yielded populations of chitosomes of the same magnitude as those obtained by homogenization with glass beads in the MSK homogenizer (S. Bartnicki-Garcia, unpublished results). Thus, it seems improbable that chitosomes are simply fragments of larger organelles. Furthermore, the sedimentation profiles (Fig. 7.2) indicate that chitosomes comprise a distinct and rather uniform population of particles. There is little or no evidence for a continuum in particle size as might be expected if chitosomes were merely fragments of larger membranous structures. The latter sediment at, or near the bottom of these gradients where, significantly, there is little chitin synthetase activity.

Relative abundance of chitosomes

How much of the total chitin synthetase activity is in the form of chitosomes? An accurate estimation of the proportion of the total chitin synthetase of the cell found in chitosomes is plagued with technical difficulties of two major kinds: loss of enzymatic activity during fractionation and incomplete recovery of chitosomes. Differential centrifugation studies of cell-free extracts of *Mucor rouxii*, and other fungi, reveal chitin synthetase activity in all major particulate fractions, cell walls, mixed membranes, and miniorganelles (Table 7.1). We know that chitin synthetase in the miniorganelle fraction is essentially, if not entirely, in the form of chitosomes. The structural state of the chitin synthetase associated with the cell-wall fraction is not known. Roughly one-half of the chitin synthetase activity in the mixed membrane fraction is in the form of chitosomes. We estimated that about 30–50 % of the chitin synthetase activity in a cell-free extract can be recovered in the miniorganelle fraction (54 000 g supernatant) and that almost all of it is in the form of chitosomes. This, however, is a minimum estimate; many chitosomes are excluded because they sediment in the 54 000 g pellet. These can be recovered by resuspending the pellet and centrifuging it on a sucrose density gradient. Accordingly, we have estimated that *at least* 60–70 % of the total activity in yeast cells of *M. rouxii* is in the form of chitosomes. The remaining activity in the mixed membrane fraction may be in endoplasmic reticulum, multivesicular bodies, and/or in plasma membrane vesicles.

Role of chitosomes

The existence of exquisitely localized, spatial patterns of cell wall construction has been clearly demonstrated by autoradiography, e.g., in hyphal tips (Bartnicki-Garcia & Lippman, 1969; Katz & Rosenberger, 1970; Gooday, 1971) and hyphal septa (Galun, 1972; Hunsley & Gooday; 1974; van Der Valk & Wessels, 1977). Electron microscopic studies have shown repeatedly an accumulation of vesicles underlying areas of actively growing walls (e.g., hyphal tips, septa) (Girbardt, 1969; Grove & Bracker, 1970; Grove, Bracker & Morré, 1970). More than one kind of vesicle, distinguished by size and electron density, participate in the process of wall growth. With this knowledge, a unitary hypothesis of cell wall growth was proposed in which the growth of a cell wall is viewed as the summation of submicroscopic growth units; each unit corresponding to the discharge of one vesicle or a minimum combination of different vesicles (Bartnicki-Garcia, 1973). Our work on chitosomes leads us to conclude that the skeletal part of the wall, i.e., chitin microfibrils, are made by microvesicles – or, more specifically, chitosomes. A chitosome delivers to the cell surface an organized packet of chitin synthetase; the packet contains as many chitin synthetase molecules as are necessary to synthesize the many chains that comprise a single microfibril. As these chains grow, they assemble collectively into a long ribbon by the intrinsic forces of crystallization of chitin molecules.

Origin of chitosomes

The capacity of reaggregation manifested by chitosome subunits (see p. 160) suggests that chitosomes are formed independently via self-assembly. On the other hand, the higher level of structural organization, revealed by digestion with high concentrations of trypsin, suggests assembly in conjunction with other membranous components. Accordingly, we should also consider assembly within specialized areas of the endoplasmic reticulum from which isolated chitosomes may later bud off. Thirdly, if the microvesicles found inside some multivesicular bodies (Bracker *et al.*, 1976) are proven to be chitosomes, self-assembly in the lumen of the enclosing macrovesicle is another possibility.

Microfibril formation

How does the chitosome make microfibrils *in vivo*? In the chitosome, chitin synthetase is present largely, if not entirely, as zymogen. Purified chitosomes do not undergo self-activation (except after several days of storage at 22 °C; unpublished results). Presumably, activation requires contact with protease(s) as the chitosome reaches its destination at the cell surface. Exactly what happens to the chitosome when it reaches the cell surface is, at the moment, a matter of conjecture. The chitosome may simply fuse with the plasma membrane, in a similar fashion to the larger vesicles of fungal cells (Pinto Da Silva & Nogueira, 1977) and thus create an organized patch of chitin-synthetase. Alternatively, in extraordinary fashion, the chitosome may traverse the plasma membrane and operate in the so-called periplasmic space.

These different possibilities for the origin and fate of chitosomes are illustrated in Fig. 7.9. We emphasize that these possibilities are at the

Fig. 7.9. Genesis and fate of chitosomes. A diagrammatic representation of the role of chitosomes as cytoplasmic conveyors of chitin synthetase zymogen to the cell surface. Three plausible origins for chitosomes and two hypothetical modes of operation during fibrillogenesis are illustrated.

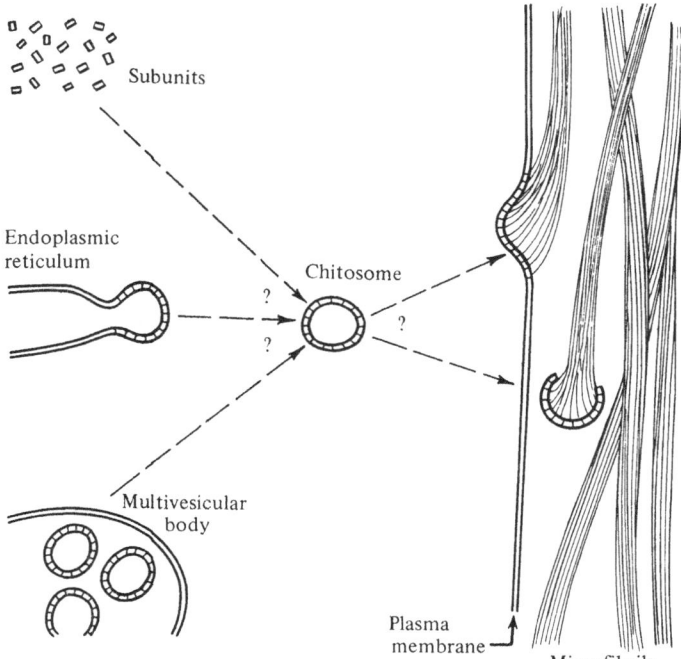

moment pure speculations, embraced with widely differing enthusiasm by the individual authors of this article.

The findings of Duran, Bowers & Cabib (1975) support the presence of chitin synthetase zymogen in the plasma membrane of *Saccharomyces cerevisiae*. On the other hand, there is a significant amount of chitin synthetase associated with mycelial walls (McMurrough *et al.*, 1971), and to a lesser extent, yeast cell walls of *Mucor rouxii* (unpublished results). This enzyme which does not wash off readily, may represent chitin synthetase, at its final site of operation affiliated with the wall itself.

Conclusions

(1) *Ubiquity.* Chitosomes as originally described for *Mucor rouxii* can be isolated from other chitinous fungi. These microvesicular structures account for a major portion of the total chitin synthetase activity

(2) *Self-assembly.* Chitosomes of *M. rouxii* can be dissociated with digitonin into subunits (16 S); upon removal of digitonin, many of the chitosome subunits reaggregate into vesiculoid structures reminiscent of chitosomes. Accordingly, we believe that self-assembly plays a role in the organization of chitin synthetase molecules into chitosomes.

(3) *Role in fibrillogenesis.* We propose that the chitosome has a key role in cell wall construction, and hence, morphogenesis: it is the microvesicular vehicle by which the cell delivers, to specific sites on the cell surface, individual, organized packets of chitin synthetase, each packet being responsible for the synthesis of one microfibril.

Acknowledgements. The experimental work reported herein was supported in part by research grants from the National Institutes of Health (AI-05540), National Science Foundation Latin America–Mexico Cooperative Program (OIP-7508378), and Consejo Nacional de Ciencia y Tecnologia of Mexico (847 and PNCB 071). Journal paper No. 7402 of the Purdue University Agricultural Experiment Station.

References

Aronson, J. M. (1965). The cell wall. In *The Fungi*, vol. 1, eds. G. C. Ainsworth & A. S. Sussman, pp. 49–76. New York, London: Academic Press.

Bartnicki-Garcia, S. (1968). Cell wall chemistry, morphogenesis, and taxonomy of fungi. *Annual Review of Microbiology*, **22**, 87–108.

Bartnicki-Garcia, S. (1973). Fundamental aspects of hyphal morphogenesis. In *Microbial Differentiation* (23rd Symposium of the Society for General Microbiology) eds.

J. M. Ashworth and J. E. Smith, pp. 245–67. Cambridge University Press.

Bartnicki-Garcia, S. & Lippman, E. (1969). Fungal morphogenesis: cell wall construction in *Mucor rouxii. Science,* **165,** 302–4.

Bartnicki-Garcia, S., Bracker, C. E. & Ruiz-Herrera, J. (1977). Reassembly of solubilized chitin synthetase into functional structures resembling chitosomes. *Abstracts of the Second International Mycological Congress.* Tampa, Florida. p. 41.

Bartnicki-Garcia, S., Bracker, C. E. & Ruiz-Herrera, J. (1978*a*). Synthesis of chitin microfibrils *in vitro* by chitin synthetase particles, chitosomes, isolated from *Mucor rouxii.* Proceedings of the First International Conference on Chitin/Chitosan, eds. R. A. A. Muzzarelli & E. R. Pariser, pp. 450–63. Massachusetts Institute of Technology, Cambridge, Massachusetts.

Bartnicki-Garcia, S., Bracker, C. E., Reyes, E. & Ruiz-Herrera, J. (1978*b*). Isolation of chitosomes from taxonomically diverse fungi and synthesis of chitin microfibrils *in vitro. Experimental Mycology,* **2,** 173–92.

Bracker, C. E., Ruiz-Herrera, J. & Bartnicki-Garcia, S. (1976). Structure and transformation of chitin synthetase particles (chitosomes) during microfibrils synthesis *in vitro. Proceedings of the National Academy of Sciences, USA,* **73,** 4570–4.

Cabib, E. & Farkas, V. (1971). The control of morphogenesis: an enzymatic mechanism for the initiation of septum formation in yeast. *Proceedings of the National Academy of Sciences, USA,* **68,** 2052–6.

Camargo, E. P., Dietrich, C. P., Sonneborn, E. & Strominger, J. L. (1967). Biosynthesis of chitin in spores and growing cells of *Blastocladiella emersonii. Journal of Biological Chemistry,* **242,** 3121–8.

Duran, A., Bowers, B. & Cabib, E. (1975). The chitin synthetase zymogen is attached to the yeast plasma membrane. *Proceedings of the National Academy of Sciences, USA,* **72,** 3952–5.

Duran, A., Bowers, B. & Cabib, E. (1976). The chitin synthetase zymogen and its participation in the formation of yeast primary septum. *Federation Proceedings,* **35,** 1584.

Frey-Wyssling, A. & Mühlethaler, K. (1950). Der submikroskopische Feinbau von chitinzellwanden. *Vierteljahresschrift Naturforschende Gesellschaft in Zürich,* **95,** 45–52.

Galun, E. (1972). Morphogenesis of *Trichoderma:* Autoradiography of intact colonies labeled by [^{3}H]N-acetylglucosamine as a marker of new cell wall biosynthesis. *Archiv für Mikrobiologie,* **86,** 305–14.

Girbardt, M. (1969). Die Ultrastruktur der Apikalregion von Pilzhyphen. *Protoplasma,* **67,** 413–41.

Glaser, L. & Brown, D. H. (1957). The synthesis of chitin in cell-free extracts of *Neurospora crassa. Journal of Biological Chemistry,* **228,** 729–42.

Gooday, G. W. (1971). An autoradiographic study of hyphal growth of some fungi. *Journal of General Microbiology* **67,** 125–33.

Gooday, G. W. & de Rousset-Hall, A. (1975). Properties of chitin synthetase from *Coprinus cinereus. Journal of General Microbiology,* **89,** 137–45.

Grove, S. N. & Bracker, C. E. (1970). Protoplasmic organization of hyphal tips among fungi: vesicles and Spitzenkörper. *Journal of Bacteriology,* **104,** 989–1009.

Grove, S. N., Bracker, C. E. & Morré, D. J. (1970). An ultrastructural basis for hyphal tip growth in *Pythium ultimum. American Journal of Botany,* **57,** 245–66.

Hunsley, D. & Gooday, G. W. (1974). The structure and development of septa in *Neurospora crassa. Protoplasma,* **82,** 125–46.

Jan, Y. N. (1974). Properties and cellular localization of chitin synthetase in *Phycomyces blakesleeanus. Journal of Biological Chemistry,* **249,** 1973–9.

Katz, D. & Rosenberger, R. F. (1970). The utilization of galactose by an *Aspergillus nidulans* mutant lacking galactose phosphate-UDP glucose transferase and its relation to cell wall synthesis. *Archiv für Mikrobiologie,* **74,** 41–51.

Keller, F. A. & Cabib, E. (1971). Chitin and yeast budding. Properties of chitin synthetase from *Saccharomyces carlsbergensis. Journal of Biological Chemistry,* **246,** 160–6.

Lopez-Romero, E. & Ruiz-Herrera, J. (1976). Synthesis of chitin by particulate preparations from *Aspergillus flavus. Antonie van Leeuwenhoek. Journal of Microbiology and Serology,* **42,** 261–76.

Lopez-Romero, E., Ruiz-Herrera, J. & Bartnicki-Garcia, S. (1978). Purification and properties of an inhibitory protein of chitin synthetase from *Mucor rouxii. Biochimica et Biophysica Acta,* **525,** 338–45.

McMurrough, I. & Bartnicki-Garcia, S. (1971). Properties of a particulate chitin synthetase from *Mucor rouxii. Journal of Biological Chemistry,* **246,** 4008–16.

McMurrough, I. & Bartnicki-Garcia, S. (1973). Inhibition and activation of chitin synthesis by *Mucor rouxii* cell extracts. *Archives of Biochemistry and Biophysics,* **158,** 812–16.

McMurrough, I., Flores-Carreón, A. & Bartnicki-Garcia, S. (1971). Pathway of chitin synthesis and cellular localization of chitin synthetase in *Mucor rouxii. Journal of Biological Chemistry,* **246,** 3999–4007.

Peberdy, J. F. & Moore. P. M. (1975). Chitin synthetase in *Mortierella vinacea:* properties, cellular location and synthesis in growing cultures. *Journal of General Microbiology,* **90,** 228–36.

Pinto Da Silva, P. & Nogueira, M. L. (1977). Membrane fusion during secretion. A hypothesis based on electron microscope observation of *Phytophthora palmivora* zoospores during encystment. *Journal of Cell Biology,* **73,** 161–86.

Porter, C. A. & Jaworski, E. G. (1966). The synthesis of chitin by particulate preparations of *Allomyces macrogynus. Biochemistry,* **5,** 1149–54.

Ruiz-Herrera, J. & Bartnicki-Garcia, S. (1974). Synthesis of cell wall microfibrils *in vitro* by a 'soluble' chitin synthetase from *Mucor rouxii. Science,* **186,** 357–9.

Ruiz-Herrera, J. & Bartnicki-Garcia, S. (1976). Proteolytic activation and inactivation of chitin synthetase from *Mucor rouxii. Journal of General Microbiology,* **97,** 241–9.

Ruiz-Herrera, J., Sing, V. O., Van Der Woude, W. J. & Bartnicki-Garcia, S. (1975). Microfibril assembly by granules of chitin synthetase. *Proceedings of the National Academy of Sciences, USA,* **72,** 2706–10.

Ruiz-Herrera, J., Bartnicki-Garcia, S. & Bracker, C. E. (1976). Solubilizacion de la quitina sintetasa a partir de quitosomas de *Mucor rouxii. Abstracts of the XI Annual Meeting of the Mexican Society of Biochemistry.* Mazatlan, Mexico. p. 176.

Ruiz-Herrera, J., Lopez-Romero, E., & Bartnicki-Garcia, S. (1977). Properties of chitin synthetase in isolated chitosomes from yeast cells of *Mucor rouxii. Journal of Biological Chemistry,* **252,** 3338–43.

van Der Valk, P. & Wessels, J. G. H. (1977). Light and electron microscopic autoradiography of cell-wall regeneration by *Schizophyllum commune* protoplasts. *Acta Botanica Neerlandica,* **26,** 43–52.

8
Chitin and its degradation

J. L. STIRLING, G. A. COOK AND A. M. S. POPE*

Department of Biochemistry, Queen Elizabeth College, Campden Hill, London W8 7AH, UK
*Research Division, G. D. Searle and Company, Lane End Road, High Wycombe, Bucks HP12 4HL, UK

Introduction

The disposition of the major macromolecular components in fungal walls has been analysed in very few species. However, the gross analysis of a wide variety of fungal walls has shown that chitin has a very widespread occurrence (Bartnicki-Garcia, 1968). In the most intensively studied wall, that of *Neurospora crassa* Shear & Dodge, chitin is found in the innermost layer or primary wall as a continuous network of microfibrils (Hunsley & Burnett, 1970; Hunsley & Kay, 1976). A similar arrangement is thought to occur in *Aspergillus oryzae* (Ahlburg) Cohn and *Fusarium solani* (Martius) Saccardo (Skujins, Potgieter & Alexander, 1965), *Phycomyces blakesleeanus* Burgeff and *Aspergillus nidulans* (Eidam) Winter (Trinci, 1978) and *Schizophyllum commune* Fries (Hunsley & Burnett, 1970), and the occurrence of chitin in the primary wall of filamentous fungi may be a general phenomenon, except in the Oomycetes. For this reason, knowledge of the structure and metabolism of chitin is important for an understanding of the biogenesis of fungal walls and the modifications that they undergo during morphogenesis.

The biosynthesis of chitin is a subject of active research (see Chapters 7, 9 and 10), but the enzymes that modify and degrade chitin are also of interest to mycologists, since spore dispersal by dissolution of the fruiting bodies of certain basidiomycetes involves developmentally regulated production of lytic enzymes including chitinase (Iten & Matile, 1970). Moreover, it has been suggested that chitinases might be involved in modifying chitin structure during wall growth (Bartnicki-Garcia, 1973) and this has received active consideration by Polacheck & Rosenberger (1975) (also see Chapter 12). Purified chitinases from various sources have been used as specific reagents for structural analysis of fungal walls (Skujins *et al.*, 1965; Hunsley & Burnett, 1970; Hunsley & Kay, 1976).

Chitin is also a major structural component of arthropod cuticles, and consequently its biosynthesis and degradation is of enormous quantitative significance in the biosphere. Chitinolytic enzymes undoubtedly have an important function in nature, and one might expect that the structures and modes of action of these enzymes would be well understood. However, surprisingly little is known about their detailed biochemistry. In part this might be due to difficulties encountered in assaying the enzymes. Some difficulties are due to the insolubility of chitin in aqueous solutions, others arise because chitin is difficult to prepare in a chemically pure form and consequently there is variation in the structure and physical state of the substrate used to assay chitinases.

Chitinases from fungi have not as yet been studied in detail, but it is likely that they will share many properties of chitinases from other sources. In this chapter we have made a broad survey of the biochemistry of chitinases and the methods that have been developed for their assay. But first, in view of the importance of using chemically defined substrates for the assay of chitinolytic enzymes, a brief survey of the biochemistry of chitin is presented.

Chitin
Extraction and purification
Crustacean shells are the most commonly available source of chitin. Drastic methods are required for the removal of protein and inorganic constituents and involve protracted extractions with 2 M HCl at room temperature and 1 M NaOH at 100 °C. Chitin obtained in this way is frequently impure, judging from its nitrogen content, and further treatment with 0.5 % potassium permanganate at 60 °C, half-saturated sodium metabisulphite, hot water and alcohol (Foster & Webber, 1960; Jeuniaux, 1966) is necessary to remove contaminants. Further purification can be achieved by precipitating chitin from its solution in concentrated H_2SO_4 or HCl by adding it to cold water or 50 % aqueous ethanol with continuous stirring. The precipitate of colloidal chitin is then washed until neutral (Jeuniaux, 1966; Skujins, Pukite & McLaren, 1970).

Variations of this procedure have been described, but in general all methods for the purification of chitin are based on the progressive removal of contaminants, leaving chitin as an insoluble residue, rather than the specific extraction of chitin.

Chitin is reported to exist in a pure form in the fibres secreted by the diatom *Thalassiosira fluviatilis* (McLachlan, McInnes & Falk, 1965) but this has not been used as a substrate for chitinases. A relatively pure form of chitin is also found in the internal skeleton or 'pen' of the squid *Loligo*. A suspension of chitin can be prepared conveniently by decalcifying dried cuttlefish bones in 0.5 M HCl at 100 °C followed by treatment with 0.5 M NaOH at 100 °C and washing to neutrality (Jeuniaux, 1966).

Chitin prepared from cuttlefish seems to be more easily degraded by chitinase than chitin prepared from crab and lobster shells (Nord & Wadström, 1972). Fungal chitin may be prepared by the acid and alkali extraction methods described above, but alkali-resistant glucans must be removed by selective oxidative degradation in order to obtain purified chitin (Foster & Webber, 1960). The widely used strategy devised by Mahadevan & Tatum (1965) for the fractionation of fungal walls involves extractions with 2 M NaOH at room temperature for 16 h, 0.5 M H_2SO_4 at 90 °C for 16 h and 2 M NaOH at room temperature for 30 min to leave chitin as an insoluble residue. Despite the difficulties of extracting and purifying fungal chitin, it seems that it is identical to chitin prepared from animals (Michell & Scurfield, 1967; Rudall, 1969). The use of enzymes for the systematic removal of glucans, mannans and proteins from fungal walls to leave the network of microfibrils has been described by Hunsley & Burnett (1970) and Hunsley & Kay (1976), but this technique has not been used generally for the purification of chitin. Presently, chitin extracted from crustaceans is preferred to fungal chitin as a substrate for chitinases since it is obtained more readily and in a less degraded state.

Composition and covalent structure of chitin

There is good evidence that chitin is a homogeneous, unbranched polymer of 2-acetamido-2-deoxy-D-glucopyranose (*N*-acetyl-D-glucosamine) linked by $\beta(1–4)$ glycosidic bonds (Fig. 8.1); for a review see Foster & Webber (1960). However, chitin extracted from several sources (Hackman & Goldberg, 1974; Rudall & Kenchington, 1973) is partially deacetylated. This might result from the harsh extraction procedures used in the purification of chitin, but it is possible that the polysaccharide exists *in situ* in a partly deacetylated form. Extensively deacetylated chitin, chitosan, (Brimacombe & Webber,

1964) has been extracted from the walls of *Phycomyces blakesleeanus* (Kreger, 1954) and *Mucor rouxii* (Calmette) Wehm. (Bartnicki-Garcia & Nickerson, 1962).

It is important to establish whether chitin *in situ* is partially deacetylated, since the free amino groups of glucosamine in the polymer could be involved in the formation of covalent linkages to polypeptides or proteins, and thus play an important part in the interactions between chitin and other macromolecules. Evidence for such covalent linkages will probably come from the use of enzymes for controlled lysis of walls or the use of specific chemical probes, rather than the harsh chemical treatments already outlined.

Configuration of chitin molecules

X-ray diffraction analysis of chitin has shown that it contains highly ordered regions having one of three (a, β or γ) crystallographic forms. In a-chitin, the form that is found in fungi and arthropods, antiparallel poly-N-acetylglucosamine chains are stabilized by intrachain and interchain hydrogen bonds. The most satisfactory interpretation of the crystal structure of a-chitin was proposed by Carlström (1962) and refined by Ramakrishnan & Prasad (1972) and is largely consistent with the polarized infrared spectrum (Pearson, Marchessault & Liang, 1960). In Carlström's model, the polysaccharide chains have a two-fold screw axis (two sugar residues per turn) and are in the 'bent' configuration, with intramolecular hydrogen bonding between the C-3 hydroxyl group of each sugar and the pyranose ring oxygen atoms of the following sugar residues. Thus the sterochemical repeating unit in a-chitin is the disaccharide, N,N'-diacetylchitobiose, a feature of the chitin molecule that might be significant for the substrate specificity of chitinases. Interaction between chitin chains in a-chitin involves full $C=O \cdots HN$ interchain hydrogen bonding between the

Fig. 8.1. The covalent structure of chitin.

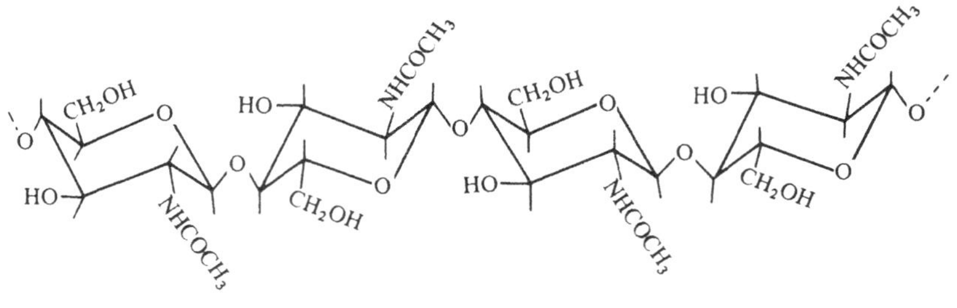

acetamido groups (Fig. 8.2) and is fully discussed by Minke & Blackwell (1978).

β-Chitin is found in *Loligo*, in tubes of Pogonophores and the chitinous spines of certain diatoms, and consists of parallel sheets of chitin molecules (Blackwell, 1969). γ-Chitin (two parallel chains and one antiparallel chain) is the form found in peritrophic membranes of insects (Rudall & Kenchington, 1973). Both β- and γ-chitin can be converted to the more stable α-structure (Rudall & Kenchington, 1973).

Fig. 8.2. Proposed structure for α-chitin. (a) *bc* projection; (b) *ab* projection. The shaded circles correspond to half-oxygens at the 2-positions on each residue. The proposed structure contains a statistical mixture of these two conformations for the CH_2OH group. Reproduced by kind permission of J. Blackwell.

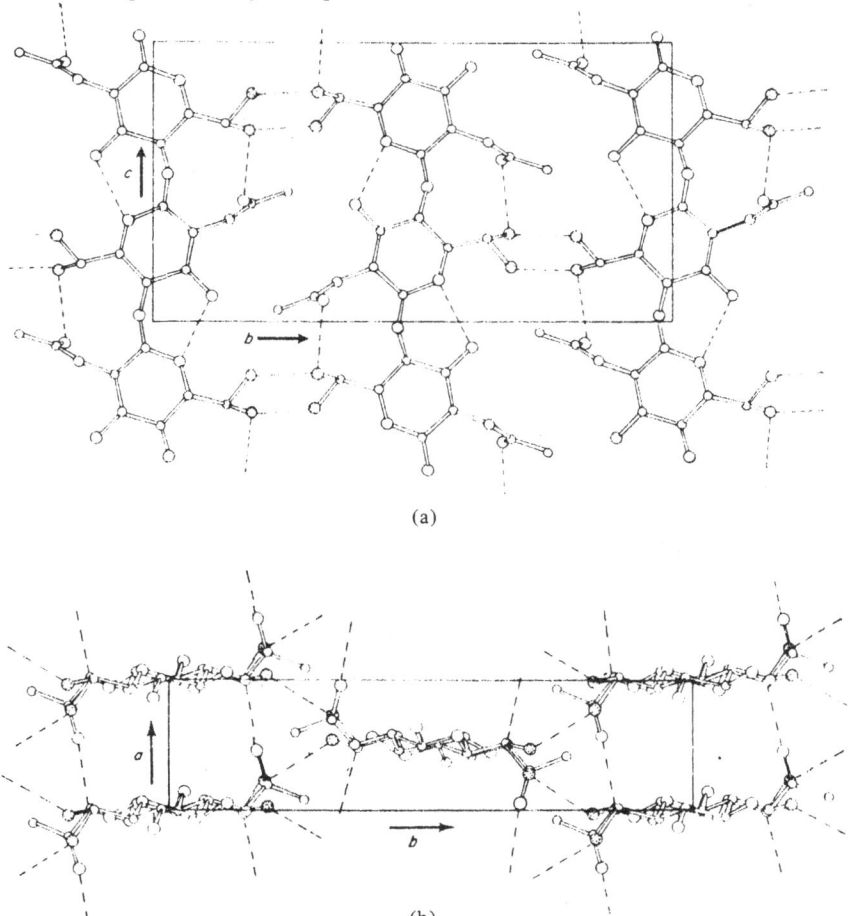

(a)

(b)

Interactions between chitin and other macromolecules

Chitin in insect cuticle exists in the form of chitin–protein complexes. X-ray analysis of insect cuticles has revealed an axial periodicity in the chitin fibrils of 3.1 nm. This period corresponds to the length of a hexasaccharide, and is completely lost on removing protein by treatment with 5 % KOH (Rudall, 1969; Rudall & Kenchington, 1973). Electron microscopy of negatively stained strips of locust cuticle has revealed that this repeating structure is due to 'corpuscles' which pack in longitudinal rows along the chitin fibrils, their axial separation being close to 3.1 nm. The nature of the interaction between chitin and these proteins is not known, but may be analogous to the non-covalent interaction between lysozyme and chitin oligosaccharides (Rudall, 1969). The major proteins extracted from insect cuticles have molecular weights of less than 18 000 (Hackman, 1972). In fungi, glucans are the principal substances thought to be closely associated with chitin (Rudall, 1969). Datema, Wessels & van den Ende (1977) have produced evidence that is consistent with the existence of covalent links between *N*-deacetylated residues of chitin and proteins in the walls of *Mucor mucedo* Auct. Similar linkages exist in the walls of *Schizophyllum commune* (see Chapter 2). The demonstration of chitin–protein linkages is clearly of importance for a full understanding of the relations between chitin and other wall components and it will be interesting to see whether such linkages are of widespread occurrence in fungal walls.

Chitinases

The major and best-documented pathway for the degradation of chitin involves the sequential action of chitinase, an endo-*N*-acetyl-*β*-glucosaminidase, and chitin oligosaccharase, an exo-*N*-acetyl-*β*-glucosaminidase.

Distribution of chitinases

Chitinases (poly-*β*1–4-(2-acetamido-2-deoxy)-D-glucoside glucanohydrolase, EC. 3.2.1.14) are widely distributed in nature and have been detected in bacteria, fungi and plants, and in the digestive systems of coelenterates, nematodes, polychaetes, oligochaetes, molluscs and arthropods. In the vertebrates, chitinases are secreted by the pancreas and digestive mucosa of insectivorous fishes, amphibians and reptiles as well as by the gastric mucosa of some insectivorous birds and mammals (Jeuniaux, 1963). Chitinases are secreted by nematodes during hatching (Ward & Fairbairn, 1972) and by the integument of insects during moulting (Zielkowski & Spindler, 1978).

Sera from several species of mammals have chitinase, as distinct from muramidase, activity (Lunblad *et al.*, 1974). Chitinase has also been found in the gastric mucosa of the prosimian *Perodicticus potto* (Cornelius, Dandrifosse & Jeuniaux, 1976).

The chitinolytic system of streptomycetes has been studied most intensively, but chitinase activity can also be induced in many other micro-organisms by growing them on a medium containing chitin as a carbon source. A survey of seventy fungi and thirty bacteria, including streptomycetes, for chitinase production was made by Monreal & Reese (1969) who found that the bacteria *Serratia marcescens* Bizio and *Enterobacter liquefaciens* produced almost ten times the activity of the next highest producers *Aspergillus fumigatus* Fresenius and a streptomycete.

Specificity of chitinases

Chitinases are sometimes compared with cellulases, since they both act on $\beta(1-4)$-linked hexose polymers. Two types of cellulases, C_1 and C_x are distinguished by their activities towards highly ordered cellulose and the derivative, carboxymethylcellulose. Cellulase C_1, a $\beta(1-4)$-glucan cellobiosidase from *Trichoderma koningii* Oudemans (Halliwell & Griffin, 1973) and *Trichoderma viride* Persoon ex S. F. Gray (Berghem & Pettersson, 1973) degrades microcrystalline cellulose and is inactive towards CM-cellulose while cellulase C_x, an endo$\beta(1-4)$glucanase is active towards CM-cellulose and also participates in the hydrolysis of microcrystalline cellulose along with cellobiase (Berghem, Pettersson & Axiö-Fredriksson, 1975).

The complexities of chitinase attack on chitin have not been investigated in the same detail as the substrate specificities and modes of action of cellulases, but there are indications that chitinases, too, may exist as two types distinguished by their activities towards crystalline and swollen chitin. Monreal & Reese (1969) reported that the extracellular chitinolytic system of *Serratia marcescens* is composed of an endochitinase, a chitobiase and a factor Ch_1 required for the hydrolysis of crystalline chitin, although these components were not purified. Tiunova *et al.* (1976*b*), using ground demineralized crab shells and colloidal chitin as substrates, found that the ratio of chitinase activities varied by a factor of two in culture filtrates from nine strains of Actinomycetes. They interpreted this as evidence for the existence of Ch_1 and Ch_x chitinases analogous to the C_1 and C_x enzymes in the cellulase system, but the putative activities were not isolated. Chitinase from *Actinomyces kurssanovii* Preobrazhenskaya, Kudrina, Ryabova & Blinov in Gauze *et al.* was selected for further investigation; this organism was thought to secrete both Ch_1 and Ch_x activities into the medium (Tiunova, Pirieva & Feniksova, 1976a). Such direct comparison of cellulases and chitinases may be premature since there is no evidence that the Ch_1 chitinase component from these sources is an exochitobiosidase. Indeed there has been no convincing demonstration of the participation of an exochitobiosidase from any source in the degradation of chitin.

Generally, chitinases are inactive towards cellulose or the deacetylated portions of chitosan, but they are active towards a variety of other chitin derivatives. It should be emphasized, however, that there have been few systematic studies of the substrate specificities of highly purified chitinases. Chitinases have a low specificity for the nature of the substituent at C-6 of the *N*-acetylglucosamine residues. The derivative, 6-*O*-hydroxyethylchitin (glycolchitin) is hydrolysed by chitinases from *Aspergillus niger* Van Tieghem (Otakara, 1961a), *Streptomyces* sp. (Nord & Wadström, 1972; Tominaga & Tsujisaka, 1976) and serum from goats and other animals (Lundblad *et al.*, 1974); another derivative, 6-*O*-hydroxypropyl chitin has been used for the assay of chitinase from *Pseudomonas* sp. (Takiguchi, Nagahata & Shimahara, 1976). In contrast, chitinases are highly specific for the nature of the *N*-acyl group at C-2. The snail chitinase studied by Karrer & White (1930) was found to be inactive towards chitosan and its *N*-formyl, *N*-propionyl, *N*-butyryl and *N*-benzoyl derivatives but was active towards *N*-acetylated chitosan. Chitinase from *A. niger* did not hydrolyse either chitosan or glycolchitosan (Otakara, 1964). Peptidoglycan, a polysaccharide from bacterial cell walls, in which β-linked residues of *N*-acetylglucosamine and muramic acid (3-*O*-lactyl-*N*-acetylglucosamine) alternate was not hydrolysed by a *Streptomyces* chitinase (Nord & Wadström, 1972).

The major end product of chitinase attack on native or colloidal chitin is the disaccharide *N*,*N'*-diacetylchitobiose (Jeuniaux, 1966). However there are many reports that appear to show that the monosaccharide *N*-acetylglucosamine is the end-product, although this is rarely observed with highly purified chitinases. It seems likely that in many cases the appearance of mono- and trisaccharides as products is due to contaminating chitobiase activity, but an exception to this may be the chitinase from *Ascaris suum* purified by Ward & Fairbairn (1972).

Kinetic properties of chitinases

The kinetics of chitinase attack on chitin have been investigated in some cases and the shape of the progress curve for the reaction has been found to depend not only on the source of the enzyme but also on the type of assay. Using colloidal chitin as a substrate, Skujins *et al.* (1970) found that the rate of release of *N*-acetylglucosamine by a *Streptomyces* chitinase was almost constant up to two hours of incubation. In fact the progress curves were slightly sigmoid, as might be expected for the activity of an endoglycosidase assayed in this way. Since *N*-acetylglucosamine was the end-product in this reaction, it is likely that chitobiase was present in their chitinase preparation. In the nephelometric assay described by Jeuniaux (1958) the *Streptomyces* chitinase gave linear progress curves. Recently, however, Molano, Duran & Cabib (1977), using tritiated chitin as a substrate for *Streptomyces* chitinase, found a marked departure from linearity in the progress curve that could not be explained by product inhibition or the denaturation of the enzymes under the conditions of assay.

Despite the difficulties of interpreting the non-ideal kinetic behaviour of chitinases, K_m values have been estimated. Chitinase from *Streptomyces antibioticus* (Waksman & Woodruff) Waksman & Henrici has a K_m value of 1.05 mg ml^{-1} (Jeuniaux, 1966) or 0.5 mM calculated as the concentration of polymeric *N*-acetylglucosamine. Molano *et al.* (1977) have reported a K_m value of about 5 mM for the chitinase from *Streptomyces griseus* (Krainsky) Waksman & Henrici when calculated as *N*-acetylglucosamine. The significance of these values is doubtful however, since the Michaelis–Menten equation is not strictly applicable to enzymes acting on insoluble substrates (Skujins, Pukite & McLaren, 1973).

The pH optimum for chitinases is usually in the range pH 4.5 to 6.5 but an exception to this is the enzyme from *Serratia marcescens* which has an optimal activity between pH 8.5 and 9.0 (Lysenko, 1976). Buffer constituents are known to inhibit the activity of some glycosidases, and in view of the requirement for *N*-acetylation of the substrate it may be a sensible precaution to avoid the use of acetate buffers unless it has been established that they do not exert inhibitory effects.

Chitinases do not require cofactors, and in this respect they resemble most other glycosidases. Several microbial chitinases are inhibited by

Na^+, Ca^{2+}, Mg^{2+}, Cu^{2+} and Hg^{2+} (Jeuniaux, 1966; Skujins *et al.*, 1970; Reisert, 1972; Tominaga & Tsujisaka, 1976) but *p*-chloromercuribenzoate has no effect on the chitinases from *Streptomyces antibioticus* (Jeuniaux, 1966) or *Streptomyces orientalis* McCormick, Stark, Pittenger (R.C.), Pittenger (J.M.) & McGuire (Tominaga & Tsujisaka, 1976). In contrast, the chitinases from *Locusta* and *Drosophila* are not inhibited by $CuSO_4$ or $HgCl_2$ (Zielkowski & Spindler, 1978). *N,N'*-diacetyl chitobiose (2 mM) inhibited the chitinase from *Streptomyces griseus* by 59 % (Molano *et al.*, 1977) but free *N*-acetylglucosamine (1.5 mM) had no effect on the activity of chitinase from *Chytriomyces halinus* Karling (Reisert, 1972). Melanins have been reported to inhibit the action of chitinases on fungal cell walls (Kuo & Alexander, 1967; Bull, 1970). The chitinase from *Manduca* larvae (Bade & Stinson, 1976) is probably activated by limited proteolysis of an inactive zymogen, but it is not known whether this type of activation is restricted to insects or is of more general occurrence.

Physical properties of chitinases

The molecular weight of chitinases from a variety of sources has been estimated using gel filtration or ultracentrifugation. Values of 30 000 (Jeuniaux, 1959), 29 000 (Skujins *et al.*, 1970), 33 000 for form I and 25 000 for form II (Tominaga & Tsujisaka 1976) have been obtained for *Streptomyces* chitinases. Chitinase from *Serratia marcescens* has a molecular weight of 36 000 (Lysenko, 1976), and the enzyme from *Phycomyces blakesleeanus* has a molecular weight of 30 000 (Cohen, 1974). It is not known whether these chitinases are monomeric or composed of subunits.

The charge characteristics have been investigated by studying their behaviour on ion-exchange resins, by electrophoresis and isoelectric focussing. Chitinases from a Streptomycete have been separated into three forms which migrated towards the cathode on electrophoresis in agar at pH 8.2. These forms each had the same specific activity but had a synergistic effect on chitin hydrolysis when combined (Jeuniaux 1957, 1959). Two chitinases from *Streptomyces griseus* with different electrophoretic mobilities but the same specificity were identified by electrophoresis in starch at pH 7.0 (Berger & Reynolds, 1958). One component migrated towards the anode and one towards the cathode, and this latter form also migrated towards the cathode at pH 8.5. Chitinases I and II from *Streptomyces orientalis* have pI values of pH 8.8 and 8.65 respectively, as determined by isoelectric focussing (Tominaga

& Tsujisaka, 1976). There was only one band of chitinase activity from the *Streptomyces* sp. investigated by Skujins *et al.* (1970) when electrophoresis was performed on cellulose acetate. This enzyme was not retained on DEAE-cellulose at pH 8.9 indicating that it, too, is a basic protein.

The existence of multiple forms of chitinase differing in their molecular weights (Tominaga & Tsujisaka, 1976) or net charge (Berger & Reynolds, 1958; Jeuniaux, 1959) have been best documented in streptomycetes, but they have also been found in beans (Powning & Irzykiewicz, 1965) where two forms with similar activities have been separated by chromatography on DEAE-cellulose. The structural relationships and biological significance of these multiple forms of chitinases has not been established.

Chitinases seem to be fairly stable at temperatures below 40 °C unless they are incubated at extremes of pH. Tominaga & Tsujisaka (1976) have shown that both chitinases (I and II) from *Streptomyces orientalis* were stable when heated at 40 °C for 3 h at pH values between 5.5 and 8.0, but outside this range there was progressive loss of activity towards extremes of pH. Chitinase from a *Streptomyces* sp. (Skujins *et al.* 1970) was rapidly inactivated by heating it at 65 °C. The half-life of *Lycoperdon pyriforme* (Schaeff.) Pers. chitinase was 15 min when heated at 51 °C (Tracey, 1955).

Purification of chitinases

A variety of techniques have been used in the purification of *Streptomyces* chitinases; $(NH_4)_2SO_4$ precipitation was used by Jeuniaux (1959), Lloyd, Noveroske & Lockwood (1965), Skujins *et al.* (1970) and Cabib & Bowers (1971); alcohol precipitation by Lloyd *et al.* (1965); hydroxyapatite chromatography by Skujins *et al.* (1970); DEAE-cellulose chromatography by Skujins *et al.* (1970) and SP-Sephadex chromatography by Tominaga & Tsujisaka (1976); and gel filtration by Skujins *et al.* (1970) and Tominaga & Tsujisaka (1976). Affinity chromatography is potentially the most powerful method for chitinase purification and this has been used with varying degrees of success by Jeuniaux (1959) and Lloyd *et al.* (1965), Noble & Sturgeon (1968) and Cabib & Bowers (1971). An improved method using regenerated chitin has been developed by Molano *et al.* (1977). It seems to be possible to obtain pure chitinases from streptomycetes without much difficulty. Jeuniaux (1959, 1966) obtained a preparation which he judged to be pure by ultracentrifugation and electrophoresis after a

9-fold purification; Lloyd *et al.* (1965) purified the enzyme 7.5-fold. Tominaga & Tsujisaka (1976) purified chitinases I and II from *Streptomyces orientales* by 7-fold and 11-fold respectively and Molano *et al.* (1977) purified chitinase from *Streptomyces griseus* by at least 20-fold. Chitinase from *Aspergillus niger* was purified 40-fold by a combination of $(NH_4)_2SO_4$ precipitation, adsorption and elution from calcium phosphate gel and column chromatography on hydroxylapatite and was judged to be nearly homogeneous by ultracentrifugation (Otakara, 1961*b*). Chitinase from *Serratia marcescens* was purified 5-fold by a combination of $(NH_4)_2SO_4$ precipitation, DEAE-cellulose chromatography and gel filtration on Sephadex G-75 (Lysenko, 1976).

Assay methods for chitinases
The methods used for the assay of chitinases can be classified into four main groups. These are viscometric, turbidometric, radiometric and those depending on the chemical detection of end-products.

Viscometric assays. Although native and colloidal chitin are insoluble in aqueous solutions, several soluble derivatives have been prepared that form viscous solutions. Random depolymerization by chitinases rapidly decreases the viscosity of the substrate solution and this can be monitored conveniently using a viscometer. The rate of decrease in viscosity is a function of enzyme concentration (Jeuniaux, 1966).

Chitosan acetate (Tracey, 1955), carboxymethylchitin (Hultin, 1955) glycolchitin (Otakara, 1961*a*; Kooiman, 1964, Takeda *et al.*, 1966; Nord & Wadström, 1972) and 6-hydroxypropylchitin (Takiguchi, *et al.*, 1976) have been used as substrates for the viscometric assay of chitinases. These assays have the advantage that they are very sensitive monitors of the primary attack of chitinases on the polymer, but they have several disadvantages. Viscometric assays have to be performed individually and consequently are unsuitable for screening large numbers of samples such as fractions of column eluates; assays based on some of the soluble derivatives of chitin are very sensitive to ionic composition (Hultin, 1955; Tracey, 1955); glycolchitin is also a substrate for lysozyme (Takeda *et al.*, 1966; Nord & Wadström, 1972). In addition it is possible that these chitin derivatives are not substrates for all chitinases, particularly those that are more specific for crystalline chitin.

Turbidometric assays. These assays depend on measurement of the rate of decrease in light-scattering that accompanies depolymerization of a suspension of colloidal chitin. The method is suitable only for estimation of relatively high activities (Jeuniaux, 1966). Nord & Wadström (1972) found that colloidal chitin was more easily degraded than chitin from crab and lobster shells but reported that the method was not very reproducible for different batches of substrate. This method has been used for the assay of *Streptomyces* chitinase (Berger & Reynolds, 1958; Jeuniaux, 1958; Lloyd, *et al.*, 1965) and the chitinase from goat serum (Lundblad *et al.*, 1974).

Assays based on the measurement of end-products. Two types of assays have been devised that depend on the measurement of end-products. The increase in reducing activity that results from the depolymerization of chitin has been used as an assay for chitinases. Iten & Matile (1970) used the Nelson method for detection of reducing sugars while Monreal & Reese (1969) and Tiunova, *et al.* (1976) used the dinitrosalicylic acid method.

Measurement of the rate of liberation of *N*-acetylglucosamine from chitin is probably the most widely used method for the assay of chitinases. Colloidal chitin is generally used as a substrate, and free *N*-acetylglucosamine released from the polymer by the sequential attack of chitinase and chitobiase is usually detected by the Reissig, Strominger & Leloir (1955) adaptation of the Morgan–Elson reaction. This assay detects *N*-acetylhexosamines that are unsubstituted at both the C-1 and C-4 positions. The end-product of chitinase activity, the disaccharide *N,N′*diacetylchitobiose has a $\beta(1-4)$ linkage and consequently reacts poorly in the Morgan–Elson reaction. For full colour development the disaccharide has to be hydrolysed to *N*-acetylglucosamine by a chitobiase.

Chitinase from many sources is accompanied by chitobiase activity, but a reliable estimation of chitinase activity requires the presence of chitobiase in sufficient amount not to be rate limiting (Jeuniaux, 1966). Where necessary chitobiase, free from chitinase, has to be added either to the reaction mixture (Jeuniaux, 1966) or to the filtered chitinase digest after termination of the reaction (Cabib & Bowers, 1971). This method has been used for the assay of chitinases from a variety of sources (Tracey, 1955; Toyama, 1967; Iten & Matile, 1970; Skujins *et al.*, 1970, 1973; Amagase, Mori & Nakayama, 1972; Ward & Fairbairn, 1972; Cohen, 1974; Cornelius *et al.*, 1976), but in most cases it is not clear whether sufficient chitobiase was present to ensure reliable assay of chitinase. Preparations of colloidal chitin used in this type of assay should be free from chitin oligosaccharides, since these may be hydrolysed to *N*-acetylglucosamine by the chitobiase and *N*-acetyl-β-glucosaminidase activities that usually accompany chitinase in crude extracts.

Colorimetric assay of chitinase. The chromogenic substrate 3,4 dinitrophenyl-tetra-*N*-acetyl-chitotetraose was developed for the assay of lysozyme, (Ballardie & Capon, 1972) but has recently been used for the assay of chitinase from *Vibrio alginolyticus* (Aribisala & Gooday, 1978). Hydrolysis of the substrate releases the aglycone which, as the 3,4-dinitrophenolate ion, is highly coloured and may be estimated spectrophotometrically. Assays using this substrate are rapid and sensitive but are not specific for chitinase, since the glycoside is hydrolysed by lysozyme and by chitin oligosaccharase from some sources (Cook, Pope & Stirling, unpublished work).

Radiochemical assays. These assays involve the radioactive counting of water-soluble oligosaccharides released from ^{14}C-labelled colloidal chitin (Reisert, 1972; Spindler, 1976; Zielkowski & Spindler, 1978) or ^{3}H-labelled reconstituted chitin (Molano, *et al.*, 1977). In the method of Molano *et al.*, (1977) reconstituted chitin is prepared using [^{3}H]acetic anhydride of very high specific activity to acetylate chitosan, and this substrate forms the basis of a rapid and sensitive assay for chitinase. This assay is not completely specific for chitinase since chitin labelled in the acetyl groups is also a substrate for chitin deacetylase. However, it is relatively easy to distinguish between these activities by identifying the end-products of the reaction.

Other enzymes involved in chitin degradation
N,N'-Diacetylchitobiase

Comparatively little is known about the biochemistry of the enzymes that hydrolyse chitobiose and chitinodextrins. They frequently accompany chitinases but have distinct activities, and chitobiase has been separated from chitinase by Zechmeister, Tóth & Balint (1938), Zechmeister & Tóth (1939), Berger & Reynolds (1958), Jeuniaux (1963), Otakara (1964) and Spindler (1976). There is some confusion over the identity of chitobiase and *N*-acetyl-β-glucosaminidase activities, this arises from the fact that the enzymes sometimes have overlapping specificities. It has been possible to isolate a chitobiase that lacks activity towards aryl-*N*-acetyl-β-glucosaminidase (Stirling, 1974) and one of the N-acetyl-β-glucos-aminidases from *Physarum polycephalum* Schw. is inactive towards chitobiose (Cook, Pope & Stirling, unpublished work). The chitobiase from *Aspergillus niger* has aryl-N-acetyl-β-glucosaminidase activity but is more active towards the disaccharide (Otakara, 1961). One of the best-characterized microbial exo-β-N-acetylglucosaminidases, the enzyme from *Bacillus subtilis* B Cohn emend. Prazmowski, hydrolyses aryl-N-acetyl-β-glucosaminides and chitobiose, but the primary substrate is O-2-acetamido-2-deoxy-β-D-glucopyranosyl-β(1–4)-2-acetamido-3-O(D-1-carboxyethyl)-2-deoxy-D-glucose, consistent with a role in the hydrolysis of bacterial cell walls (Berkely *et al.*, 1972). In contrast to the chitinases, which seem in general to be low molecular weight basic enzymes, N-acetyl-β-glucosaminidase (chitobiase) from *Sclerotina fructigena* Aderhold & Ruhland has a molecular weight of 140 000 and is acidic (Reyes & Byrde, 1973), while the enzyme from *Aspergillus oryzae* has a molecular weight of 141 000 (Mega, Ikenaka & Matsushima, 1970).

Chitin deacetylase

In addition to the major pathway of chitin degradation there exists in some organisms another route of breakdown, involving initial deacetylation to chitosan by chitin deacetylase. This enzyme has been detected in *Mucor rouxii* by its activity towards glycol ([³H]acetyl) chitin (Araki & Ito, 1974) and has been purified 140-fold (Araki & Ito, 1975). The high activity of chitin deacetylase in *M. rouxii* is considered by Araki & Ito (1975) to be consistent with a role for the enzyme in the formation of chitosan, which is a major wall component of this fungus.

Chitosanase

An enzyme, distinct from chitinase, that depolymerizes chitosan was detected in the culture broths of several organisms when they were grown on dead hyphae of *Rhizopus rhizopodiformis* (Cohn) Zopf. (Monaghan *et al.*, 1973). The chitosanase from *Myxobacter* AL-1 was purified to homogeneity, as judged by ultracentrifugation and electrophoresis (Hedges & Wolfe, 1974), and was found to copurify with β(1–4) glucanase. On denaturation both activities were lost at the same rate, they both had the same pH optima and the same molecular weight of 30 000, indicating that both activities were carried by the same enzyme. The chitosanase from *Bacillus* R-4 was purified to homogeneity by Tominaga & Tsujisaka (1975) who found that the enzyme had a molecular weight of 31 000, an isoelectric point of pH 8.30 and a pH optimum of 5.6, but this enzyme did not have β(1–4)glucanase activity.

Chitin was first discovered in fungi by Braconnot in 1811 and since then there has been much published on chitin and the enzymes involved in its degradation (Muzzarelli, 1977). However, with the application of the rapid and sensitive assay methods recently introduced there are excellent prospects for future developments in the biochemistry of chitinolytic enzymes.

References

Amagase, S., Mori, M. & Nakayama, S. (1972). Digestive enzymes in Insectivorous plants. IV. Enzymatic digestion of insects by *Nepenthes* secretion and *Drosera peltata* extract: Proteolytic and chitinolytic activities. *Journal of Biochemistry, Tokyo*, **72**, 765–7.

Araki, Y. & Ito, E. (1974). A pathway of chitosan formation in *Mucor rouxii*: enzymatic deacetylation of chitin. *Biochemical and Biophysical Research Communications*, **56**, 669–75.

Araki, Y. & Ito, E. (1975). A pathway of chitosan formation in *Mucor rouxii*. Enzymatic deacetylation of chitin. *European Journal of Biochemistry*, **55**, 71–8.

Aribisala, O. A. & Gooday, G. W. (1978). Properties of chitinase from *Vibrio alginolyticus* as assayed with the chromogenic substrate 3,4 dinitrophenyltetra-*N*-acetylchitotetraoside. *Biochemical Society Transactions*, **6**, 568–9.

Bade, M. L. & Stinson, A. (1976). Activation of cuticle chitinase. A probable new instance of activation by partial proteolysis. *Proteolysis and Physiological Regulation* (Miami Winter Symposium), vol. 2, eds. D. W. Ribbons & K. Brew, p. 391. New York, London: Academic Press.

Ballardie, F. W. & Capon, B. (1972). 3,4-Dinitrophenyl tetra-*N*-Acetyl-β-chitotetraoside a good chromophoric substrate for Hens Egg-White lysozyme. *Journal of the Chemical Society, Chemical Communications*, 828–9.

Bartnicki-Garcia, S. (1968). Cell wall chemistry, morphogenesis and taxonomy of fungi. *Annual Review of Microbiology*, **22**, 87–108.

Bartnicki-Garcia, S. (1973). Fundamental aspects of hyphal morphogenesis. In *Microbial Differentiation* (23rd Symposium of the Society for General Microbiology, 1973), eds. J. M. Ashworth & J. E. Smith, pp. 245–68. Cambridge University Press.

Bartnicki-Garcia, S. & Nickerson, W. J. (1962). Isolation, composition and structure of cell walls of filamentous and yeast-like forms of *Mucor rouxii*. *Biochimica et Biophysica Acta*, **58**, 102–19.

Berger, L. R. & Reynolds, D. M. (1958). The chitinase system of a strain of *Streptomyces griseus*. *Biochimica et Biophysica Acta*, **29**, 522–34.

Berghem, L. E. R. & Pettersson, L. G. (1973). The mechanism of enzymatic cellulose degradation. Purification of a cellulolytic enzyme from *Trichoderma viride* active on highly ordered cellulose. *European Journal of Biochemistry*, **37**, 21–30.

Berghem, L. E. R., Pettersson, L. G. & Axiö-Fredriksson, U.-B. (1975). The mechanism of enzymatic cellulose degradation. Characterisation and enzymatic properties of a β-1,4-glucan cellobiohydrolase from *Trichoderma viride*. *European Journal of Biochemistry*, **53**, 55–62.

Berkley, R. C. W., Brewer, S. J., Ortiz, J. M. & Gillespie, J. B. (1972). An exo-β-N-acetylglucosaminidase from *Bacillus subtilis* B, characterisation. *Biochimica et Biophysica Acta*, **309**, 157–68.

Blackwell, T. (1969). Structure of β-chitin or parallel chain systems of Poly-β-(1–4)N-acetyl-D-glucosamine. *Biopolymers*, **7**, 281–98.

Braconnot, H. (1811). Recherches analytiques sur la nature des champignons. *Annales de chimie, Paris*, **79**, 265–304.

Brimacombe, J. S. & Webber, J. M. (1964). *Mucopolysaccharides*. Amsterdam: Elsevier.

Bull, A. T. (1970). Inhibition of polysaccharases by melanin: Enzyme inhibition in relation to mycolysis. *Archives of Biochemistry and Biophysics*, **137**, 345–56.

Cabib, E. & Bowers, B. (1971). Chitin and yeast budding. Location of chitin in yeast bud scars. *Journal of Biological Chemistry*, **246**, 152–9.

Carlström, D. (1962). The polysaccharide chain of chitin. *Biochimica et Biophysica Acta*, **59**, 361–4.

Cohen, R. J. (1974). Some properties of chitinase from *Phycomyces blakesleeanus*. *Life Sciences*, **15**, 289–300.

Cornelius, C., Dandrifosse, G. & Jeuniaux, C. (1976). Chitinolytic enzymes of the Gastric mucosa of *Periodicticus potto* (primate prosimian): Purification and enzyme specificity. *International Journal of Biochemistry*, **7**, 445–8.

Datema, R., Wessels, J. G. H. & van den Ende, H. (1977). The hyphal wall of *Mucor mucedo*. 2. Hexosamine containing polymers. *European Journal of Biochemistry*, **80**, 621–6.

Foster, A. B. & Webber, J. M. (1960). Chitin. *Advances in Carbohydrate Chemistry*, **15**, 371–93.

Hackman, R. H. (1972). Gel electrophoresis and Sephadex thin-layer studies of protein from an insect cuticle. *Insect Biochemistry*, **2**, 235–42.

Hackman, R. H. & Goldberg, M. (1974). Light Scattering and infrared-spectrophoto-metric studies of chitin and chitin derivatives. *Carbohydrate Research*, **38**, 35–45.

Halliwell, G. D. & Griffin, M. (1973). The nature and mode of action of the cellulolytic component C_1 of *Trichoderma koningii* on native cellulose. *Biochemical Journal*, **135**, 587–94.

Hedges, A. & Wolfe, R. S. (1974). Extracellular enzyme from *Myxobacter ALI* that exhibits both β1,4 glucanase and chitosanase activities. *Journal of Bacteriology*, **120**, 844–53.

Hultin, E. (1955). Carboxymethyl chitin, a new substance available for the determination of chitinase activity. *Acta Chemica Scandanavica*, **9**, 192.

Hunsley, D. & Burnett, J. H. (1970). The ultrastructural architecture of the walls of some hyphal fungi. *Journal of General Microbiology*, **62**, 203–18.

Hunsley, D. & Kay, D. (1976). Wall structure of *Neurospora* hyphal apex: Immunofluorescent localization of wall surface antigens. *Journal of General Microbiology*, **95**, 233–48.

Iten, W. & Matile, P. (1970). Role of chitinase and other lysosomal enzymes of *Coprinus lagopus* in the autolysis of fruiting bodies. *Journal of General Microbiology*, **61**, 301–9.

Jeuniaux, C. (1957). Purification of a *Streptomyces* chitinase. *Biochemical Journal*, **66**, 29.

Jeuniaux, C. (1958). Recherches sur les chitinases. I. Dosage néphélométrique et production de chitinase par des Streptomycètes. *Archives Internationales de Physiologie et Biochimie*, **66**, 408–27.

Jeuniaux, C. (1959). Recherches sur les chitinases. II. Purification de la chitinase d'un streptomycète, et séparation électrophorétique de principes chitinolytiques distincts. *Archives Internationales de Physiologie et Biochimie*, **67**, 597–617.

Jeuniaux, C. (1963). *Chitine et Chitinolyse, un Chapitre de la Biologique Moleculaire.* Paris: Masson.

Jeuniaux, C. (1966). Chitinases. In *Methods in Enzymology,* vol. 8 eds. E. F. Neufeld & V. Ginsburg, pp. 644–50. New York, London: Academic Press.

Karrer, P. & White, S. M. (1930). Chitin. *Helvetica Chimica Acta,* **13,** 1105–13.

Kooiman, P. (1964). The occurrence of carbohydrases in digestive juice and in hepatopancreas of *Astacus fluviatilis* Fabr. and of *Homarus vulgaris* M.-E. *Journal of Cellular and Comparative Physiology,* **63,** 197–201.

Kreger, D. R. (1954). Observations on cell walls of yeasts and some other fungi by X-ray diffraction and solubility test. *Biochimica et Biophysica Acta,* **13,** 1–9.

Kuo, M. J. & Alexander, M. (1967). Inhibition of lysis of fungi by melanins. *Journal of Bacteriology,* **94,** 624–9.

Lloyd, A. B., Noveroske, R. L. & Lockwood, J. L. (1965). Lysis of fungal mycelium by *Streptomyces* spp. and their chitinase systems. *Phytopathology* **55,** 871–5.

Lundblad, G., Hederstedt, B., Lind, J. & Steby, M. (1974). Chitinase in goat serum. Preliminary purification and characterization. *European Journal of Biochemistry,* **46,** 367–76.

Lysenko, O. (1976). Chitinase of *Serratia marcescens* and its toxicity to insects. *Journal of Invertebrate Pathology* **27,** 385–6.

McLachlan, J., McInnes, A. G. & Falk, M. (1965). Studies on chitan (chitin: poly-*N*-acetylglucosamine) fibres of the Diatom *Thalassiosiva fluviatilis* Histedt. Production and isolation of chitin fibres. *Canadian Journal of Botany,* **43,** 707–13.

Mahadevan, P. R. & Tatum, E. S. (1965). Relationship of the major constituents of the *Neurospora crassa* cell wall to wild type and colonial morphology. *Journal of Bacteriology,* **90,** 1073–81.

Mega, T., Ikenaka, T. & Matsushima, Y. (1970). Studies on *N*-acetyl-β-glucosaminidase of *Aspergillus oxyzae.* I. Purification and characterisation of *N*-acetyl-β-glucosaminidase obtained from Takadiastase. *Journal of Biochemistry,* **68,** 109–17.

Michell, A. J. & Scurfield, G. (1967). Composition of extracted fungal cell walls as indicated by infrared spectroscopy. *Archives of Biochemistry and Biophysics,* **120,** 628–37.

Minke, R. & Blackwell, J. (1978). The structure of *a*-chitin *Journal of Molecular Biology,* **120,** 167–82.

Molano, J., Duran, A. & Cabib, E. (1977). A rapid and sensitive assay for chitinase using tritiated chitin. *Analytical Biochemistry* **83,** 648–56.

Monaghan, R. L., Eveleigh, D. E., Tewari, R. P. & Reese, E. T. (1973). Chitosanase, a novel enzyme. *Nature, New Biology,* **245,** 78–80.

Monreal, J. & Reese, E. T. (1969). The chitinase of *Serratia marcescens. Canadian Journal of Microbiology,* **15,** 689–96.

Muzzarelli, R. A. A. (1977). *Chitin.* Oxford: Pergamon Press.

Noble, D. W. & Sturgeon, R. J. (1968). The purification and properties of some glycoside hydrolases. *Biochemical Journal,* **110,** 7P.

Nord, C. E., & Wadström, T. (1972). Chitinase activity and substrate specificity of three bacteriolytic endo-β-*N*-acetyl-muramidases and one endo-β-*N*-acetylglucosaminidase. *Acta Chemica Scandanavica,* **26,** 653–60.

Otakara, A. (1961*a*). Studies on the chitinolytic enzymes of black-koji mold I. Viscometric determination of chitinase activity by application of glycol chitin as a new substrate. *Agricultural and Biological Chemistry, Japan,* **25,** 50–4.

Otakara, A. (1961*b*). Studies on the chitinolytic enzymes of black-koji mold II. Purification of chitinase. *Agricultural and Biological Chemistry, Japan,* **25,** 54–60.

Otakara, A. (1964). Studies on the chitinolytic enzymes of black-koji mold. VI. Isolation and some properties of N-acetyl-β-glucosaminidase. *Agricultural and Biological Chemistry, Japan,* **28,** 745–51.

Pearson, F. G., Marchessault, R. H. & Liang, C. Y. (1960). Infrared spectra of crystalline polysaccharides. VI. Chitin. *Journal of Polymer Science,* **43,** 101–16.

Polachek, Y. & Rosenberger, R. F. (1975). Autolytic enzymes in hyphae of *Aspergillus nidulans,* their action on old and newly formed walls. *Journal of Bacteriology,* **121,** 332–7.

Powning, R. F. & Irzykiewicz, H. (1965). Studies on the chitinase system in beans and other seeds. *Comparative Biochemistry and Physiology,* **14,** 127–33.

Ramakrishnan, C. & Prasad, N. (1972). Rigid-body refinement and conformation of α-chitin. *Biochimica et Biophysica Acta,* **261,** 123–35.

Reisert, P. S. (1972). Studies on the chitinase system of *Chytriomyces hyalinus* using a C^{14}-chitin assay. *Mycologia,* **64,** 288–97.

Reissig, J. L., Strominger, J. L. & Leloir, L. F. (1955). A modified colorimetric method for the estimation of N-acetylamino sugars. *Journal of Biological Chemistry,* **217,** 959–66.

Reyes, F. & Byrde, R. J. W. (1973). Partial Purification and Properties of β-N-Acetylglucosaminidase from the fungus *Sclerotinia fructigena. Biochemical Journal,* **131,** 381–8.

Rudall, K. M. (1969). Chitin and its association with other molecules. *Journal of Polymer Science,* Part C **28,** 83–102.

Rudall, K. M. & Kenchington, W. (1973). The chitin system. *Biological Reviews,* **49,** 597–636.

Skujins, J. J., Potgieter, H. J. & Alexander, M. (1965). Dissolution of fungal cell walls by a streptomycete chitinase and β-(1→3)glucanase. *Archives of Biochemistry and Biophysics,* **111,** 358–64.

Skujins, J. J., Pukite, A. & McLaren, A. D. (1970). Chitinase of *Streptomyces* spp., purification and properties. *Enzymologia,* **39,** 353–70.

Skujins, J. J., Pukite, A. & McLaren, A. D. (1973). Absorbtion and reactions of chitinase and lysozyme on chitin. *Molecular and Cellular Biochemistry,* **2,** 221–8.

Spindler, K. D. (1976). Initial characterization of chitinase and chitobiase from the integument of *Drosophila hydei. Insect Biochemistry,* **6,** 663–7.

Stirling, J. L. (1974). Human N-acetyl-β-hexosaminidases: Hydrolysis of N,N′-Diacetylchitobiose by a low molecular weight enzyme. *F.E.B.S. Letters,* **39,** 171–5.

Takeda, H., Strasdine, G. A., Whitaker. D. R. & Roy, C. (1966). Lytic enzymes in the digestive juices of *Helix pomatia. Canadian Journal of Biochemistry,* **44,** 509–18.

Takiguchi, Y., Nagahata, N. & Shimahara, K. (1976). A new method of chitinase assay using 6-O-hydroxypropyl-chitin. *Journal of the Agricultural Chemical Society, Japan,* **50,** 243–4.

Tiunova, N. A., Pirieva, D. A. & Feniksova, R. V. (1976a). Formation and properties of chitinase *Actinomyces kursanovii. Mikrobiologiya,* **45,** 543–6.

Tiunova, N. A., Pirieva, D. A., Feniksova, R. V. & Kuznetsov, V. D. (1976b). Formation of Chitinase by *Actinomycetes* in submerged cultures. *Mikrobiologiya,* **45,** 280–3.

Tominaga, Y. & Tsujisaka, Y. (1975). Purification and some enzymatic properties of the chitosanase from *Bacillus* R-4 which lyses *Rhizopus* cell walls. *Biochimica et Biophysica Acta,* **410,** 145–55.

Tominaga, Y. & Tsujisaka, Y. (1976). Purifications and some properties of two chitinases from *Streptomyces orientalis* which lyse *Rhizopus* cell walls. *Agricultural and Biological Chemistry,* **40,** 2325–33.

Toyama, N. (1967). Mycolytic and cellulolytic enzymes of *Trichoderma viride*. *Journal of Fermentation Technology*, **45**, 663–70.

Tracey, M. V. (1955). Chitinase in Some Basidiomycetes. *Biochemical Journal*, **61**, 579–86.

Trinci, A. P. J. (1978). Wall and hyphal growth. *Science Progress*, **65**, 75–99.

Wadström, T. (1971). Chitinase activity and substrate specificity of endo-β-N-acetylglucosaminidase of *Staphylococcus aureus* strain M18. *Acta Chemica Scandanavica*, **25**, 107–12.

Ward, K. A. & Fairbairn, D. (1972). Chitinase in developing eggs of *Ascaris suum* (Nematoda). *Journal of Parasitology*, **58**, 546–9.

Zechmeister, L. & Tóth, G. (1939). Chromatographic analysis of chitinase. *Naturwissenschaften*, **27**, 367–71.

Zechmeister, L., Tóth, G. & Balint, M. (1938). Chromatographic separation of some of the enzymes from emulsin. *Enzymologia*, **5**, 302–6.

Zielkowski, R. & Spindler, K. D. (1978). Chitinase and chitobiase from the integument of *Locusta migratoria*: characterization and their titer during the fifth larval instar. *Insect Biochemistry*, **8**, 67–71.

9
Localized activation of chitin synthetase in the initiation of yeast septum formation

E.CABIB*, A.DURAN* AND B.BOWERS†

*National Institute of Arthritis and Metabolic Diseases, and †National Heart, Lung, and Blood Institute, National Institutes of Health, Bethesda, Maryland 20014, USA

Morphology of yeast septum formation

The process of morphogenesis, i.e. the formation of structures with defined shapes during growth and differentiation of living organisms, requires underlying molecular mechanisms endowed with a sense of direction. One simple way of obtaining this goal is to synthesize molecules which are capable of polymerization. If the interacting domains of such molecules are limited to specific regions, the coupling of either identical or different monomers to each other can create relatively complicated structures, such as ribosomes, microtubules or viruses. Here the only directional element is the recognition by a specific portion of a molecule of the complementary portion of another molecule. An input of energy, perhaps in the form of an appropriate nucleotide, may be necessary in some cases to complete the process but the pattern of growth is dictated by the structure of the polymerizing molecules.

Although self-assembling systems are quite common in nature, it is very unlikely that they are at the origin of morphogenetic events connected with growth and division of a complex structure, such as a whole cell or some of its organelles. One senses here the need for a spatial and temporal coordination of different elements if an ordered process of growth is to ensue. How is it possible to investigate these more complex cases? In an ideal system the chemical nature of the structure created by the cell would be simple enough to study at the molecular level; yet its synthesis should be integrated into the general process of cell growth, to permit observation of the interaction between the different directional and regulatory elements. One system that approaches these requirements is the formation of the primary septum during cell division of *Saccharomyces cerevisiae* Hansen. The formation of this septum is illustrated in Fig. 9.1; this diagram is based on observations made by electron microscopy (Bowers, Levin & Cabib, 1974) and fluorescence microscopy (Cabib & Bowers, 1975). The primary septum is initiated at the beginning of budding as a ring at the base of the incipient bud (Fig. 9.1*a*). Before cytokinesis occurs, the septum extends centripetally to fill the neck between mother and daughter cell; at this point the ring has now become a disk (Fig. 9.1*d*). Completion of the primary septum is followed by deposition of two secondary septa, each facing one of the dividing cells (Fig. 9.1*e*); finally an asymmetric separation takes place, whereby the primary septum remains embedded in the 'bud scar' of the mother cell (Fig. 9.1*f*).

Chitin as the main constituent of the primary septum. Molecular mechanism of its synthesis

By chemical and enzymatic digestion of the cell, it was possible to isolate the primary septum disks embedded in the bud scars, and to show that they were composed of chitin (Cabib & Bowers, 1971). The establishment of the chemical composition of the septum permitted an investigation of its biosynthesis. A particulate preparation obtained after lysis of yeast protoplasts was found to catalyse the biosynthesis of chitin from uridine diphosphate acetyglucosamine (Keller & Cabib, 1971). The crucial discovery, however, was that chitin synthetase could be obtained in an inactive, or zymogenic, state; its activation was achieved by incubation either with an extract from the same yeast, which contains

Fig. 9.1. Scheme of formation of yeast septum. The dotted line represents the plasma membrane. SP, septal primordia; PS, primary septum; SS, secondary septa; Bi Sc, birth scar; Bu Sc, bud scar. (Reprinted from Trends in Biochemical Sciences, with permission of the publisher.)

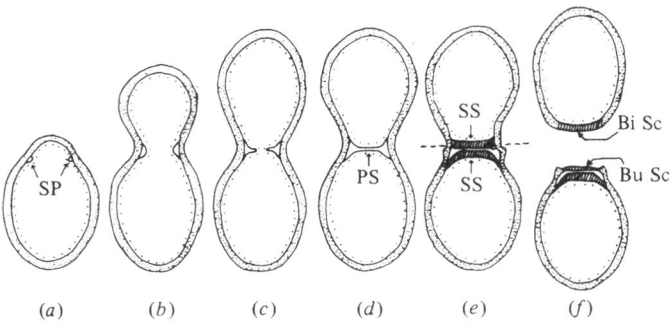

(*a*)   (*b*)   (*c*)   (*d*)   (*e*)   (*f*)

an 'activating factor', or with a protease, such as trypsin (Cabib & Farkas, 1971). This finding provided a possible explanation for the initiation of chitin synthesis at the budding site (Fig. 9.1*a*), an event that is precisely determined with respect to spatial location and temporal position in the cell cycle. It seemed probable that polysaccharide synthesis could be triggered by contact between chitin synthetase, zymogen and activating factor, although it was not clear how the localization and timing of this event could be achieved. An explanation of this process was suggested by Cabib & Farkas (1971) and is shown in Fig. 9.2; an inhibitor of the activating factor, which was found in yeast (Cabib & Keller, 1971; Ulane & Cabib, 1974) is also incorporated into this hypothesis. According to this scheme, chitin synthetase zymogen is uniformly distributed on the plasma membrane (Fig. 9.2*a*), and is activated at specific sites by the impact of vesicles containing activating factor (Fig. 9.2*b*). In our hypothesis, the inhibitor would function as a safety valve, to trap and neutralize any activating factor that might be released into the cytoplasm (Fig. 9.2*c*) and thus prevent activation of zymogen elsewhere in the plasma membrane. The activating factor and inhibitor have been purified to homogeneity (Ulane & Cabib, 1974, 1976). The activator was identified as a protease (Ulane & Cabib, 1976), which was identical with protease B described by Lenney (1956) and others (Hata, Hayashi & Doi, 1967). The inhibitor was shown to be

Fig. 9.2. Hypothesis for initiation of chitin synthesis (from Cabib & Farkas, 1971).

▲ Chitin synthetase zymogen
△ Active chitin synthetase
⟱ Activating factor
▐ Inhibitor of activating
   factor

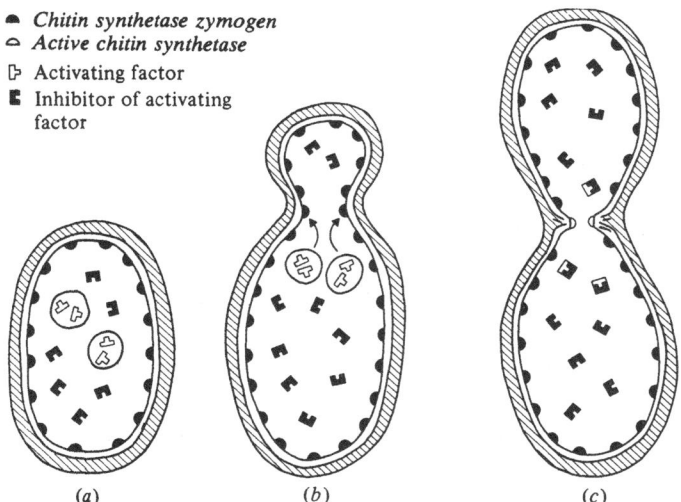

(*a*)   (*b*)   (*c*)

a small protein of molecular weight 8500, which formed a tight complex with the protease in a stoichiometric ratio of one to one (Ulane & Cabib, 1974, 1976).

Subcellular distribution of the chitin synthetase system

In our hypothesis it was deemed essential for the activating factor to be contained within vesicles, since if it were distributed uniformly in the cell sap it is difficult to understand how it could be directed to a specific site. Proteinase B was indeed found to be associated with a purified preparation of yeast vacuoles (Cabib, Ulane & Bowers, 1973). Mild sonication opened the vesicles and released the protease in soluble form.

Our hypothesis also postulated that the chitin synthetase zymogen would be located on the plasma membrane; verification of this assumption required the isolation of purified plasma membranes which were not significantly contaminated by other membranes. Because of the difficulty in distinguishing different types of membranes we decided to isolate plasma membranes in an intact, unfragmented state. We did this by adapting a procedure developed by Scarborough (1975) for *Neurospora crassa* Shear & Dodge. In this method yeast protoplasts are coated with Concanavalin A to reinforce the plasma membranes and then lysed by a combination of osmotic shock and mechanical shearing. This treatment opens the plasma membranes without breaking them into fragments. The resulting lysate is fractionated by centrifugation on a Renografin density gradient. The intact plasma membranes collect in a well defined band, with little contamination from other components. In agreement with our expectation, the bulk of the chitin synthetase activity was found in the zymogen state in the plasma membrane fraction (Duran, Bowers & Cabib, 1975).

Although this result provided strong evidence for the association of the chitin synthetase zymogen with the plasma membrane, we felt it desirable to confirm association by cytochemical and autoradiographic methods. This was all the more important, because Bracker, Ruiz-Herrera & Bartnicki-Garcia (1976) have reported that chitin synthetase from *Mucor rouxii* (Calmette) Whem. is present in small particles which they call 'chitosomes'.

Since the synthetase could not be tagged specifically, an indirect approach was used. It had been observed earlier that the reaction product, chitin, appeared to remain associated with the particulate material that catalysed its formation. Thus, when purifed plasma membranes were first activated with protease, then allowed to react with ^{14}C-labelled UDP-*N*-acetylglucosamine, and finally re-isolated in another Renografin gradient, most of the radioactivity incorporated into chitin was found in the repurified membrane fraction. This result suggested that the nascent chitin chains might remain attached to the synthetase, and that the location of the enzyme could perhaps be inferred from the position of the chitin whose formation it catalysed. In order to visualize chitin, we took advantage of the fluorescence produced by its association with 'brighteners', in this case Calcofluor White M2R (Cabib & Bowers, 1975). Addition of Calcofluor to plasma membranes that had been allowed to synthesize some chitin resulted in the appearance of bright fluorescent spots on the membranes (Fig. 9.3*a* and *b*). Fluorescence was not observed between groups of plasma membranes, therefore the preparation does not appear to contain smaller particles, such as 'chitosomes', able to synthesize chitin. If such particles exist, they must be associated with the plasma membranes. Furthermore, it was observed that on each plasma membrane there were several bright spots. However, because of the impossibility of focussing the entire surface of a plasma membrane simultaneously, and the relatively large grain of the film, this cannot be shown in the photographs.

In order to obtain more information about the site of synthesis of chitin, we tried to locate the chitin by autoradiography. In this method it was essential to eliminate the unreacted radioactive substrate. This required some manipulation of the plasma membranes, such as centrifugation into a pellet or on a new gradient. It was observed, however, that during this manipulation the synthesized chitin tended to coalesce into relatively large lumps, thus losing the localization of the nascent state. In order to prevent this movement of chitin, the membranes, after proteolytic activation of the chitin synthetase zymogen, were immobilized inside small agar blocks which were suspended in the normal reaction mixture for chitin synthetase. It was found that the substrate could easily diffuse in and out of the blocks, whereas the chitin remained trapped inside, presumably near the place where it had been formed. An autoradiogram prepared from a thin section of a fixed and embedded agar block is shown in Fig. 9.3*c*. The silver grains are in close association with the plasma membrane, usually localized in one or more spots along a membrane section. This uneven localization of chitin is similar to that observed by fluoresence. When polyoxin D, a specific inhibitor of chitin synthetase, was included in the incubation mixture, no silver grains were seen (Keller & Cabib, 1971).

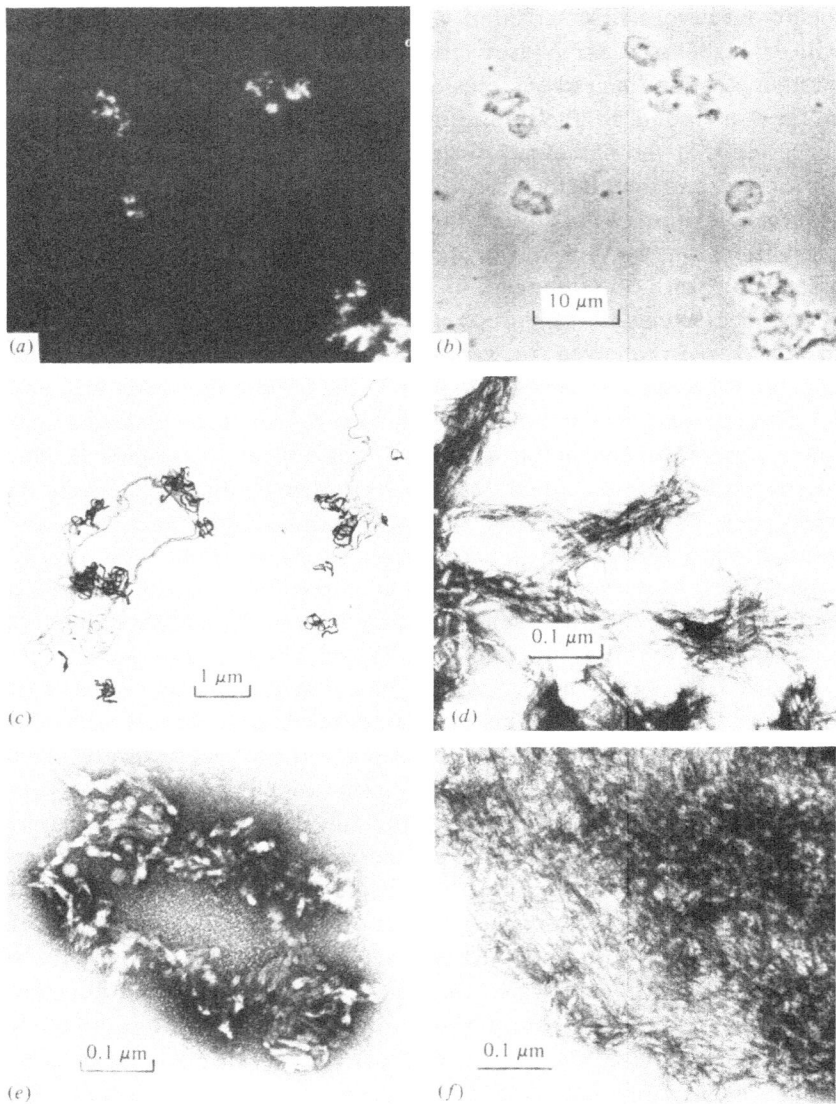

Fig. 9.3. (a) Appearance of yeast plasma membrane in the fluorescent microscope, after incubation with UDP-N-acetylglucosamine and in the presence of Calcofluor White M2R. Membranes were obtained as described by Duran *et al.* (1975), treated with trypsin and incubated for 30 min at 30 °C with non-radioactive UDP-N-acetylglucosamine, as already reported (Cabib, 1972). After incubation, Calcofluor was added to a final concentration of 0.007 %(w/v).

The zymogenic nature of chitin synthetase

The combined evidence from the fluorescence and autoradiography experiments confirmed the presence on chitin synthetase of the plasma membrane. The enzyme associated with purified plasma membranes also required proteolytic activation (Duran, *et al.*, 1975), in agreement with the zymogen hypothesis. Nevertheless, since the chitin synthetase preparations were particulate, it could be argued that their apparent inactivity was due to a masking effect by other proteins and lipids; the protease would then be acting by destroying this barrier, rather than by partial proteolysis of a zymogen. In order to resolve this question it was necessary to solubilize the enzyme. This was finally achieved with digitonin, which Gooday & Rousset-Hall (1975) had shown could solubilize the chitin synthetase of *Coprinus cinereus* (Schaeff.) Fr. About 50–60 % of the enzymatic activity was found in the $200\,000\,g$ supernatant of a digitonin extract. A further indication that the enzyme was no longer particulate was provided by chromatography on a Sepharose 6B column, where the synthetase emerged considerably retarded with respect to the void volume and slightly ahead of ferritin

---

(*b*) Phase contrast image of (*a*).

(*c*) Autoradiography of chitin synthesized by a plasma membrane preparation. Membranes were prepared and treated with trypsin as specified under (*a*), then embedded in 1.5 % (w/v) agar. Blocks of membrane-containing agar (1 mm^3) were incubated for 20 min in the reaction mixture for chitin synthetase (Cabib, 1972), containing UDP-[^{3}H]*N*-acetylglucosamine. The blocks were fixed in 3 % (v/v) glutaraldehyde, washed in phosphate buffer to remove unincorporated substrate, post-fixed in osmium tetroxide and embedded in Epon 812. Thin sections were prepared for autoradiography by the flat-substrate method (Salpeter & Bachman, 1972), coated with Ilford L4 emulsion and exposed for 3 weeks before development in Microdol X.

(*d*) Chitin synthesized by solubilized enzyme. A preparation of solubilized chitin synthetase, obtained as outlined in the legend of Fig. 9.4, was activated with trypsin and incubated with UDP-*N*-acetylglucosamine for 16 h under the conditions previously described (Cabib, 1972). The chitin that precipitated out during incubation was negatively stained with 1 % uranyl acetate.

(*e*) Chitin synthesized by plasma membranes. For preparation of plasma membranes and incubation conditions see under (*a*). After incubation, the plasma membranes were sedimented by centrifugation at 12 000 *g* for 10 min and the pellet was extracted with boiling 1% sodium dodecyl sulphate. The residue was negatively stained with 0.5% uranyl acetate.

(*f*) Purified bud scar preparation (Cabib & Bowers, 1971) negatively stained with 0.5 % uranyl acetate.

(molecular weight *c.* 500 000) (Duran & Cabib, 1978). The solubilized enzyme was almost completely inactive before proteolysis (Fig. 9.4). It can be concluded that chitin synthetase is a true zymogen, and is not simply masked inside a membrane or vesicle. The results described so far indicate that chitin synthetase, in the zymogen state, is bound to the plasma membrane at multiple sites. It follows almost by necessity that initiation of chitin synthesis at a particular site must occur by localized activation of the enzyme, as predicted by our hypothesis.

On the other hand, our results are not consistent with the hypothesis that localized formation of chitin is achieved by directing particles containing the synthetase (such as chitosomes (Bracker *et al.*, 1976)) to a specific site. If this were the case, the enzyme would presumably be found only in one restricted area on the plasma membrane, and most of it would be in the active rather than in the zymogen state. The reason for finding most of the synthetase in a precursor form is, in our view, that only the small portion of it which is found at the location of the growing septum is functioning at any given time, and this occurs only in that portion of the cell population which is in the appropriate phase of the cell cycle.

Fig. 9.4. Activation of solubilized enzyme by trypsin. A particulate preparation obtained by lysing yeast protoplasts in 1 mM EDTA and centrifuging at 113 000 *g* was extracted with 10 mg ml^{-1} digitonin, containing 25 mM Tris chloride, pH 7.5, 5 mM MgSO$_4$ and 0.2 M NaCl. The extract was centrifuged for 20 min at 165 000 *g*. The chitin synthetase zymogen present in the supernatant was activated with different amounts of trypsin, and the active enzyme was assayed as previously described (Cabib, 1972).

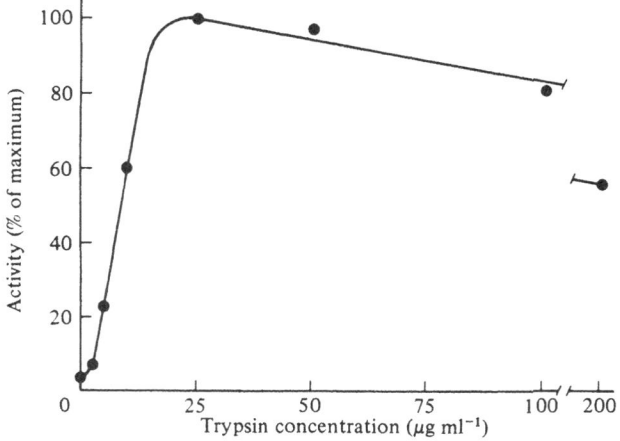

The activation of chitin synthetase *in vivo* need not be by action of protease B, although this is a very likely candidate. It is conceivable that the zymogen might be transformed into an active form by some other post-translational modification. We would suggest that even in this case the activating agent would probably be carried inside a vesicle, in order to permit localization of the effect to the target area.

Activation of chitin synthetase by protease is not confined to *Saccharomyces cerevisiae*. It has been also reported in five of the twelve different genera of fungi in which chitin synthetase has been investigated (see Table 9.1). The finding that in *S. cerevisiae* the enzyme is a true zymogen suggests that this may be the case in other organisms (however, see Chapter 10). One wonders about the physiological function of the chitin synthetases zymogen in a filamentous fungus such as *Aspergillus nidulans* (Eidam) Wint. By analogy with yeast, it is possible that chitin synthesis at the hyphal tip is continuously stimulated by activation of the zymogen present in the plasma membrane. Some of the numerous vesicles observed at the hyphal tip (Girbardt, 1969; Grove & Bracker, 1970) might be the carrier of such an activator.

Formation of insoluble chitin by the solubilized synthetase

Some properties of the solubilized chitin synthetase are worth mentioning. In the first place the enzyme purified by chromatography on Sephadex showed a requirement for a phospholipid. Only acidic phospholipids could fulfil this requirement; phosphatidyl serine and lysophosphatidylserine were the most active. Secondly, the soluble enzyme catalyses the formation of insoluble chitin in the absence of an added primer. The presence of an endogenous primer in the enzyme preparation cannot be discounted although certain limitations can be put on its nature. Thus, since the enzyme has been centrifuged at $200\,000\,g$, the primer cannot be a long chitin chain, which would have sedimented under those conditions. Similarly, the chromatography through Sepharose 6B would have resolved a small primer, such as an acetylglucosamine oligosaccharide, unless it was tightly bound to the enzyme.

One major problem in wall morphogenesis is how the main constituents of the wall, i.e. the polysaccharide chains, acquire a specific structure and orientation in space. It is not known to what extent this requires a spatial organization of the enzymes which catalyze their synthesis and perhaps of other 'helper' proteins. To what extent, on the other hand, does the conformation of the polysaccharides arise by mere

Table 9.1. *Activation of chitin synthetase by partial proteolysis in different fungi*

Species	Class	Activation	Reference
Saccharomyces cerevisiae	Ascomycetes	+	Cabib & Farkas (1971)
Aspergillus nidulans	Ascomycetes	+	Ryder & Peberdy (1977)
			Lopez-Romero & Ruiz-Herrera (1976)
Neurospora crassa	Ascomycetes	?	Glaser & Brown (1957)
Venturia inaequalis (Cooke) Winter	Ascomycetes	?	Jaworski, Wang & Carpenter (1965)
Mucor rouxii yeast form	Zygomycetes	+	Ruiz-Herrera &
Mucor rouxii, mycelial form	Zygomycetes	+	Bartnicki-Garcia (1976)
Phycomyces sp.	Zygomycetes	+	Thomson & Fischer (1976)
Mortierella vinacea Dixon Stewart	Zygomycetes	?	Peberdy & Moore (1975)
Cunninghamella elegans Lendner	Zygomycetes	?	Moore & Peberdy (1975)
Candida albicans (Robin) Berkhout, yeast form	Deuteromycetes	+	Braun & Calderone (1978)
Candida albicans, mycelial form	Deuteromycetes	+	Plessman Camargo *et al.* (1967)
Blastocladiella emersonii Cantino & Hyatt	Chytridiomycetes	?	Porter & Jaworski (1966)
Allomyces macrogynus Emerson & Wilson	Chytridiomycetes	?	Gooday & deRousset-Hall (1975),
Coprinus cinereus	Basidiomycetes	–	deRousset-Hall & Gooday (1975)

physicochemical interactions between their component sugars? An insight into this problem was provided by the possibility of synthesizing chitin from yeast with three systems of increasing complexity: soluble synthetase, isolated plasma membranes and the intact yeast cell. Electron-microscope observation with negative staining of the chitin obtained with soluble enzyme revealed that the chitin was in the form of flake-like particles, roughly $9 \times 4 \times 55$ nm in size, which tended to cluster in clumps (Fig. 9.3d). Thus, despite the lack of orientation in the enzyme molecules, the product displayed a surprising degree of organization. Furthermore, particles of similar aspect were found in electron micrographs of chitin obtained with membrane (Fig. 9.3e) or in the chitin of bud scars (Fig. 9.3f). Although these observations are very preliminary, they suggest that perhaps for the chitin of *Saccharomyces cerevisiae*, physicochemical forces may be sufficient to give rise to the organization of the polysaccharide that is found *in vivo*.

References

Bowers, B., Levin, G. & Cabib, E. (1974). Effect of polyoxin D on chitin synthesis and septum formation in *Saccharomyces cerevisiae*. *Journal of Bacteriology*, **119**, 564–75.

Bracker, C. E., Ruiz-Herrera, J. & Bartnicki-Garcia, S. (1976). Structure and transformation of chitin synthetase particles (chitosomes) during microfibril synthesis *in vitro*. *Proceedings of the National Academy of Sciences, USA*, **73**, 4570–4.

Braun, P. C. & Calderone, R. A. (1978). Chitin synthesis in *Candida albicans*. Comparison of yeast and hyphal forms. *Journal of Bacteriology*, in press.

Cabib, E. (1972). Chitin synthetase system from yeast. *Methods in Enzymology*, vol. 28, ed. V. Ginsburg, pp. 572–80. New York, London: Academic Press.

Cabib, E. & Bowers, B. (1971). Chitin and yeast budding. Localization of chitin in yeast bud scars. *Journal of Biological Chemistry*, **246**, 152–9.

Cabib, E. & Bowers, B. (1975). Timing and function of chitin synthesis in yeast. *Journal of Bacteriology*, **124**, 1586–93.

Cabib, E. & Farkas, V. (1971). The control of morphogenesis: an enzymatic mechanism for the initiation of septum formation in yeast. *Proceedings of the National Academy of Sciences, USA*, **68**, 2052–6.

Cabib, E. & Keller, F. A. (1971). Chitin and yeast budding. Allosteric inhibition of chitin synthetase by a heat-stable protein from yeast. *Journal of Biological Chemistry*, **246**, 167–73.

Cabib, E., Ulane, R. & Bowers, B. (1973). Yeast chitin synthetase. Separation of the zymogen from its activating factor and recovery of the latter in the vacuole fraction. *Journal of Biological Chemistry*, **248**, 1451–8.

Duran, A. & Cabib, E. (1978). Solubilization and partial purification of yeast chitin synthetase. Confirmation of the zymogenic nature of the enzyme. *Journal of Biological Chemistry*, in press.

Duran, A., Bowers, B. & Cabib, E. (1975). Chitin synthetase zymogen is attached to the yeast plasma membrane. *Proceedings of the National Academy of Sciences, USA,* **72,** 3952–5.

Girbardt, M. (1969). Die Ultrastruktur der apikalregion von Pilzhyphen. *Protoplasma,* **67,** 413–41.

Glaser, L. & Brown, D. H. (1957). The synthesis of chitin in cell-free extracts of *Neurospora crassa. Journal of Biological Chemistry,* **228,** 729–42.

Gooday, G. W. & de Rousset-Hall, A. (1975). Properties of chitin synthetase from *Coprinus cinereus. Journal of General Microbiology,* **89,** 137–45.

Grove, S. N. & Bracker, C. E. (1970). Protoplasmic organization of hyphal tips among fungi: vesicles and Spitzenkörper. *Journal of Bacteriology,* **104,** 989–1009.

Hata, T., Hayashi, R. & Doi, E. (1967). Purification of yeast proteinases. I. Fractionation and some properties of the proteinases. *Agricultural and Biological Chemistry,* **31,** 150–9.

Jaworski, E. G., Wang, L. C. & Carpenter, W. D. (1965). Biosynthesis of chitin in cell-free extracts of *Venturia inaequalis. Phytopathology,* **55,** 1309–12.

Keller, F. A. & Cabib, E. (1971). Chitin and yeast budding. Properties of chitin synthetase from *Saccharomyces carlsbergensis. Journal of Biological Chemistry,* **246,** 160–6.

Lenney, J. F. (1956). A study of two yeast proteinases. *Journal of Biological Chemistry,* **221,** 919–30.

Lopez-Romero, E. & Ruiz-Herrera, J. (1976). Synthesis of chitin by particulate preparations from *Aspergillus flavus. Antonie van Leeuwenhoek, Journal of Microbiology and Serology,* **42,** 261–76.

Moore, P. M. & Peberdy, J. F. (1975). Biosynthesis of chitin by particulate fractions from *Cunninghamella elegans. Microbios,* **12,** 29–39.

Peberdy, J. F. & Moore, P. M. (1975). Chitin synthetase in *Mortierella vinacea:* properties, cellular location and synthesis in growing cultures. *Journal of General Microbiology,* **90,** 228–36.

Plessmann Camargo, E., Dietrich, C. P., Sonneborn, D. & Strominger, J. L. (1967). Chitin synthesis in spores and vegetative cells of *Blastocladiella emersonii. Journal of Biological Chemistry,* **242,** 3121–8.

Porter, C. A. & Jaworski, E. G. (1966). The synthesis of chitin by particulate preparations of *Allomyces macrogynus. Biochemistry,* **5,** 1149–54.

de Rousset-Hall, A. & Gooday, G. W. (1975). A kinetic study of a solubilized chitin synthetase preparation from *Coprinus cinereus. Journal of General Microbiology,* **89,** 146–54.

Ruiz-Herrera, J. & Bartnicki-Garica, S. (1976). Proteolytic activation and inactivation of chitin synthetase from *Mucor rouxii. Journal of General Microbiology,* **97,** 241–9.

Ryder, N. S. & Peberdy, J. F. (1977). Chitin synthetase in *Aspergillus nidulans:* properties and proteolytic activation. *Journal of General Microbiology,* **99,** 69–76.

Salpeter, M. M. & Bachman, L. (1972). Autoradiography. *Principles and Techniques of electron microscopy,* vol. 2, ed. M. A. Hayat, pp. 221–78. New York: Van Nostrand Reinhold.

Scarborough, G. A. (1975). Isolation and characterization of *Neurospora crassa* plasma membranes. *Journal of Biological Chemistry,* **250,** 1106–11.

Thomson, K. S. & Fischer, E. P. (1976). Purification of two inhibitors of the chitin synthase-activating protease in *Phycomyces. Abstracts, 10th International Congress of Biochemistry* (Hamburg) p. 420.

Ulane, R. E. & Cabib, E. (1974). The activating system of chitin synthetase from *Saccharomyces cerevisiae.* Purification and properties of an inhibitor of the activating factor. *Journal of Biological Chemistry,* **249,** 3418–3422.

Ulane, R. E. & Cabib, E. (1976). The activating system of chitin synthetase from *Saccharomyces cerevisiae*. Purification and properties of the activating factor. *Journal of Biological Chemistry*, **251**, 3367–74.

10
Chitin synthesis and differentiation in *Coprinus cinereus*

G.W.GOODAY

Department of Microbiology, Marischal College, University of Aberdeen, Aberdeen AB9 1AS, UK

Introduction

This account describes the biosynthesis of chitin during elongation of the stipe of the toadstool *Coprinus cinereus* (Schaeff.) Fr. The stipe has only one function – that of raising the cap into a position suitable for the release and dispersal of the basidiospores; to this end it is positively phototropic and strongly negatively geotropic. In *Coprinus lagopus* Fr. the stages of fruit body morphogenesis have been described in detail by Borriss (1934). In the early stages the stipe is dependent on the presence of the cap for its continued development, as is common for the elongation of the stipe in other species of agarics (Gruen, 1969), but it outgrows this dependency so that the phase of rapid elongation can occur in the absence of the cap. Excised stipes of *C. cinereus,* freed of traces of cap tissue and basal vegetative mycelium, will elongate in the absence of exogenous water and nutrients to an extent which is dependent on their size – larger stipes elongate more than smaller ones (Gooday, 1974). This autonomous endotrophic phenomenon *in vitro* provides a useful experimental system to determine which processes are most important in controlling elongation. One such process, which appears to be a necessary accompaniment to elongation, is chitin biosynthesis.

The structure and site of chitin

Chitin is a homopolymer of *N*-acetylglucosamine units, linked by $\beta(1–4)$glycosidic bonds. The sugar units are in the chair form, and the resultant conformation is a linear chain, with each monomer inverted 180° with respect to its neighbour, i.e. it is a linear helix with two residues per turn. Rudall & Kenchington (1973) and Neville (1975) discuss the possibility that single chitin chains might have a slight right-handed helicity, but conclude that crysalline chain packing would force them into an exact two-fold screw axis.

The chain is given rigidity by hydrogen bonding, $O \cdots O$ between the O group of C-3 of one unit and the ring-O of its neighbour (Fig. 10.1). Chains can then stack on top of each other, with interchain hydrogen bonding, $CO \cdots NH$, above and below between the acetyl groups and the secondary amine groups at C-2.

The crystalline form found in fungi is a-chitin, in which the accepted structure is of adjacent piles of chains lying beside each other in an antiparallel form, i.e. facing alternately in opposite directions (Rudall, 1969). This structure is the unavoidable conclusion of X-ray data showing that the unit cell has only two chains of a repeating unit of diacetylchitobiose with two fold screw axes for both its a and b axes. This contrasts with β-chitin, in which the chains are parallel, but which can be converted to a-chitin by swelling with a concomitant contraction, representing a folding back of the chains on themselves (Preston, 1974). Rudall (1969) suggests that the conformation at the fold regions might

Fig. 10.1. The structure of the diacetylchitobiose unit of chitin. The dashed lines indicate hydrogen bonds. ●, C; ○, O; ◉, N. The unit cell of a-chitin is two such units, alongside one another but pointing in opposite directions.

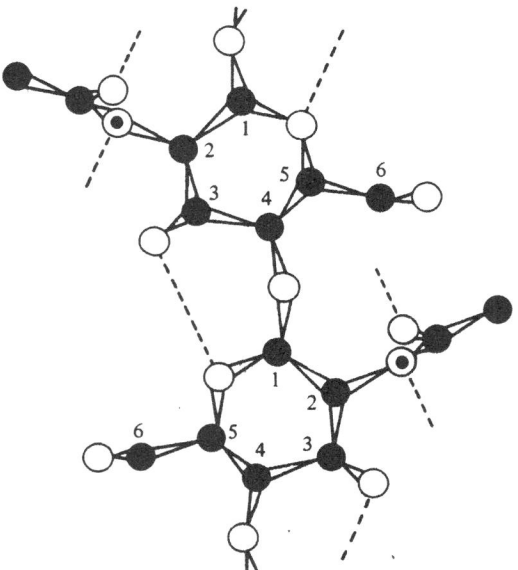

be assisted by a change of a sugar conformation from the chair form, or by a deacetylation.

Some deacetylation is found in nearly all native chitins, but, except in the case of the massive deacetylation to chitosan found in the Mucorales (Araki & Ito, 1974), it is not clear to what extent this is a controlled post-synthetic modification as opposed to a random uncontrolled process. In addition there is some evidence for cross-linking of proteins and other polysaccharides to chitin in fungal walls (see Burnett, Chapter 1; Wessels and Sietsma, Chapter 2) which may have significant consequences for the final conformation of the chitin.

The resultant chitin chains are assembled into structures visible in the electron microscope, the microfibrils (see Burnett, Chapter 1). In all systems examined in detail, the chitin microfibrils lie in the innermost layer of the fungal wall, adjacent to the plasmalemma (e.g. *Neurospora crassa* Shear & Dodge and *Schizophyllum commune* Fries; Hunsley & Burnett, 1970).

The mechanical strength of the chitin microfibrils suggests that they play a major role in the determination of the shape of the fungal cell, whilst withstanding the turgor pressure exerted by the cytoplasm. Thus chemical extraction of stipe cells by the method of Hunsley & Burnett (1970), to remove all other components, leaves chitinous shells in the shape of the original cell (Fig. 10.2), and treatment of vegetative hyphae of *Coprinus cinereus* (Schaeff.) Fr. with chitinase in an isotonic medium results in the release of protoplasts (Moore, 1975). However, the importance of turgor pressure in supporting the structure of the stipe of *C. cinereus* is shown by the collapse of the tissue after freezing and thawing.

Chitin synthesis during stipe elongation

Analysis of stipes of different lengths has shown that chitin synthesis takes place throughout elongation (Fig. 10.3). In terms of dry weight, the chitin content of these 54 stipes was constant at $10.9 \pm 2.3\%$ from start to finish. The results shown in Fig. 10.3 indicate an increment of about $32\,\mu g$ of chitin per mm increase in stipe length. Elongation of excised stipes also involves an increase in chitin content (Fig. 10.4). The final chitin content of excised stipes was less than those elongating *in vivo*, and during the elongation *in vitro* the increment was about $16\,\mu g$ of chitin per mm increase in stipe length, giving a final chitin content of about $22\,\mu g$ per mm length. However, as no exogenous nutrients were

Fig. 10.2. Metal shadowcast electron micrographs of chitin microfibrillar residues of stipe cells of *Coprinus cinereus*. Preparation (successive treatments with alkali, oxidizing agent, acid; shadowing with gold/palladium) as described by Hunsley & Burnett (1970). (*a*) A cell from the apical half of a stipe before rapid elongation (20 mm long). (*b*) A cell from the apical half of a stipe after rapid elongation (70 mm long). (*c*) As for (*b*) but cell has been torn during preparation so that the chitin in the wall has 'unwound' in a strip. Two portions of the strip are seen, each with an inner face, I_1 and I_2, and an outer face, O_1 and O_2. Note how the microfibrils visible on the torn edges face in the direction consistent with a Z-helix.

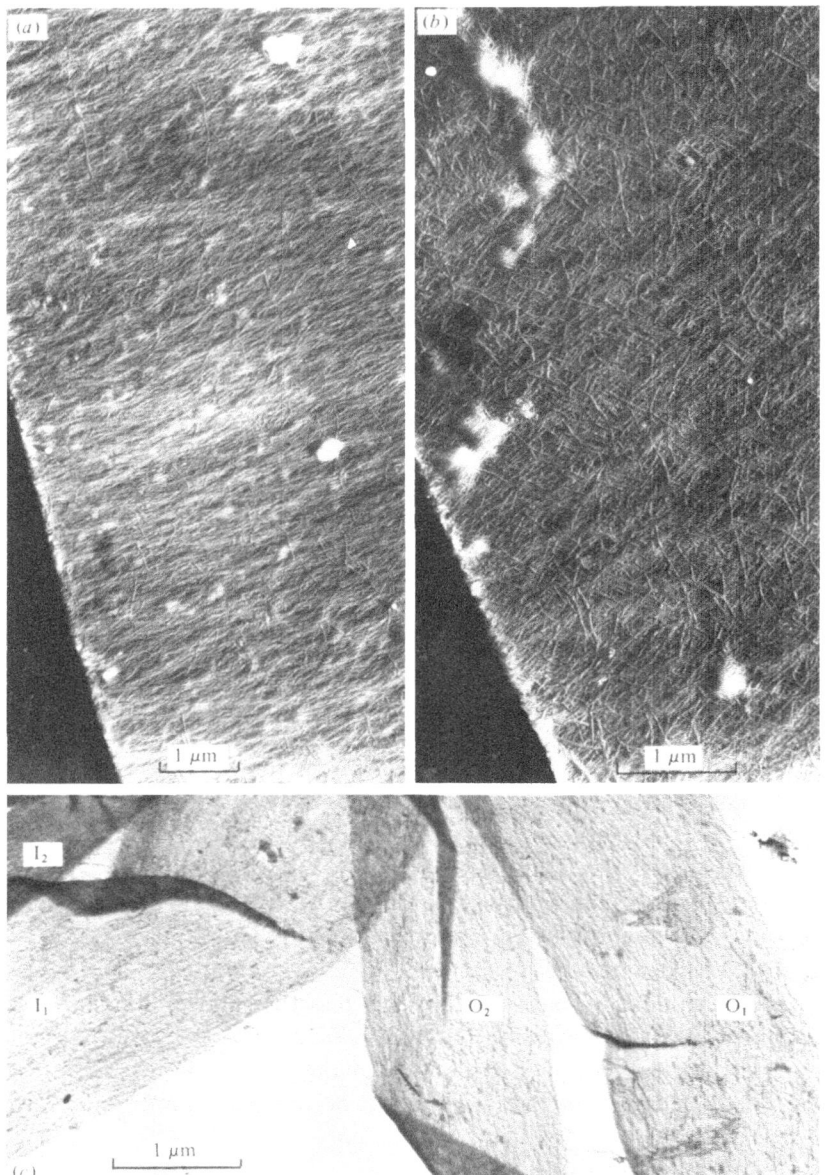

available, this increase in chitin (from 10.9 % to 16.1 ± 1.4 % of stipe dry weight) must have been at the expense of other cell constituents.

The techniques of light-microscopic and electron-microscopic auto-radiography allow the sites of chitin deposition during elongation to be visualized. Tritiated *N*-acetylglucosamine is the most suitable substrate for such studies. It is readily taken up by fungi and is phosphorylated by *N*-acetylglucosamine kinase to *N*-acetylglucosamine 6-phosphate which is converted to the immediate precursor of chitin, UDP-*N*-acetylglucosamine (see Fig. 10.8). It has been confirmed that exogen-ously supplied *N*-acetylglucosamine is efficiently incorporated into

Fig. 10.3. Chitin contents of stipes of *Coprinus cinereus*. Cultures as grown by Gooday (1974). All stipes were used and there was no selection of 'typical' specimens. Stipes were measured, freeze-dried, ground to a powder, extracted thrice with 2 ml methanol: water (4:1, v/v) for 15 h at 45°C, then thrice with 1.5 ml 0.24 M NaOH for 15 min at 100 °C. The residue was washed with methanol, dried, soaked overnight with 0.75 ml 6 M HCl, and hydrolysed at 100 °C for 8 h. Glucosamine was measured in the hydrolysate by the Elson–Morgan colorimetric method as described by Tracey (1955).

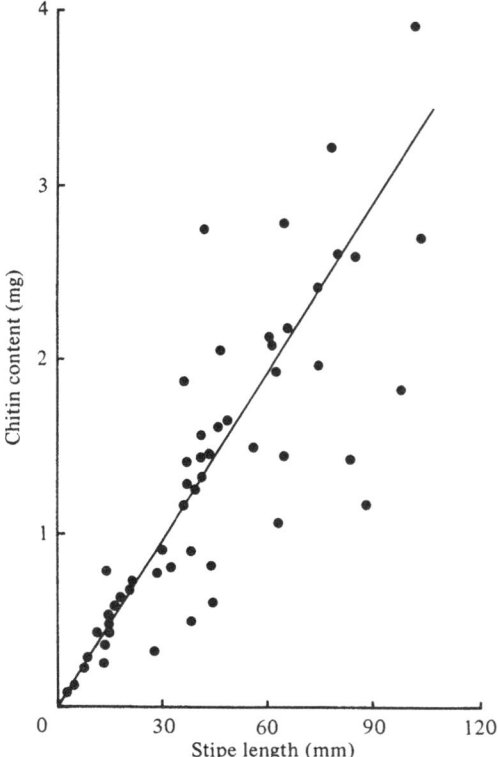

chitin in elongating stipes of *Coprinus cinereus*, and light-microscopic autoradiographs have shown that this incorporation is uniform along the length of the cells (Gooday, 1975). Only a few hyphae showed the pattern of apical labelling associated with 'normal' hyphal extension, and these probably represented secondary hyphae growing through the stipe tissue. This uniform incorporation of chitin by the stipe cells is consistent with intercalary elongation being a major factor involved in the elongation of the stipe, as has been described for *Agaricus bisporus* by Craig, Gull & Wood (1977). Electron-microscopic autoradiography has also shown a uniform deposition of chitin along the walls (Gooday, 1975).

Chitin microfibril orientation during elongation

In the walls of higher plants and algae, the structural microfibrils, particularly of cellulose but also of mannans and xylans, very often show very strong preferred orientations, which can be related to their structural roles (Preston, 1974). In fungi this is rarely the case, although closer analysis can reveal that there is more order than first appears in the arrangements of chitin microfibrils in hyphal walls (see Chapter 1).

Polarization microscopy is a powerful technique for determining crystallinity in biological material. It has been used to study the arrangement of chitin microfibrils in the walls of stipe cells of *Coprinus*

Fig. 10.4. Chitin contents of excised stipes allowed to elongate *in vitro*, as described by Gooday (1974). Lines are extrapolated from the points back to the chitin content corresponding to the length of the stipe at excision (dashed line taken from Fig. 10.3).

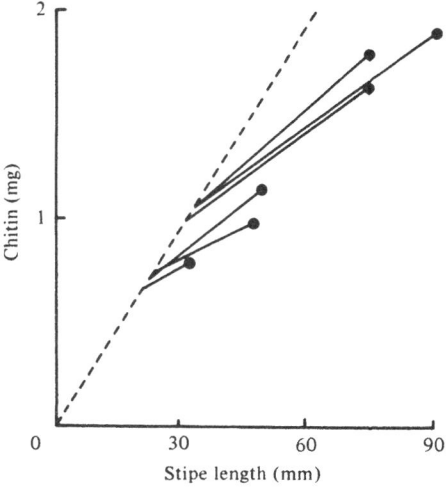

cinereus. The material has been examined unstained or stained with saturated aqueous Congo red. Observations have been made on fresh stipe material, squashed or teased out to individual cells; on fixed and sectioned material; and on material that has been chemically extracted to leave all cell components other than chitin (described in the legend to Fig. 10.2). Results have been essentially the same in all treatments.

Although chitin itself will be only weakly birefringent, oriented chitin shows a marked positive rodlet birefringence, due to its submicroscopic interstices (Roelofsen, 1959), i.e. the microfibrils are embedded in an orientated way in a medium with a different refractive index. These interstices can be stained by a colloidal dye such as Congo red, which results in increased birefringence as well as dichroitic staining.

With crossed polarizer and analyser, fungal tissues and cells are dark when parallel or transverse to the initial plane of polarization, but bright at diagonal positions, indicating either predominantly transverse or longitudinal orientations. This is true of stipes before rapid elongation (i.e. apical halves of stipes less than 20 mm long) and during and after elongation (i.e. apical halves of stipes 60 mm to 100 mm long), but stipes after elongation are more brightly birefringent. Material stained with Congo red and examined in polarized light is bright red when the plane of polarization is transverse to the long axes of the cells, and pale when the plane of polarization is along the length of the cells. This dichroism indicates predominantly transverse orientations. When squashed slices of tissue are examined, all hyphae show this phenomenon. When chemically extracted chitinous residues of walls or fresh specimens of tissue are stained with Congo red and examined with crossed polarizer and analyser, an increase in birefringence is observed. Accompanying this increase there is a resolution into different colours at the two different extinction positions (these are orange and bright green for chemically purified walls; blue and green-yellow for squashed tissue and teased-out cells; and bright green and gold for these viewed with a Red I compensator plate). This phenomenon shows that the microfibrils are in a shallow helix rather than being transverse. Thus, the weak birefringence seen in unstained cells results from an approximate averaging of that of their lower walls with that of their upper walls. With the stronger birefringence of the Congo red staining, the result is not an average, but addition or subtraction colours due to the order in which the polarized light meets the different layers.

Metal shadowcast electron micrographs of chemically purified walls of stipe cells of *Coprinus cinereus* (Fig. 10.2) show that the chitin is present as microfibrils. Preston (1974) warns of the dangers of changes in microfibril orientations that might accompany the acid, alkali and oxidizing treatments and consequent shrinkage during preparations, but if these possibilities are borne in mind, examination of such micrographs can be instructive (see Chapter 1). When preparations are made from stipes before (less than 20 mm long) and after (60 to 100 mm long) rapid elongation, the microfibrillar orientations of the chitin can be judged to be predominantly 'transverse' in both types of preparation, unequivocally so in about 60 % and 70 % of the cells respectively. The surface microfibrils of the remaining 40 % and 30 % of cells appear to have 'random' orientations. As nearly all cells appear birefringent by polarizing microscopy, presumably due to chitin microfibril alignments, these random orientations may have occurred during preparation. Alternatively, they may represent a superficial loss of parallel orientations by outer chitin microfibrils penetrating matrix materials such as glucans. When visible, torn inner faces of cells nearly always show a predominantly transverse orientation of microfibrils.

Analysis of electron micrographs shows that the apparent transverse orientations are really shallow helices. Thus Fig. 10.5 shows frequency

Fig. 10.5. Distribution of angles of 100 chitin microfibrils each in two cells from stipes of *C. cinereus*. (*a*) A cell from the apical half of a stipe 20 mm long, before rapid elongation; (*b*) a cell from the apical half of a stipe 70 mm long, after rapid elongation. Measurements made of angles with respect to the longitudinal axes of the cells on metal shadowcast electron micrographs of chemically extracted cells (as in Fig. 9.2*a* and *b*). Radius of semicircular scale represents the point at which 25 % of the fibrils fall into the particular 10° category (175–4°, 5–14°, etc.).

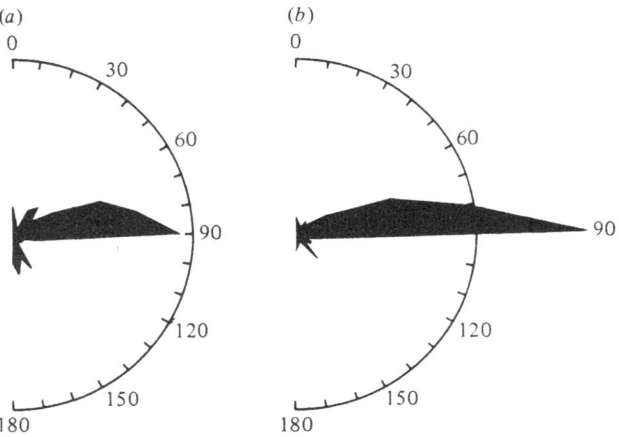

diagrams for two cells, one from a young stipe, one from an old stipe. The average angles in these two particular cases are respectively 82° and 86° with respect to the longitudinal axes of the cells. In the analyses performed to date, no differences in gross patterns of orientations have been observed between cells before and after elongation. The helix is right-handed, i.e. a Z-helix, as opposed to a left-handed S-helix (use of Z and S as in Preston, 1974). Although not taken into account by this analysis, closer observation of these preparations indicates that microfibrils not showing the predominant orientation tend to be those lying on the surface, which (as suggested above) might have been disturbed during preparation or might *in vivo* have been raised away from the main mass of chitin. Occasionally, chitinous residues are seen that have been 'unwound' during preparation, like a spring stretched, and closer examination shows that the chitinous wall has been pulled apart in the fashion to be expected from a Z-helix (Fig. 10.2*c*).

Stipe elongation requires elongation and division of cells, and these processes require the uniform intussusception of new chitin to increase the length of the walls at constant cell diameter. Given that (as seems likely) microfibrils are a finite length, and that any lateral bridges between microfibrils can readily be broken, then it would seem quite possible for increases in surface area of the walls to occur by insertion of new microfibrils between existing ones while maintaining the shallow Z-helicity.

Chitin synthase from *Coprinus* stipes

Chitin synthase can be obtained in very high activity from stipes of *Coprinus cinereus*. There is sufficient activity to account for the observed rate of chitin synthesis (Gooday, 1973). Results from cell fractionations and solubilizations with digitonin treatment are consistent with its being localized in the cell membrane (Gooday, 1977). Electron-microscopic autoradiography shows no evidence for any intracellular polymerization and transport of chitin, nearly all of the silver grains being associated with the walls and plasma membranes of the hyphae (the resolution of the technique does not allow these to be readily distinguished) (Table 10.1).

The chitinous product of the enzyme preparations from *Coprinus cinereus* has many properties consistent with its being macromolecular crystalline chitin (Gooday & de Rousset-Hall, 1975), as have the products from enzymes from other fungi (see Chapters 3, 7 & 9). The infrared spectrum is nearly identical to that of chemically purified arthropod a-chitin (Fig. 10.6; Table 10.2). The formation of the product can be visualized by light microscopy, as it forms as birefringent precipitates which stain fluorescently with FITC-labelled wheat germ agglutinin.

Control of chitin synthase

In yeast cells and in filamentous fungi there is now evidence (discussed in detail in Chapters 7 and 9) for a spatial and temporal sequence resulting in chitin synthesis; this is summarized in Fig. 10.7. It can be seen that there is considerable scope for control of chitin synthase production and activity in these seven stages.

Unlike the enzyme extracted from yeast cells or from apically growing hyphae, there is negligible zymogen activity in our preparations from stipe tissue of *Coprinus cinereus*. This is unlikely to be due to a straightforward difference in our preparation technique, as *Candida albicans* (Robin) Berkhout, subjected to an identical technique, yields considerable quantities of chitin synthase zymogen (J. Hardy & G. W. Gooday, unpublished results). It may be due to a rapid specific proteolysis of zymogen by specially active proteases in *C. cinereus* stipes, but this does not appear likely, as a preparation made of equal amounts of stipe tissue of *C. cinereus* and mycelium of *Mucor mucedo* Auct. had properties that were an average between those of the two prepared separately, i.e. enzymes from each tissue had no noticeable effect on those from the other tissue (W. H. Leith & G. W. Gooday, unpublished results).

Table 10.1 *Electron-microscopic autoradiography of chitin formation in elongating stipes of* Coprinus cinereus

Location	Relative intensity of labelling
Wall	65.8
Cytoplasm	5.0
Nuclei, mitochondria	1.7
Lomasomes	7.5
Intercellular space	3.0
Vacuoles	1.0

Results from segments of elongating stipe tissue incubated for 10 min in N-acetyl-D-[1 −^{3}H] glucosamine at 25°C. Expressed as ratio of silver grain distribution to 'random circle' distribution for each item compared to ratio for vacuoles.

Fig. 10.6. Infrared spectra of chemically purified arthropod a-chitin (above) and chitin produced by the action of chitin synthase from *Coprinus cinereus* (below). Spectra from KBr disks, by courtesy of Dr F. B. Williamson. (See Table 10.2.)

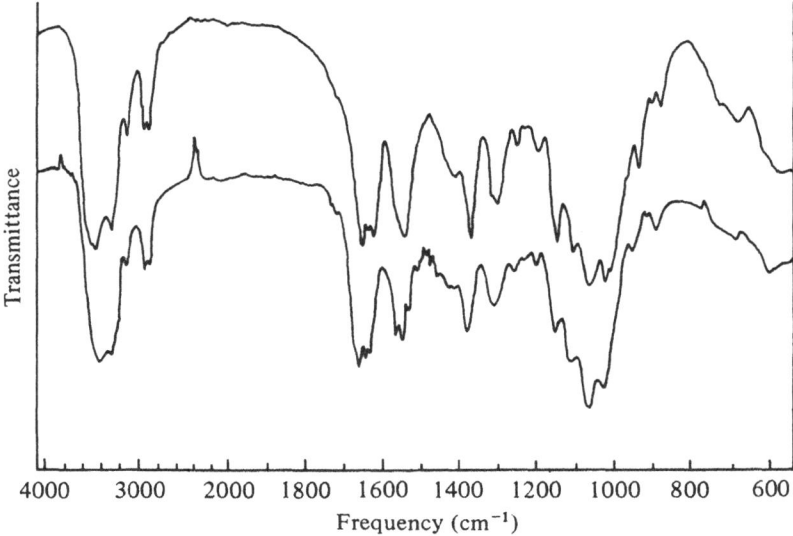

Table 10.2. *Characteristics of infrared spectrum of enzymically synthesized chitin compared to that of chemically purified arthropod a–chitin (cf. Fig. 10.6)*

Enzyme product (cm^{-1})	Purified chitin (cm^{-1})	Possible assignment[a]
895	895	Axial H of β-linkage
1030, 1070, 113	1025, 1070, 1115	Stretching in ring
115	1160	C_1–O–C_4 stretching
1310, 1555, 1630, ⎫	1310, 1557, 1626, ⎫	Stretchings within
1660, 3100, 3250 ⎭	1660, 3100, 3293 ⎭	amide group
2870, 2920	2870, 2929	CH axial stretching
2950	2950	CH$_2$ axial stretching

Reaction product obtained by centrifugation of several 100 μl incubations of 10 mM UDP-GlcNAc, 30 mM MgCl$_2$, 1 mM Na$_2$EDTA, 50 mM tris–HCl, pH 7.5, with solubilized enzyme preparation, for 210 min, 25°C. Product treated in a vacuum for 30 min with 350 mg KBr per mg sample for spectrometry with a Grubb Parsons Spectromaster MK3.

[a] F. B. Williamson (personal communication).

Fig. 10.7. The suggested temporal and spatial organization of chitin synthase in fungi. All steps are assumed to be reversible.

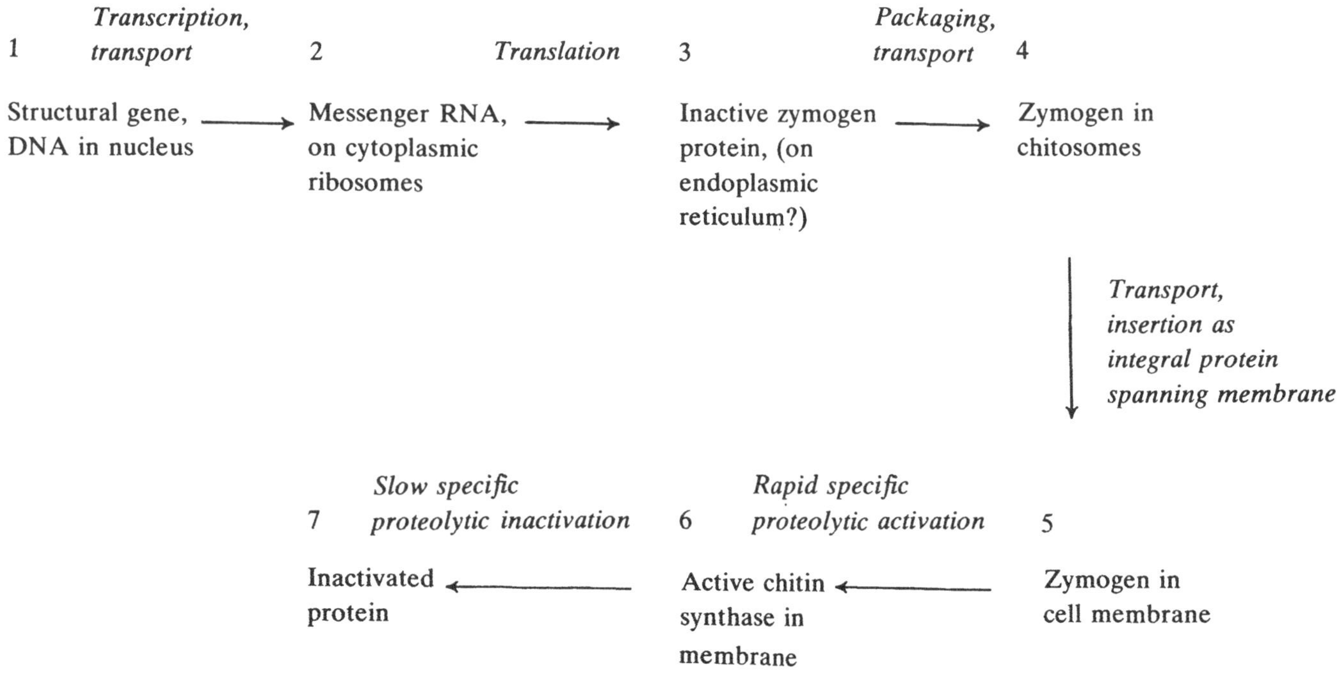

	Transcription, transport		*Translation*		*Packaging, transport*	
1		2		3		4

Structural gene, DNA in nucleus $\longrightarrow$ Messenger RNA, on cytoplasmic ribosomes $\longrightarrow$ Inactive zymogen protein, (on endoplasmic reticulum?) $\longrightarrow$ Zymogen in chitosomes

Transport, insertion as integral protein spanning membrane $\longrightarrow$

	Slow specific proteolytic inactivation		*Rapid specific proteolytic activation*	
7		6		5

Inactivated protein $\longleftarrow$ Active chitin synthase in membrane $\longleftarrow$ Zymogen in cell membrane

Rather, it appears that chitin synthase in the stipes of *C. cinereus* is substantially present as active enzyme in the plasma membrane, i.e. it is in state 6 of Fig. 10.7. If this is so, it may reflect the fact that chitin synthesis in this tissue is not localized, but occurs uniformly along the cell surfaces; perhaps it reflects also that the tissue is ephemeral, with autolysis rapidly following elongation. A consequence will be that in the stipe cells the time and place of chitin synthesis may be controlled much more by availability of substrates and effectors – as compared to other fungal systems, where control of activation of zymogen appears to play a major role.

Chitin synthase is allosterically activated by its substrate UDP-*N*-acetylglucosamine (de Rousset-Hall & Gooday, 1975). This will ensure that the enzyme is only active when there is sufficient substrate to support chitin synthesis. However, there is very little information with which to assess whether the enzyme is regulated by substrate availability. Slayman (1973) estimates the concentration of UDP-*N*-acetyl-glucosamine in exponentially growing cells of *Neurospora crassa* Shear & Dodge as 1.45 mM, which is very close to the K_m value for chitin synthase from this fungus of between 1 and 2 mM (Endo, Kakiki & Misato, 1970), and so in this fungus the substrate is unlikely to be limiting.

Chitinolysis and the significance of activation by N-acetylglucosamine

When stipe cells in *Coprinus cinereus* are elongating, chitin synthase must be responsible for the production of the chitin which must be inserted as helically orientated microfibrils in the wall to give an increase in surface area. However, at any time, the chitinous framework of the wall is a coherent structure, as seen after chemical extraction of other materials (e.g. Fig. 10.2*a, b*), and only very occasionally does one see instances where the microfibrillar structure has been torn apart (Fig. 10.2*c*). This suggests that the microfibrils are held together, either just by their interweavings, or (more probably in *C. cinereus*, where there is often a high degree of alignment and little apparent interweaving) by chemical or physical cross-bridging between adjacent microfibrils. Whatever the mechanism, it seems inevitable that a loosening of bonds between or within microfibrils is a prerequisite for insertion of new microfibrils in the elongating wall. Such a loosening of the microfibrillar network could be achieved by the action of the lytic enzyme, chitinase (see Chapters 8 and 12 by Stirling and Rosenberger). Chitinase activity is detectable in elongating stipes (Gooday, unpublished), but reliable estimates are still lacking, due to its high affinity for chitinous debris in cell homogenates.

Figure 10.8. suggests an '*N*-acetylglucosamine cycle', invoking the sequential action of chitinase and chitin synthase. Chitinase, in lysosomal vesicles (see Iten & Matile, 1970), would be released through the plasma membrane into the wall, to lyse accessible chitin chains. The result would be a wall with localized areas where the remaining microfibrils had lost their interconnections, and the product would be diacetylchitobiose. This perhaps could enter the cell intact, or would be broken down extracellularly by *N*-acetylglucosaminidase/*N,N*-diacetylchitobiase to yield *N*-acetylglucosamine, which would be taken up by a specific permease. Some of the *N*-acetylglucosamine would be phosphorylated and recycled to give more UDP-*N*-acetylglucosamine to supplement the supply of this precursor for the synthesis of the new chitin.

Fig. 10.8. Proposed *N*-acetylglucosamine cycle associated with chitin synthesis in *C. cinereus*. The cycle postulates sequential action of chitinase, *N*-acetylglucosaminidase (diacetylchitobiase), *N*-acetylglucosamine kinase, *N*-acetylglucosamine phosphomutase, uridine diphosphate *N*-acetylglucosamine pyrophosphorylase and chitin synthase.

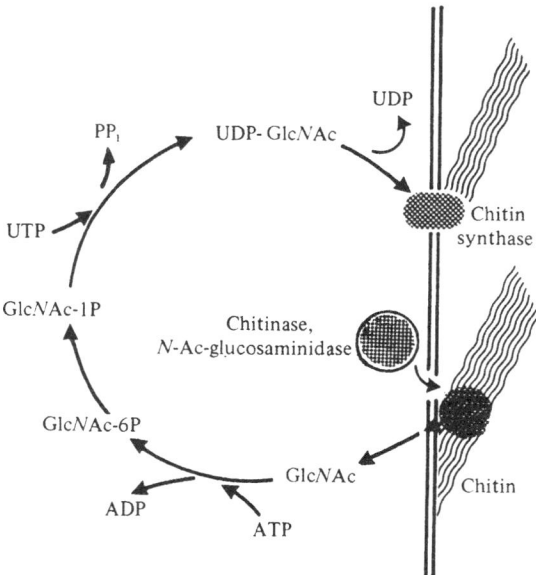

However, in common with enzyme from all other fungi reported, the chitin synthase of *Coprinus cinereus* is activated by both the monomer and dimer of chitin, *N*-acetylglucosamine and diacetylchitobiose (de Rousset-Hall & Gooday, 1975; Gooday, 1977). The kinetics of activation are consistent with these sugars acting as allosteric effectors; there is no evidence for their being incorporated into the chitin product. Thus the localized arrival of *N*-acetylglucosamine could then allosterically activate chitin synthase precisely at that site – i.e. where the chitinase had just 'loosened' the wall, so that new microfibrils could be inserted at this predetermined spot.

Requirement for metal cations

Microsomal preparations of chitin synthase from *Coprinus cinereus* that have been washed with magnesium-free buffer and solubilized with magnesium-free digitonin medium show an absolute requirement for a metal cation for activity. Magnesium chloride most efficiently fills this requirement (Fig. 10.9): there is no activity in its absence;

Fig. 10.9. Effect of concentration of magnesium chloride on enzyme activities from *C. cinereus*. Symbols ●, chitin synthase, 1 mM UDP-GlcNAc; ■ , chitin synthase, 0.1 mM UDP-GlcNAc; ▲, uridine diphosphatase, 2.0 mM UDP. Results are from separate enzyme preparations; assays are as described by Gooday & de Rousset-Hall (1975) and de Rousset-Hall & Gooday (1975). One unit on the ordinate scale is equivalent to enzyme reaction velocities (nmol substrate utilized min^{-1} (mg protein)$^{-1}$) of: ●, 40; ■ , 10; ▲ , 26.7.

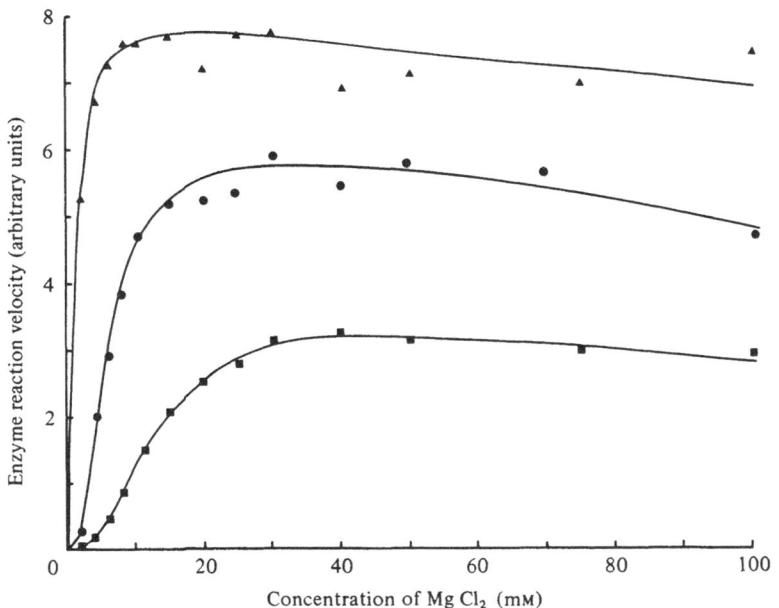

increasing its concentration gives a sigmoidal increase in activity up to a maximal activity at about 30 mM $MgCl_2$ at concentrations of 0.1 and 1.0 mM UDP-GlcNAc; higher Mg^{2+} concentrations are inhibitory, with 250 mM $MgCl_2$ giving less than 50 % of maximal activity. A Hill plot ($\log v/(V - v)$ against $\log [Mg^{2+}]$ (Atkinson, Hathaway & Smith, 1965)) drawn from results at 0.1 mM UDP-GlcNAc gave a straight line, with slope of 2.4 and a $[S]_{0.5}$ value of 12.0 mM $MgCl_2$. Magnesium must clearly be thought of as a cosubstrate.

Assays in which $MgCl_2$ is replaced by a range of other metal chlorides show that only Mn^{2+} can completely replace Mg^{2+}, indeed it gives slightly higher activities. Cu^{2+} and Cd^{2+} are strongly inhibitory when added together with Mg^{2+}. Thus 1.5 mM $CuCl_2$ gives 50 % inhibition of chitin synthase activity in the presence of 25 mM $MgCl_2$.

The finding of the requirement for Mg^{2+}, and the marked sigmoidicity of its activity, leads to the question as to what extent its availability might be a controlling factor for enzyme activity *in vivo*. There are no estimates of the free Mg^{2+} concentration in stipe cells (considerable amounts would be bound to nucleic acids and to proteins) and the concentrations could well be different in the cytoplasm and in the vacuoles which form a large part of these cells. Estimates for mammalian tissues indicate that the free Mg^{2+} concentration is usually 1 mM or less (Heaton, 1973). This is considerably less than the optimum shown in Fig. 10.9, and indeed is close to the 'turn-up' point of the sigmoidal curve. If similar levels pertain in *Coprinus cinereus* and other fungi, Mg^{2+} availability could clearly play a major role in regulating the enzyme's rate of reaction, and there might be a need for localized mechanisms for transporting or reversibly binding the Mg^{2+}.

A uridine nucleotide cycle?

Uridine disphosphate, the second product of chitin synthase activity, is a powerful inhibitor of enzyme activity. A concentration of UDP of 0.8 mM produces 50 % inhibition of both the particulate and solubilized enzyme preparations at a substrate concentration of 1.0 mM UDP-GlcNAc. A Dixon plot (Dixon, 1953) of its effect on enzyme activity at two substrate concentrations gives two straight lines intersecting to give an inhibition constant of about 0.5 mM, and indicates that the UDP is acting as a competitive inhibitor (Fig. 10.10). Hill plots ($\log (v_0 - v)/v$ against $\log [UDP]$ where v_0 is the enzyme reaction velocity in the absence of UDP) for different substrate concentrations give straight lines with slopes of 1.0, and estimates of $[M]_{0.5}$ (equivalent to K_i) of about 0.6 mM. Hill plots ($\log v/(V - v)$ against $\log$ [UDP-GlcNAc]) of assay results in the presence of 0.5 and 1.0 mM UDP give straight lines with unchanged slope (1.7) but increased $[S]_{0.5}$ (from 0.9 to 1.3 and 2.5 mM UDP-GlcNAc respectively), while V (maximal enzyme reaction velocity) is unchanged. These results are also indicative of UDP acting as a competitive inhibitor.

However, UDP does not accumulate as the sole nucleotide product during incubations of chitin synthase from *Coprinus cinereus*. Thin-layer chromatograms show that it is broken down to UMP by a uridine diphosphatase activity in the enzyme preparations (de Rousset-Hall & Gooday, 1975). The uridine diphosphatase shows classical Michaelis–Menten enzyme kinetics, with a K_m of 0.35 mM UDP as estimated by a Lineweaver–Burk double reciprocal plot (Fig. 10.11). A Hill plot of the same results gives, as expected, a slope of 1.0, and a value of $[S]_{0.5}$ of 0.3 mM UDP. This is smaller than the $[S]_{0.5}$ for chitin

Fig. 10.10. A Dixon plot of the reciprocal of enzyme reaction velocity against uridine diphosphate concentration for the solubilized chitin synthase from *C. cinereus* at two substrate concentrations: 1.0 mM UDP-GlcNAc (■), giving a regression line of $y = 20.4x + 11.2$; 0.4 mM UDP-GlcNAc (●), giving a regression line of $y = 6.1x + 4.6$.

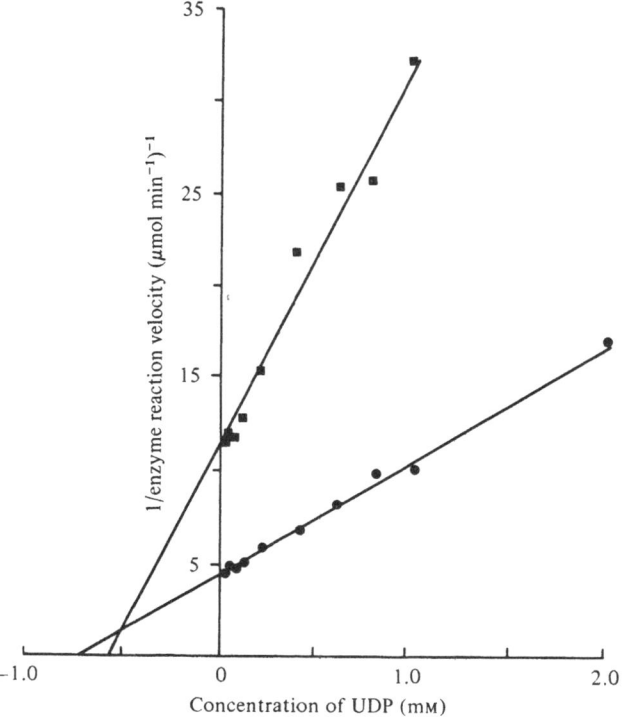

synthase of 0.9 mM UDP-GlcNAc, and thus it is to be expected that the UDP produced by the action of chitin synthase will be efficiently hydrolysed to UMP.

The UMP is much less inhibitory to the action of chitin synthase than UDP, and so the activity of this UDPase will favour activity of chitin synthase. This suggests that the UDPase may exert a direct regulatory role on chitin synthesis, especially as it is a membrane-bound enzyme, appearing in the same subcellular fractions as chitin synthase. However, continued synthesis of chitin requires a continued supply of UDP-GlcNAc, and this requires a recycling of the uridine moiety. Such a uridine nucleotide cycle is suggested in Fig. 10.12, where the membrane-bound UDPase breaks down the UDP to allow chitin synthesis to proceed, but the UMP must be rephosphorylated in the cytoplasm, to UDP, to UTP and to UDP-GlcNAc. Thus the stoichiometry of chitin synthesis during the operation of this cycle requires the participation of an extra molecule of ATP.

This suggested involvement of UDPase in the control of chitin synthesis *in vivo* has the consequence that effectors of its activity will vicariously affect chitin synthesis. Risking the danger that this argument

Fig. 10.11. A Lineweaver–Burk plot for uridine diphosphatase activity from *C. cinereus*. K_m value from this plot is 0.35 mM. Assays were as described by de Rousset-Hall & Gooday (1975).

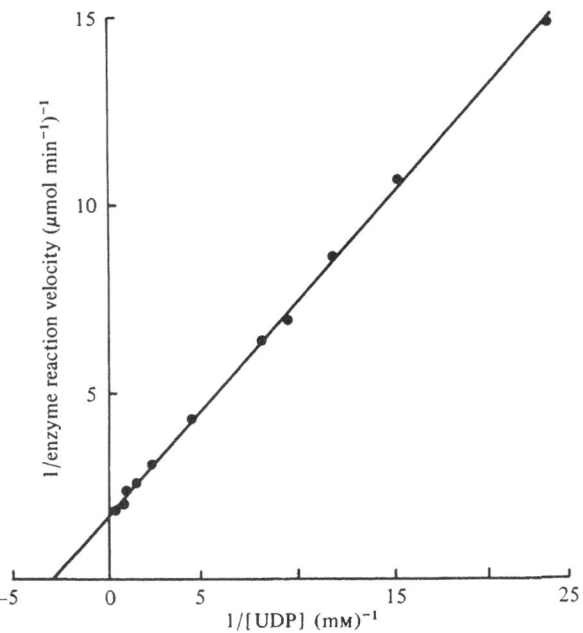

will become an insoluble Chinese puzzle – boxes within boxes – it can be pointed out that, for example, the UDPase also requires Mg^{2+} for activity (Fig. 10.9). Unlike chitin synthase, the curve for activation is not sigmoidal, and the maximal activity is rapidly achieved by a much lower concentration. Another likely effector of this UDPase is inorganic phosphate, which is an inhibitor of nucleoside diphosphatases from other sources.

Such a uridine nucleotide cycle has been proposed by Kuhn & White (1977) as a regulatory mechanism for lactose synthesis within the Golgi apparatus of mammary glands.

Slayman (1973) gives estimates of apparent intracellular concentrations of nucleotides, showing that of UDP in exponentially growing cells of *Neurospora crassa* as much lower, at 0.23 mM, than that of UDP-*N*-acetylglucosamine, at 1.45 mM. A calculation from his result of the level of a nucleotide with chromatographic mobility of UMP gives its tentative concentration at 3.4 mM. These relative concentrations might indicate that there is also active UDPase in *N. crassa*.

Fig. 10.12. Proposed uridine nucleotide cycle associated with chitin synthesis in *C. cinereus*. The cycle postulates sequential action of chitin synthase, uridine diphosphatase, uridine monophosphate kinase, uridine diphosphate kinase, uridine diphosphate *N*-acetylglucosamine pyrophosphorylase and a phosphatase hydrolysing inorganic pyrophosphate.

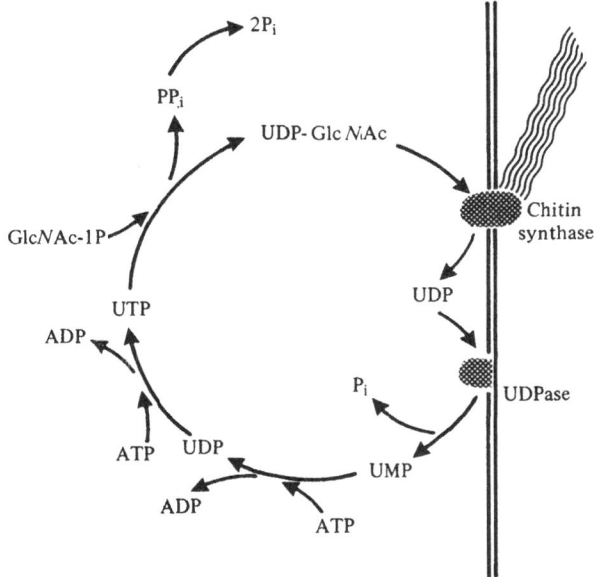

Conclusions

Chitin is a major constituent of the stipe of *Coprinus cinereus*, throughout its elongation. Its relative amount, and its disposition as nearly transverse microfibrils, remain unchanged. The elongation of individual cells must be 'driven' by their turgor pressure, but must be 'allowed' by the expansion of the cell surface by the synthesis of new chitin microfibrils. However, the mechanism for controlling the rate of synthesis, i.e. the activity of chitin synthase, remains obscure. If there is ample substrate, then activation by *N*-acetylglucosamine will have little effect, and the action of UDPase will serve to maintain the rate of utilization of UDP-Glc*N*Ac. But what factors decide what that rate should be? As the enzyme reaction velocities recovered *in vitro* closely approximate the required rate of synthesis of chitin *in vivo*, the enzyme in these stipes may be working 'flat-out', unrestrained by any control processes, i.e. this might be a rare case in which amount of enzyme protein may be the limiting factor for activity *in vivo*.

Acknowledgements—I thank my colleagues, Drs D. Hunsley and A. de Roussett-Hall, for their collaborations in the work described here.

References

Araki, Y. & Ito, E. (1974). Pathway of chitosan formation in *Mucor rouxii*. Enzymic deacetylation of chitin. *Biochemical and Biophysical Research Communications*, **56**, 669–75.

Atkinson, D. E., Hathaway, J. A. & Smith, E. C. (1965). Kinetics of regulatory enzymes. Kinetic order of the yeast diphosphopyridine nucleotide isocitrate dehydrogenase reaction and a model for the reaction. *Journal of Biological Chemistry*, **240**, 2682–90.

Borriss, H. (1934). Beiträge zur Wachstums- und Entwicklungsphysiologie der Fruchtkörper von *Coprinus lagopus*. *Planta*, **22**, 28–69.

Craig, G. D., Gull, K. & Wood, D. A. (1977). Stipe elongation in *Agaricus bisporus*. *Journal of General Microbiology*, **102**, 337–47.

Dixon, M. (1953). Determination of enzyme inhibitor constants. *Biochemical Journal*, **55**, 170–1.

Endo, A., Kakiki, K. & Misato, T. (1970). Mechanism of action of the antifungal agent. polyoxin D. *Journal of Bacteriology*, **104**, 189–96.

Gooday, G. W., (1973). Activity of chitin synthetase during the development of fruit bodies of the toadstool *Coprinus cinereus*. *Biochemical Society Transactions*, **1**, 1105–7.

Gooday, G. W. (1974). Control of development of excised fruit bodies and stipes of *Coprinus cinereus*. *Transactions of the British Mycological Society*, **62**, 391–9.

Gooday, G. W. (1975). The control of differentiation in fruit bodies of *Coprinus cinereus*. *Reports of Tottori Mycological Institute (Japan)*, **12**, 151–60.

Gooday, G. W. (1977). Biosynthesis of the fungal wall – mechanisms and implications. The first Fleming Lecture. *Journal of General Microbiology*, **99**, 1–11.

Gooday, G. W. & de Rousset-Hall, A. (1975). Properties of chitin synthetase from *Coprinus cinereus*. *Journal of General Microbiology*, **89**, 137–45.

Gruen, H. E. (1969). Growth and rotation of *Flammulina velutipes* fruit bodies and the dependence of stipe elongation on the cap. *Mycologia*, **61**, 149–66.

Heaton, F. W. (1973). Magnesium requirement for enzymes and hormones. *Biochemical Society Transactions*, **1**, 67–70.

Hunsley, D. & Burnett, J. H. (1970). The ultrastructural architecture of the walls of some hyphal fungi. *Journal of General Microbiology*, **62**, 203–18.

Iten, W. & Matile, P. (1970). Role of chitinase and other lysosomal enzymes of *Coprinus lagopus* in the autolysis of fruiting bodies. *Journal of General Microbiology*, **61**, 301–9.

Kuhn, N. J. & White, A. (1977). The role of nucleoside diphosphatase in a uridine nucleotide cycle associated with lactose synthesis in rat mammary-gland Golgi apparatus. *Biochemical Journal*, **168**, 423–33.

Moore, D. (1975). Production of *Coprinus* protoplasts by use of chitinase or helicase. *Transactions of the British Mycological Society*, **65**, 134–6.

Neville, A. C. (1975). *Biology of the Arthropod Cuticle*. Berlin: Springer-Verlag.

Preston, R. D. (1974). *The Physical Biology of Plant Cell Walls*. London: Chapman and Hall.

Roelofsen, P. A. (1959). *The Plant Cell Wall*. Berlin: Gebrüder Borntraeger.

de Rousset-Hall, A. & Gooday, G. W. (1975). A kinetic study of a solubilized chitin synthetase preparation from *Coprinus cinereus*. *Journal of General Microbiology*, **89**, 146–54.

Rudall, K. M. (1969). Chitin and its association with other molecules. *Journal of Polymer Science*, Part C **28**, 83–102.

Rudall, K. M. & Kenchington, W. (1973). The chitin system. *Biological Reviews*, **48**, 597–636.

Slayman, C. L. (1973). Adenine nucleotide levels in *Neurospora*, as influenced by the conditions of growth and by metabolic inhibitors. *Journal of Bacteriology* **114**, 752–66.

Tracey, M. V. (1955). Chitin. *Modern Methods of Plant Analysis*, vol. 2, eds. K. Paech & M. V. Tracey, pp. 264–74. Berlin: Springer-Verlag.

11
Glucanases, glucan synthases and wall growth in *Saprolegnia monoica*

M.FÈVRE
Université Claude Bernard Lyon I, Laboratoire de Physiologie Végétale, Laboratoire de Mycologie associé au CNRS No. 44, 43 Boulevard du 11 Novembre 1918, 69621, Villeurbanne, France

Introduction

The structure of the cell wall (a microfibrillar network of cellulose or chitin embedded in a glucan matrix) and the plasticity of this cell wall at the growing centres, involves a loosening of the texture during cell growth and cell wall extension.

Cell wall elongation of green plant cells and lysis of mycelial apices is promoted by numerous chemical treatments (immersion in acidic pH solutions for example) (Rolland & Pilet, 1974; Robertson, 1965; Bartnicki-Garcia, 1973). In green plants, a plasmalemma-bound ATPase is involved in the acid growth effect. This enzyme, as a hydrogen ion pump, causes the uptake of protons from the cytoplasm into the wall where they may dissociate the acid-labile bonds of cell wall components (Rolland & Pilet, 1974). In fungi, interactions between Ca^{2+} and H^+ can control the hyphal bursting of *Mucor rouxii* (Calmette) Wehm. (Dow & Rubery, 1975).

However, numerous observations show the association of lytic enzymes, in particular glucanase, with various morphogenetic processes involving fungal cell walls (Table 11.1). The balance between wall synthesis, wall lysis and the mechanical force due to cytoplasmic turgor could explain hyphal tip growth (Robertson, 1965; Bartnicki-Garcia, 1973).

Indirect evidence, derived mostly from ultrastructural studies, suggests that the cytoplasmic vesicles found in each growing centre (apices – branching sites) play a decisive role in wall growth by transporting wall precursors and/or the necessary enzymes to the centres (Grove & Bracker, 1970; Grove, 1971; Mullins & Ellis, 1974).

As the characteristic carbohydrate components of the Oomycete cell

Table 11.1. *Hydrolytic enzymes involved in morphogenetic processes.*

Process and fungus	Enzyme	Reference
(a) Yeasts		
Growth		
Schizosaccharomyces pombe Lindner	$\beta(1-3)$glucanase Cell wall autolysis	Barras, 1972; Fleet & Phaff, 1974; Kröning & Egel, 1974
Cell elongation		
Saccharomyces cerevisiae Hansen	Protein disulphide reductase	Brown & Hough, 1966
Budding		
Saccharomyces cerevisiae	$\beta(1-3)$glucanase, mannanase	Cortat, 1971; Maddox & Hough, 1971
Conjugation		
Hansenula wingei Wickerham	$\beta(1-3)$glucanase	Brock, 1964
Ascus lysis		

Hansenula anomola (Hansen) H. et P.
   Sydow
   Fabospora fragilis (Jorgensen)
   Kudriavzev
	$\beta(1-3)$glucanase	Abd El Al & Phaff, 1968

Dimorphism (Yeast – Mycelium)		

Candida albicans (Robin) Berkhaut
   Paracoccidioides brasiliensis (Splendore)
   Almeda
	Protein disulphide reductase
	Protein disulphide reductase
		Nickerson & Falcone, 1956 *a*, *b*
		Kanetsuna *et al.*, 1972

(b) Filamentous fungi		
Macroconidial germination		
Microsporum gypseum (Bodin) Guirt Grigorakis	$\beta(1-3)$glucanase, proteases	Page & Stock, 1972, 1974

Conidial germination; germ tube growth and branching		
Neurospora crassa Shear & Dodge	β(1–3)glucanase, proteases	Mahadevan & Mahadkar, 1970 Mehta & Mahadevan, 1975
Growth and hyphal morphogenesis		
Pythium sp.	β(1–3) glucanase, cellulase	Siestma & Haskins, 1968
Saprolegnia monoica	Cellulase β(1–3)glucanase	Fèvre, 1969, 1970, 1972, 1976
	Cell wall autolysis	
Aspergillus nidulans (Eidam) Winter	Cell wall autolysis	Polachek & Rosenberger, 1975
Antheridial branching		
Achlya ambisexualis J. R. Raper	Cellulase	Thomas des S. & Mullins 1967, 1969
Septum dissolution		
Schizophyllum commune Fries	R-glucanase	Wessels & Koltin, 1972
Sporangiophore elongation		
Phycomyces blakesleeanus Burgeff	Chitinase	Cohen, 1975
Conidial liberation		
Neurospora sp.	Cell wall autolysis	Selitrennikoff, Nelson & Stegel, 1974
Carpophore development		
Schizophyllum sp.	R-glucanase	Wessels, 1965
Carpophore autolysis		
Coprinus lagopus Fries	Chitinase, protease	Iten & Matile, 1970
	β(1–3)glucanase	Bush, 1974

walls are $\beta(1\text{--}3)$-linked and $\beta(1\text{--}4)$-linked glucans, major emphasis was placed on investigations of glucanases (glucanase, cellulase) and glucan synthases in order to elucidate the structural and biochemical relations of the growth centre of *Saprolegnia monoica* Pringsheim.

Hyphal Morphogenesis and hydrolases

The localization of elongation to the hyphal tip, and the regular appearance and spatial arrangement of branches, indicate the interdependence of the cells' activities in the mycelium of filamentous fungi.

If certain species, such as *Saprolegnia monoica*, are cultivated on a nutritionally deficient medium, the development of lateral branches is slowed down or even stopped and, sometimes, branching disappears (Larpent, 1966).

Morphogenetic effects of protein synthesis inhibitors

The inhibition of growth rate by puromycin or DL-*p*-fluorophenylalanine (*p*-FPA) is proportional to the logarithm of inhibitor concentration until saturation at 54 % of the control (Fig. 11.1*a*).

Fig. 11.1. Effects of DL-*p*-fluorophenylalanine on hyphal growth (*a*) and hyphal branching (*b*) of *S. monoica*.

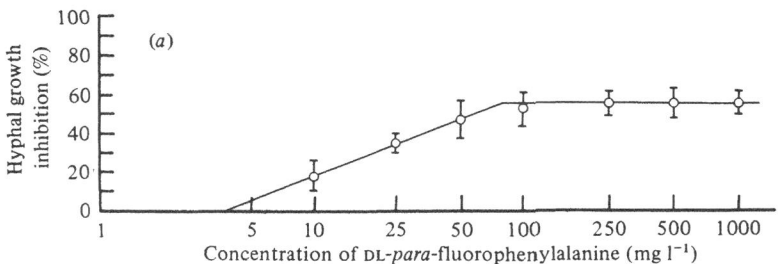

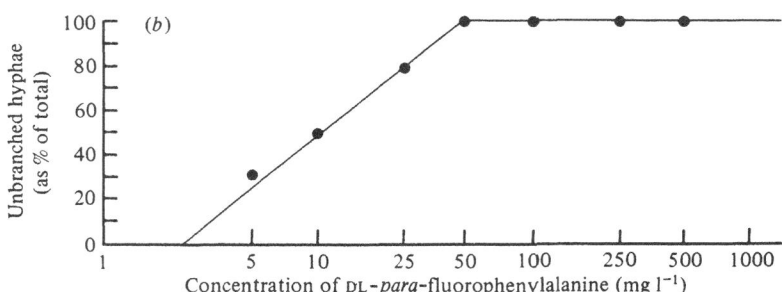

The distances separating two branches increase with increasing inhibitor concentrations, e.g. at 10 mg l⁻¹ puromycin, successive branches are separated by 380 μm instead of 110 μm as in the control. At higher concentrations of inhibitor (p-FPA at 100 mg l⁻¹) the hyphae become devoid of any lateral branches. This loss of branching capacity corresponds to the maximum of the inhibition level of growth rate (Fèvre, 1968 and Fig. 11.1b; Fig. 11.3). A relationship between elongation growth and branching capacity exists (Fèvre, 1970). A diminution of the number of branches per unit of length of the main hyphae and also a diminution of the number of branched hyphae corresponds to a reduction in the growth rate (Fig. 11.2).

Comparison of enzyme activities

Mycelia grown on cellophane in Petri dishes where analysed for $\beta(1-3)$ glucanase and cellulase activities. Colonies of the same size (but of different ages) obtained in the presence of the protein-synthesis inhibitor p-FPA, or on the control medium, contained different amounts of enzymes (Table 11.2). The loss of branching capacity corresponds to a decrease in the amounts of hydrolases able to act on cell wall polymers. Cellulase seemed to be more affected than glucanase.

Fig. 11.2. Relation between growth rate and branching capacity of the mycelium of *S. monoica* grown in the presence of different concentrations of DL-p-fluorophenylalanine.

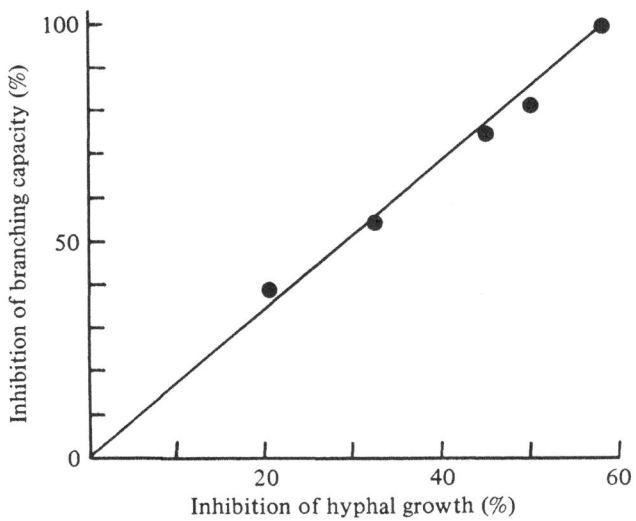

Fig. 11.3. Morphology of the mycelium of *Saprolegnia*. (*a, b*) Branched hyphae obtained on the control medium. (*c, d*) Unbranched hyphae obtained in the presence of 100 mg 1⁻¹ DL-*p*-fluorophenylalanine.

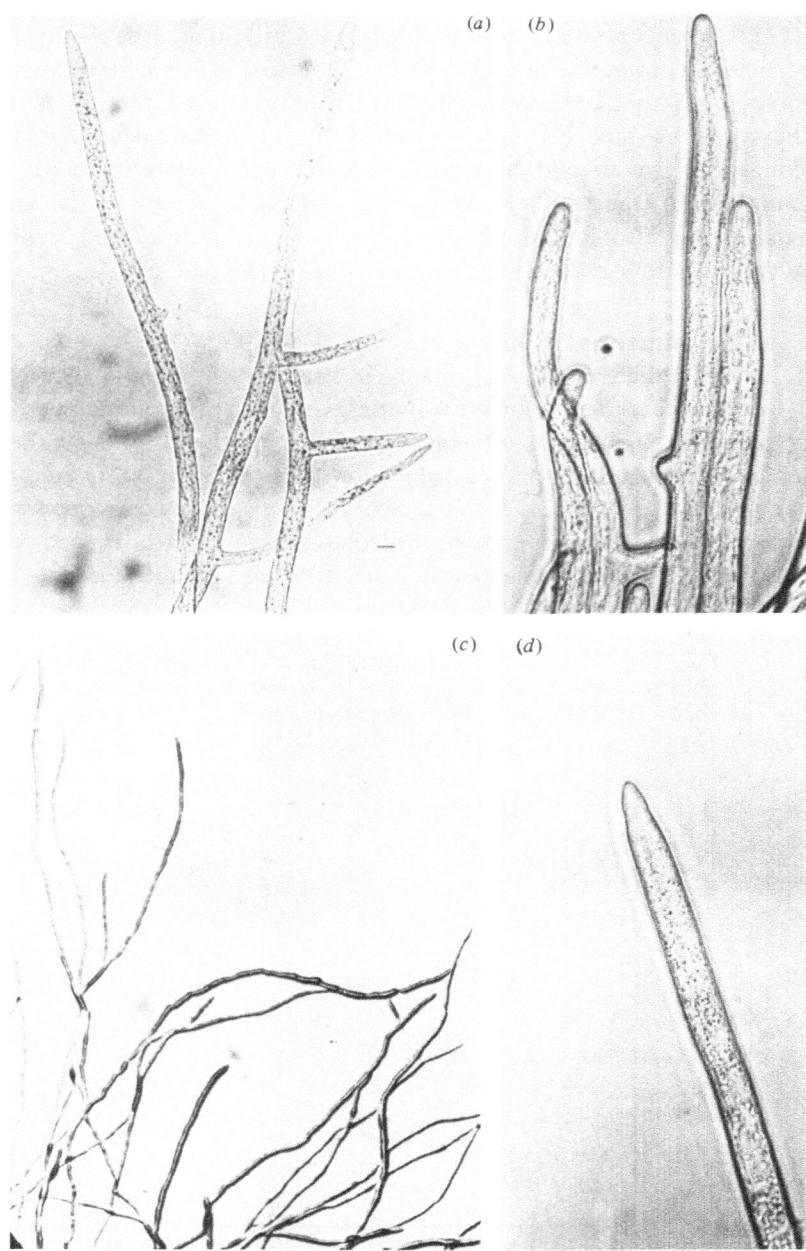

(*a*)　(*b*)

(*c*)　(*d*)

Localization of glucanase and cellulase in the fungal colony

The fungal colony obtained on a plate can, theoretically, be divided into three concentric zones. The youngest, at the margin of the colony, is the zone of extension and branching of the main hyphae. In the second zone, only the branches continue to grow, but at a lower speed than the main hyphal tips. The inner zone, the oldest, is a non-growing zone. The proportions and the relative importance of each zone are difficult to establish.

Mycelia grown on cellophane in Petri dishes were cut in concentric zones and analysed for enzyme activities. $\beta(1-3)$glucanase and cellulase were localized mainly at the edge of the colony (the extension and branching zone) and their activities decreased with increasing age of the zone. Thus, the activities (per mg freeze-dried mycelium) in the oldest zone were only 12 % (cellulase) and 40 % (glucanase) of those in the hyphal tip area (Fig. 11.4*a*). Their specific activities showed the same trend; activities in the oldest zone were 50 % (glucanase) and 20 % (cellulase) of those of the youngest zone (Fig. 11.4*b*). The distribution of other enzymes lacking any major morphogenetic role was different. For example, acid phosphatases in particular, succinic INT reductase and, to a lesser extent, β-glucosidase, were localized in the older parts of the mycelium.

Morphogenetic action of glucono-δ-lactone, a glucanase inhibitor

Gluconolactone is a competitive inhibitor of $\beta(1-3)$ glucanase (Fleet & Phaff, 1974). Purified exo-β-glucanase or β-glucosidase from

Table 11.2. *Comparison of enzyme activities of branched and unbranched mycelia*

Culture	Size (mm)	Age (h)	Activity per mg freeze-dried mycelium	
			Cellulase	$\beta(1-3)$glucanase
Branched mycelia (control)	25	41	12	480
Unbranced mycelia (treated with p-fluorophenylalanine 100 mg l⁻¹)	25	75	5	365

Colonies were grown on cellophane films and harvested when they reached the same size.
One unit of cellulase activity is expressed as a decrease of 1 % of viscosity of the initial solution (hydroxyethyl cellulose, 2 %).
$\beta(1-3)$glucanase activity is expressed as μg glucose h⁻¹ liberated from laminarin (3 %).

Saprolegnia are inhibited by low concentrations; 50 % inhibition is obtained at 2.5 μM. High concentrations are necessary to block purified endo-β(1–3) glucanase, 20 mM is required to obtain 50 % inhibition (Fèvre, unpublished).

The presence of gluconolactone in the culture medium results in a reduction of hyphal growth and a diminution of branching capacity. At a

Fig. 11.4. Mycelia of *S. monoica* grown on cellophane in Petri dishes were cut into concentric zones, freeze-dried and analysed for enzymatic activities. Activities found in the different zones are expressed as a ratio relative to those found in the apical zone (1.0). (*a*) Enzyme activities per mg freeze-dried mycelium; (*b*) specific activities. Symbols: O——O, acid phosphatase; ●——●, succinic INT reductase; ▲——▲, β-glucosidase; △——△, β(1–3)glucanase (assayed by the DB Merieux kit); ■——■, β(1–3)glucanase (assayed by dinitrosalicylic reagent); □——□, cellulase.

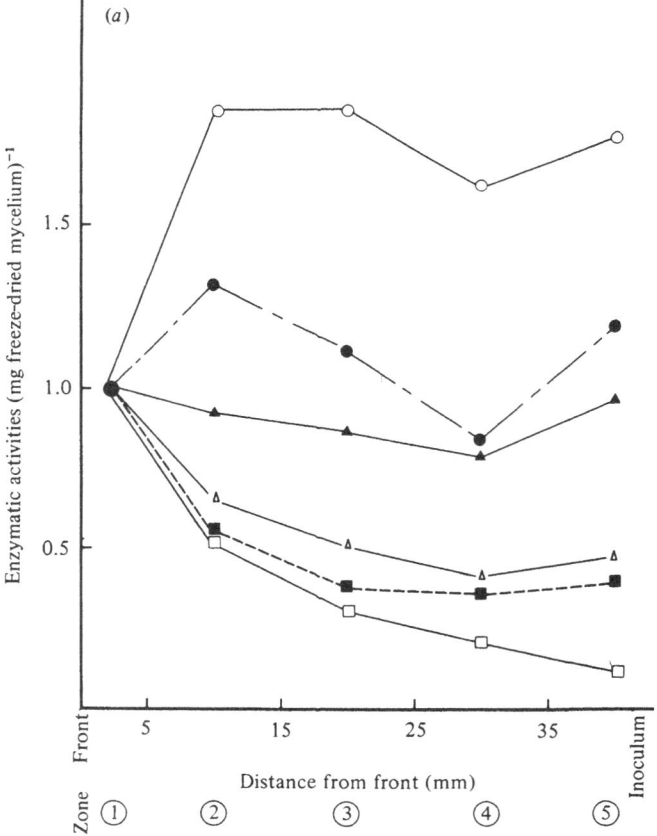

high concentration, which totally inhibits endoglucanase *in vivo*, treated hyphae were wider than the control ($27\,\mu$m instead of 14) and they grew without initiating branches (Table 11.3).

Incorporation of [6-³H]glucose in the cell walls, measured by auto-radiography, varied depending on morphogenesis. In the control hyphae, the silver grain density, very high in the apical zones, decreased rapidly in the subapical parts. On the contrary, gluconolactone-treated hyphae exhibited a more enhanced subapical incorporation, suggesting an abnormal thickening of cell walls in these zones (Fig. 11.5).

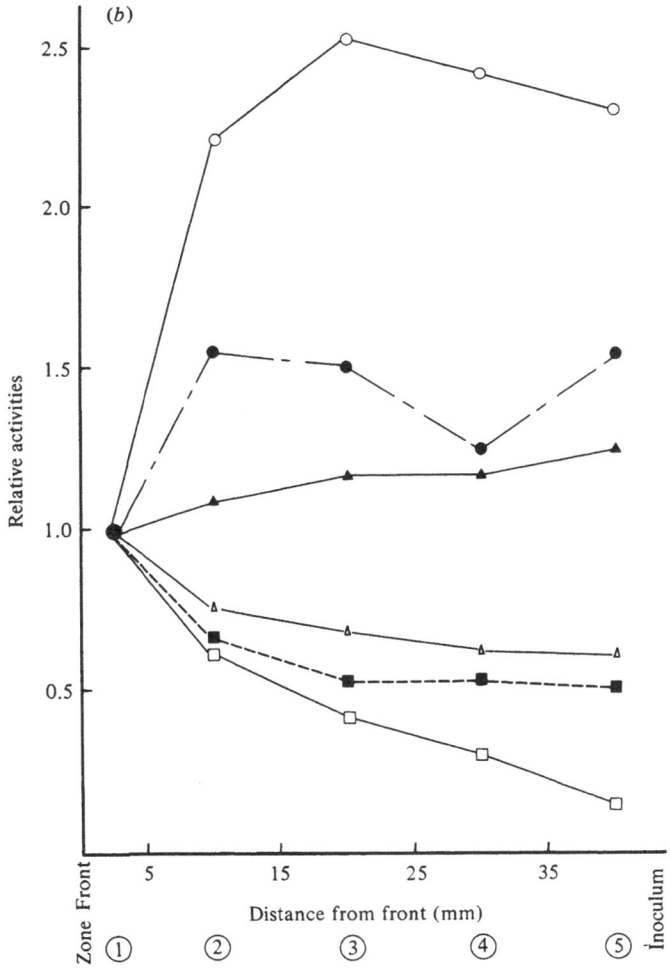

Conclusions

These results confirm the relations between hydrolases and hyphal morphogenesis. The loss of branching capacity obtained in the presence of inhibitors of protein synthesis is correlated with a diminu-

Fig. 11.5. Sites of [6–³H]glucose incorporation in branched and unbranched hyphae of *S. monoica* obtained by treatment with glucanase inhibitor (gluconolactone). Hyphae were labelled for 1 min. The cytoplasm was extracted with alkaline ethanol (1 M NaOH 95 % ethanol; 1:2 by vol.) at 100 °C for 10 min. Autoradiographs were exposed for 2 days. (*a*) silver grains per 100 μm² with standard deviation (15 tips counted), (*b*) Mean silver grain density relative to that 0 to 5 μm behind the apex. Symbols: ■——■, control (branched mycelium); O----O, gluconolactone-treated (unbranched) mycelium.

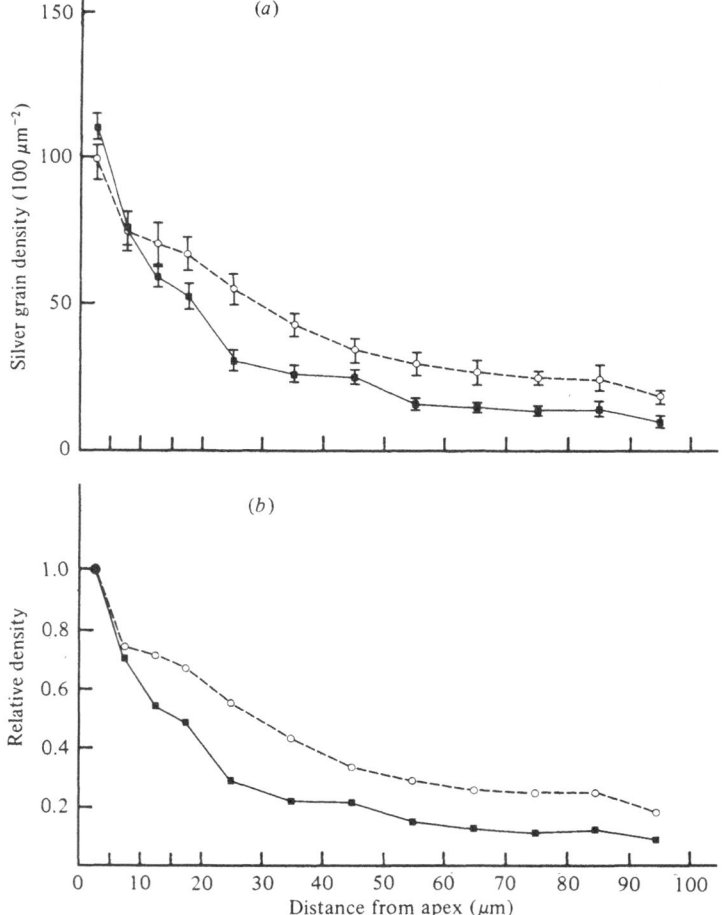

tion of glucanase and cellulase. These morphogen enzymes are mainly localized at the outer edge of the colony where growth and branching of the hyphae occur. Moreover, inhibition of glucanase leads to the suppression of branching capacity.

Subcellular localization of hydrolases

The preferential localization of hydrolases in the apical zone is correlated with the cytoplasmic differentiation of the growth centres. The presence of Golgi-derived cytoplasmic vesicles in such centres which fuse with the plasma membrane suggest that their cytoplasmic organization is orientated towards the synthesis and transport of enzymes and precursors necessary for the expression of hyphal morphogenesis.

If these hydrolases have morphogenetic roles, as I suspect, they should probably be associated with cytoplasmic structures during their transport and with cell walls at the time of their action.

Cell wall enzymes: cell wall autolysis
Differential centrifugation shows that glucanases and cellulase are bound to the cell walls and cytoplasmic particles (Fèvre, Turian & Larpent, 1974).

Cell wall bound enzymes. Wall enzymes, trehalase for example, can be liberated by the mechanical action of freezing and thawing (Chang & Trevithick, 1972; Fèvre et al., 1974). The amount of glucanase released by this treatment is related to the mycelial morphology: branched hyphae contained more enzyme than unbranched mycelia (Fig. 11.6). This treatment is ineffective in liberating β-glucosidase (Fèvre, 1976).

Table 11.3. *Effect of glucono-δ-lactone on growth and branching of* Saprolegnia monoica

D-gluconic acid-lactone Concentration (mol l^{-1})	Colony Diameter (cm)	Percentage of leading hyphae having (x) branches in the apical 800 μm					
		0	1	2	3	4	5
0	13	–	19.2	21.3	44.7	12.7	2.1
0.22	11.5	48.7	43.3	5.4	2.6	–	–
0.33	8	79	21	–	–	–	–
0.44	6	100	–	–	–	–	–

Cell wall autolysis. Purified walls incubated in McIlvaine buffer (pH 5) at 37 °C, released soluble sugars into the supernatant. Analysis of the product showed that glucose (tested by the glucostat) represented only a small part of the total sugar released (tested by the anthrone method), suggesting that the major products of autolysis were long-chain, soluble glucans, produced by the action of endoenzymes in the cell walls. At the same time, $\beta(1-3)$ glucanase was released. Comparison with mycelia grown in the presence of an inhibitor of protein synthesis showed that the loss of branching capacity corresponds to a diminution of the autolytic power of the cell walls and of the glucanase released (Fig. 11.7; Fèvre, 1977).

Fig. 11.6. *S. monoica.* Effect of freezing and thawing on glucanase release during cell wall purification. 200 mg of freeze-dried mycelium was ground with quartz sand in 0.1 M tris-HCl buffer, pH 7.2, at 4 °C. The mixture was centrifuged at 850 g for 10 min, washed twice in buffer and the supernatants pooled as the soluble fraction (S). Cell walls were washed with 0.5 M sucrose (washings 1, 2, 3, 4) then with distilled water (washings 5, 6, 7, 8). After the sixth washing (arrowed) walls were frozen then thawed. Enzymatic activities were tested in the washing solutions.
Cell walls originated from branched (●——●) or unbranched (■ --- ■) mycelia.

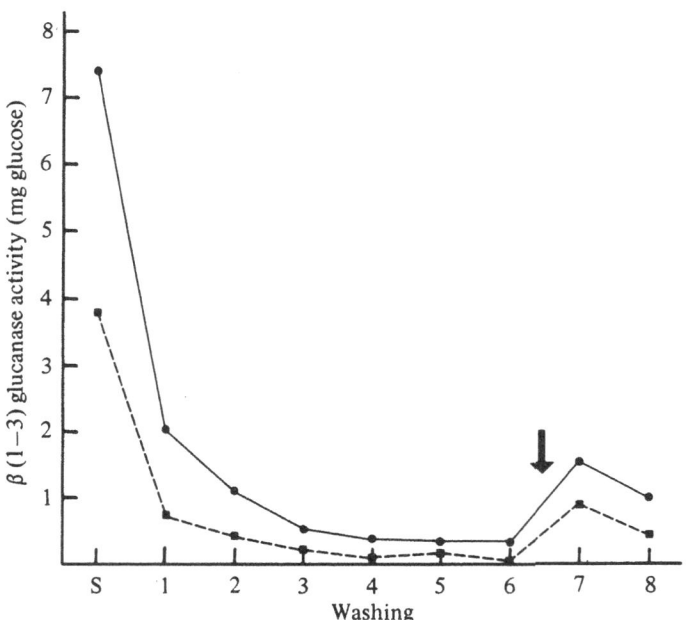

Sites of cell wall autolysis. Do the cell walls of each part of the mycelium have the same capacity for autolysis and do cell wall enzymes show a distribution comparable to those of the intracellular enzymes?

Purified walls obtained from different zones of the fungal colony were tested for their glucanase and autolytic activities (Fig. 11.8). Wall preparations were incubated with laminarin (3 %, pH 5, 37 °C). Cell walls of each zone were able to degrade this substrate, but the activity of the youngest zone was twice as great as that of the older zone. The autolytic capacity of the cell walls also differed according to their origin.

Fig. 11.7. Autolysis of purified cell walls of *S. monoica.* Comparison of the quantity and the quality of sugars released by cell walls of (*a*) unbranched (*p*-FPA) mycelia and (*b*) branched (control) mycelia. Products released were tested by anthrone for total sugar; dinitrosalicylic reagent for reducing sugar; Gluco-kit or glucostat for glucose. Activity of $\beta(1-3)$ glucanase liberated after 24 h of autolysis was tested against laminarin using dinitrosalicylic reagent. Symbols: ■——■, total sugar; ▲----▲, reducing sugar; O–·–·–O, gulcose (Gluco-kit); ●——●, glucose (glucostat); □ , $\beta(1-3)$glucanase.

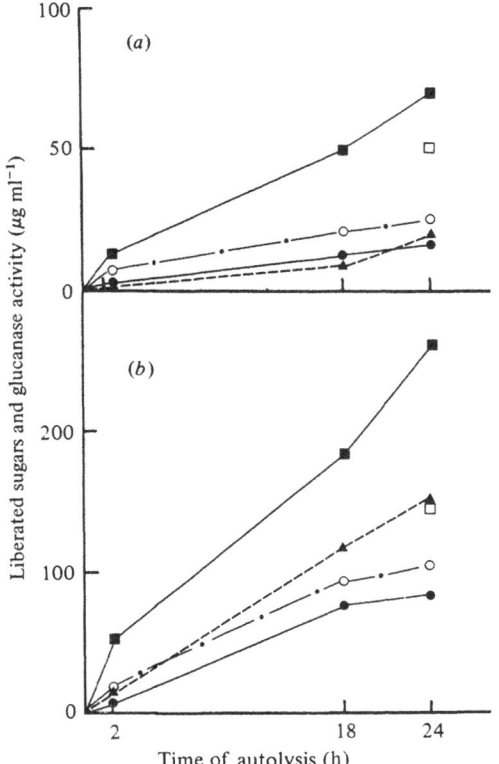

The total release of sugars from the youngest zone (growth and branching zone) was nine times greater than that from the oldest zone. In each zone, glucose was always released in small quantities, confirming the endomechanisms of the autolytic process (Fig. 11.8).

Thus, laminarin was more readily degraded than the macromolecules of the wall. Newly synthesized cell walls seemed to be attacked more easily than older cell walls because the difference in their autolytic capacity was greater (a ratio of 9:1 between zones 1 and 4) than the

Fig. 11.8. Total sugars and glucose released by autolysis and $\beta(1-3)$ glucanases of purified cell walls of different parts of the colony of *S. monoica*. Colonies grown on cellophane film were cut into concentric zones. The walls of the mycelium of each zone were purified and tested for their autolytic (200 mg walls, 4 ml buffer, pH 5, 18 h) and glucanase (100 mg walls, 1 ml 3 % laminarin, 1 h) activities. Results for the different analysis methods are expressed as glucose (μg) equivalents.
Products released were tested by anthrone (O), Gluco-kit ($\square$) and glucostat ($\blacksquare$). $\beta(1-3)$ glucanase activities ($\bullet$) were estimated by the Gluco-kit reagent.

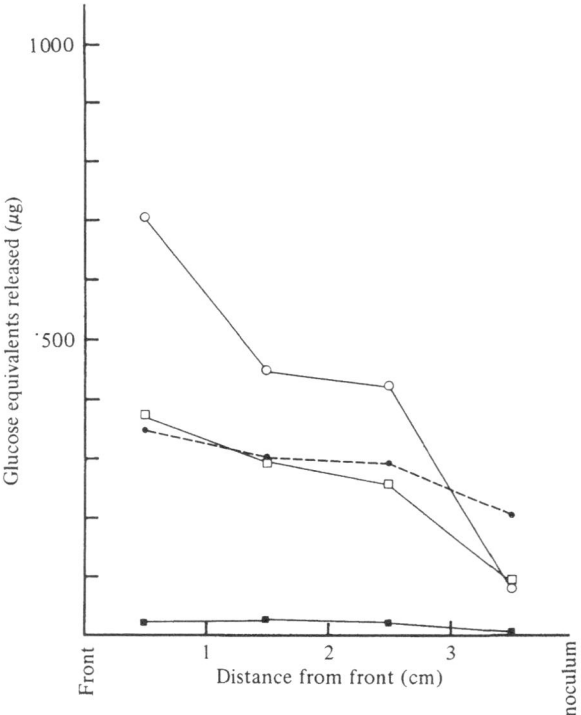

difference in their glucanase activity (a ratio of 2 : 1 between zone 1 and 4; Fèvre, 1977).

Differential centrifugation

Differential centrifugation of the supernatant of mycelial extracts from unbranched and branched colonies showed differences in their enzymatic activities (Fèvre et al., 1974; Fèvre, 1976). The amount of enzyme recovered by centrifugation at 48 000 g for 30 min represented 21 % of the total activity of branched mycelia for the glucanase and 19 % for the cellulase instead of 17 % and 12 % respectively for the unbranched hyphae (Fèvre, 1977). It is noticeable that the main part of the enzyme activity is localized in the soluble fraction (Table 11.4). These soluble enzymes could originate from exocellular enzymes, i.e. those between the plasma membrane and the cell walls, and cytoplasmic structures destroyed during the homogenization procedures.

The presence of the detergent Triton X100 stimulated the enzymatic activity of the particulate fractions (Table 11.5). The rise varied according to the fraction considered but this suggested that part of the enzyme was latent in cytoplasmic structures (Fèvre, 1976).

It was not possible to obtain a particulate fraction enriched in cellulase or glucanase by means of differential centrifugation. Comparison of the distribution of hydrolases with other enzymes which have been considered by Jan (1974) as marker enzymes for fungal cytoplasmic struc-

Table 11.4. *Cellulase and* $\beta(1-3)$*glucanase activities in particulate fractions from* S. monoica

| | $\beta(1-3)$glucanase | | Cellulase | |
	% total activity	specific activity	% total activity	specific activity
Pellet 3000 g, 10 min	1.85	347	3.3	6.45
Pellet 20 000 g, 20 min	9.8	314	7.9	5.8
Pellet 48 000 g, 30 min	8.9	166	7.6	9.05
Supernatant 48 000 g, 30 min	79.45	493	81.2	13.42

Fractions were obtained by differential centrifugation of mycelial homogenate. Cell walls were discarded after centrifugation at 1000 g for 10 min.

tures, confirmed the ineffectiveness of this isolation procedure (Fèvre, 1976).

Density gradient centrifugations

In order to obtain a better resolution between cytoplasmic organelles, the particulate fraction obtained by centrifugation at 48 000 g for 30 min was layered on density gradients.

Continuous density gradient centrifugation. In sucrose or urografin density gradients, glucanase was distributed in three different density bands corresponding to those of the plasma membrane, the Golgi apparatus and the endoplasmic reticulum (Fèvre, 1977) (Fig. 11.9).

Glucanase-rich particles from unbranched mycelium were less abundant, and the difference was marked for particles of density 1.14 g cm^{-3}–1.11 g cm^{-3}, which corresponded to those of the cytoplasmic vesicles isolated from yeast buds (Cortat, 1971) (Fig. 11.10). The difference in activity was comparable to that found during differential centrifugation. Cellulase was also distributed in three different density zones (Fig. 11.11).

Table 11.5. *Differential activation by Triton X100 of cellulase and β(1–3) glucanase fractions from* S. monoica

	Cellulase			β(1–3)glucanase		
	Total activity			Total activity		
	Without Triton X100	With Triton X100	Stimu-lation (%)	Without Triton X100	With Triton X100	Stimu-lation (%)
Pellet 3000 g, 10 min	88.7	106.7	20	3680	5000	35
Pellet 20 000 g, 20 min	213.4	364.5	70	16 000	26400	65
Pellet 48 000 g, 30 min	203.8	346.7	69	8400	24000	185

Membrane fractions were obtained by differential centrifugation of mycelial homogenates. Cell walls were discarded after centrifugation at 1000 g for 10 min. Enzyme activities were determined in the absence and in the presence of Triton X100 (0.5 %) and results are expressed as the total activity of the fraction.

According to Shore & Maclachlan (1975), a high Mg^{2+} concentration conserves the integrity of the rough endoplasmic reticulum and leads to its sedimentation at a high density. The addition of $Mg\ Cl_2$ to extraction and gradient buffers resulted in a shift of cellulase from density 1.10 to 1.20 g cm^{-3} (Fig. 11.11). β-glucosidase, which has sometimes been considered to behave as an exoglucanase (Brock, 1964; Cortat, 1971), accumulated at a density of 1.20 g cm^{-3} in the presence of this salt. The behaviour of these enzymes indicates that they are present in the rough endoplasmic reticulum.

Discontinuous density gradient centrifugations. The distribution of cellulase and glucanase activities in a discontinuous sucrose gradient is shown in Table 11.6; 34 % of the glucanase activity and 31 % of the cellulase activity of the total activities found in the gradient were blocked by the layer of density 1.16 g cm^{-3} (Fèvre, 1977).

Fig. 11.9. Distribution of enzyme activities on a continuous urografin density gradient. Cytoplasmic particles, obtained by centrifugation at 48 000 g for 30 min of a mycelial homogenate of *S. monoica* were layered on a urografin gradient and centrifuged (4 h, 26 000 rev min^{-1}, SW 27 MSE rotor). Activies of cytochrome c oxidase and β-glucosidase are expressed as change in extinction unit per fraction and glucanase activity is expressed as μg glucose released per fraction. Symbols: O—O, cytochrome c oxidase; ▲ - - - ▲, β(1–3)glucanase; ■- - -■, β-glucosidase. Fractions were 400 μl.

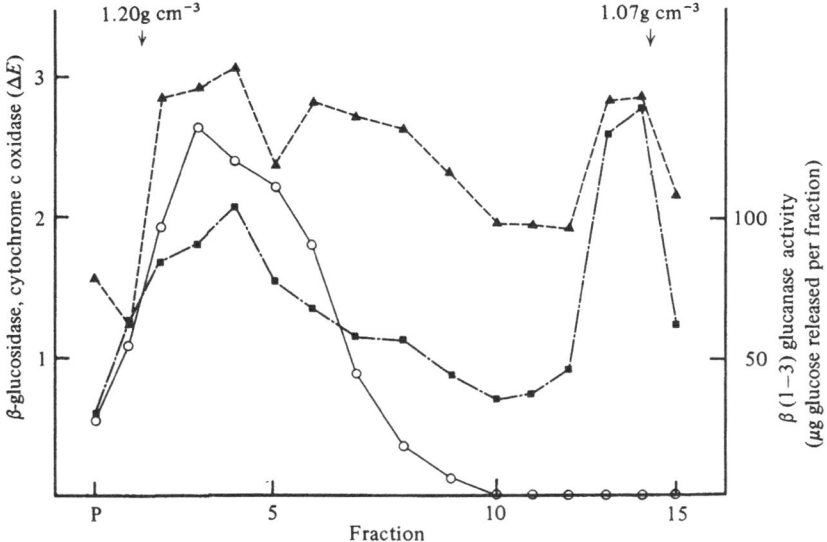

The sedimentation of these hydrolases was also compared with the distribution of enzyme markers (Fig. 11.12).

The endoplasmic reticulum, rich in glucose-6-phosphatase, was characterized by NADH-cytochrome reductase (Gardiner & Chrispeels,

Table 11.6. *Distribution of glucanase and cellulase activities amongst the membrane fraction of* S. monoica

Interface (g cm^{-3})	$\beta(1\text{–}3)$glucanase	Cellulase
soluble/1.08	1.3	10
1.08/1.10	11.8	12
1.10/1.13	30.7	22
1.13/1.16	34.6	31
1.16/1.17	16.3	15
1.17/1.20	5.3	10

Cytoplasmic particles obtained by centrifugation at 48 000 g for 30 min of a mycelial homogenate, were centrifuged on a discontinuous density gradient. For each enzyme, activities are expressed as a percentage of the total activity of that enzyme found in the gradient.

Fig. 11.10. Distribution of glucanase activity on a continuous density urografin gradient. Cytoplasmic particles, prepared from *S. monoica* branched mycelium (solid line) or unbranched mycelium grown in the presence of DL-*p*-fluorophenylalanine (broken line) were layered on the urografin gradient and centrifuged (4 h, 26 000 rev min^{-1}, SW 27 MSE rotor). $\beta(1\text{–}3)$ glucanase activities, expressed as mg glucose released in 6 h per fraction, were corrected to the same amount of protein layered on the gradients.

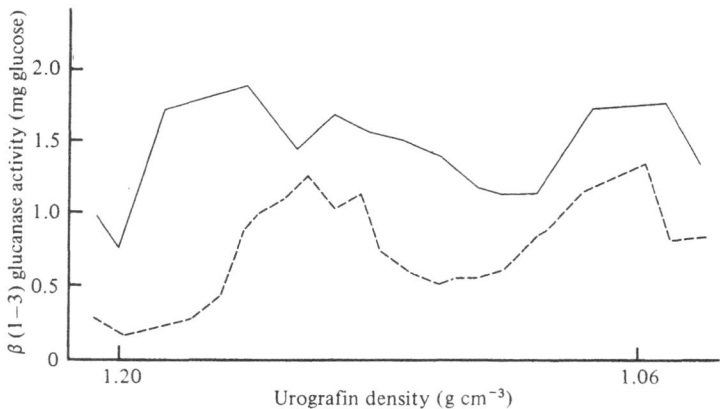

1975; Shore & Maclachlan, 1975) in the presence of KCN and antimycin (Tolbert, 1974) which inhibits the mitochondrial enzymatic activity. 5-nucleotidase, an enzymatic marker of animal cell plasma membrane (Morré, 1971) has been found in algae (Patni, Billmire & Aaronson, 1974) and in *Phycomyces* (Jan, 1974). But the distribution of this enzyme in the membrane of *Candida albicans* (Robin) Berkhout (Marriott, 1975), its weak activity in *Neurospora* (Scarborough, 1975) and even its absence in *Agaricus* (Holtz *et al.*, 1972) and *Saccharomyces* (Suomalainen & Nurminen, 1973) leads me to doubt the specificity of this marker. However, plasmalemma can be characterized by another phosphatase, Mg^{2+}-dependent ATPase stimulated by Na^+ and K^+ (Nurminen, Oura & Soumalainen, 1969; Soumalainen & Nurminen, 1973; Marriott, 1975; Scarborough, 1975; Leonard & van der Woude, 1976). However, it is necessary to inhibit the mitochondrial ATPase by the antibiotic oligomycin (Kulaev, 1973; Fuhrman, Nehrli & Boehm, 1975) when using the Mg^{2+}-dependent, ATPase.

Animal cell dictyosomes are characterized by the presence of thiamin

Fig. 11.11. Distribution of cellulase on a continuous density sucrose gradient; (*a*) Mg Cl_2 present in the extraction and gradient buffers. (*b*) Mg Cl_2 absent. Cytoplasmic particles were layered on the sucrose gradient and centrifuged (4 h, 26 000 rev min^{-1}, SW 27 Beckman rotor). Cellulase activities are expressed in arbitrary units.

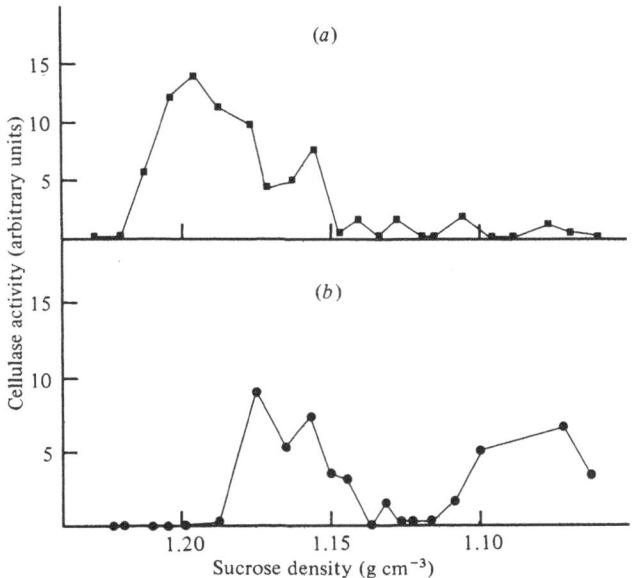

pyrophosphatase (Morré, 1971). In green plants, latent inosine diphosphatase seems to be the only specific marker (Ray, Shininger & Ray, 1969; Powell & Brew, 1974; Gardiner & Chrispeels, 1975; Shore & Maclachlan, 1975). In the particular case of *Saprolegnia monoica*, both enzymes were tested (Fèvre, 1976) but all the membrane fractions

Fig. 11.12. Distribution of marker enzymes, cellulase and exoglucanase (β-glucosidase) activities among subcellular fractions of *S. monoica* (i) NADH-cytochrome c reductase (endoplasmic reticulum marker); (ii) Mg^{2+}-dependent ATPase (plus oligomycin) (plasma membrane marker); (iii) alkaline phosphatase (Golgi marker); (iv) β-glucosidase; (v) cellulase in the presence of Triton X100 (0.5 %). For each enzyme, activities are expressed as (*a*) percentage of the total activity of that enzyme in the gradient and (*b*) as specific activities: μmol h^{-1} (mg protein)$^{-1}$ for ATPase; ΔE h^{-1} (mg protein)$^{-1}$ for alkaline phosphatase, β-glucosidase, NADH-cytochrome c reductase; Unit (20 h)$^{-1}$ (mg protein)$^{-1}$ for cellulase.

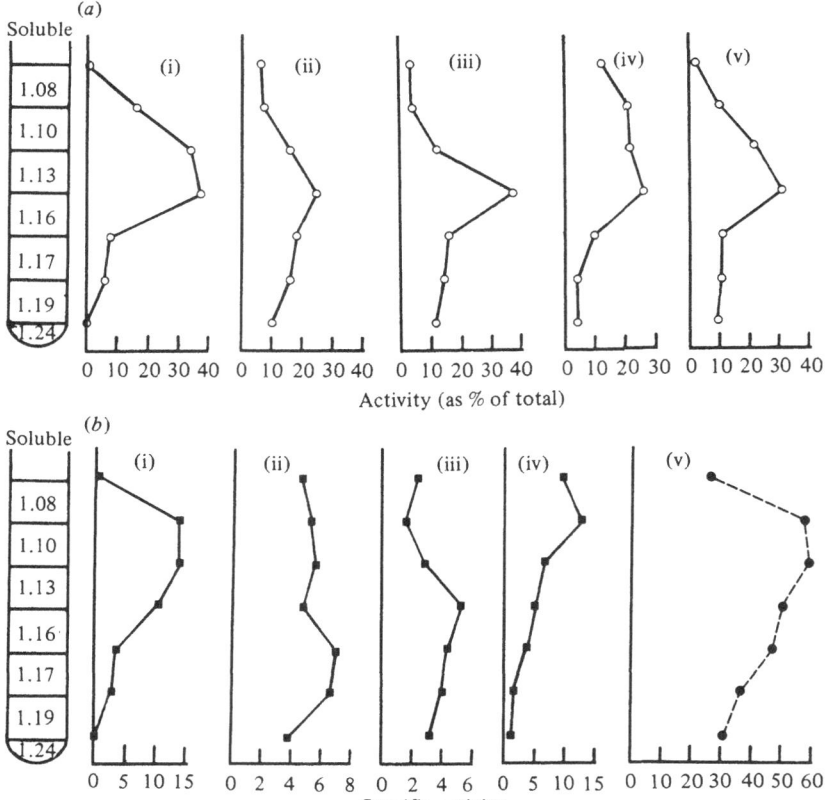

exhibited very low activities. These enzymes do not seem to be good markers of fungal dictyosomes. However, cytochemical tests have shown that they exhibit some particular, but non-specific, enzymatic activities, e.g. alkaline phosphatases (Dargent, 1975) and cellulase (Nolan & Bal, 1974) have been demonstrated in dictyosomes and in Golgi-derived vesicles.

Results are shown in Fig. 11.12. Mg^{2+}-stimulated ATPase (plasma membrane marker) showed a higher concentration and higher specific activities at the bottom than at the top of the gradients while NADH-cytochrome c reductase (Endoplasmic reticulum marker) was mainly localized in the light layers of sucrose.

Alkaline phosphatase sedimented at a medium density, which corresponded to the maximum amount of cellulase and glucanase found in the gradient. The cellulase and exoglucanase (β-glucosidase) exhibited higher specific activities in the endoplasmic reticulum zone.

Electron microscopy of the 1.16 g cm^{-3} fraction, the richest in 'morphogen' hydrolases, showed dictyosomes, some of which had been dissociated during homogenization. Cytoplasmic vesicles were also present but these were of a similar size to those found in the growing centres (Fig. 11.13). The 1.13 g cm^{-3} fraction showed larger vesicles which could be fragments of vacuole membrane. The 1.20 g cm^{-3} fraction contained vesicles with double membranes which may be

Fig. 11.13 (*a*) Electron micrographs of cytoplasmic membranes, dictyosomes and apical vesicles, obtained by centrifugation in a discontinuous density sucrose gradient and equilibrated above the layer of density 1.16 g cm^{-3}. Horizontal arrows indicate large vesicles and vertical arrows indicate small vesicles. (*b*) Dictyosomes and vesicles observed *in situ*.

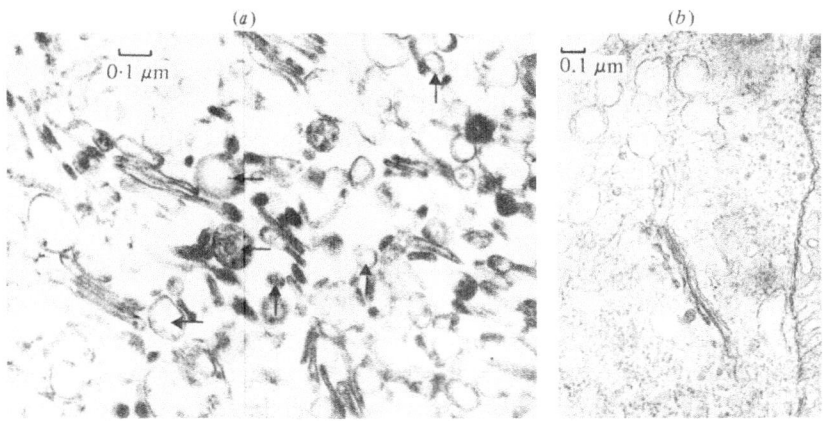

identified as plasma membrane. A few cisterna were also present (Fèvre, 1977).

Conclusions

In common with other fungal species, hyphae of *Saprolegnia* contain enzymes which can hydrolyse their own wall polymers. Autolysis occurs preferentially in newly formed walls, and branched hyphae have more glucanase and autolytic enzymes than unbranched hyphae. Differential centrifugation and density-gradient centrifugation show that the loss of branching capacity is marked by a diminution of the particulate enzymes. It is in the Golgi-rich fraction that the main quantity of these hydrolytic enzymes are found. They are also present in fractions rich in endoplasmic reticulum and plasma membrane.

The subcellular localization of the morphogen enzymes is consistent with the patterns of cytoplasmic differentiation and organization in the Oomycetes described by cytologists (Grove & Bracker, 1970; Grove, Bracker & Morré, 1970; Heath, Gay & Greenwood, 1971).

Glucan synthases

According to the accepted concept of lysis and synthesis of the cell wall during hyphal tip growth and branching, glucan synthases must play an important role. In other species, wall polymer synthases are associated with similar morphological events. Activities of chitin synthase and mannan synthase have been correlated respectively with the fission and budding of yeast (Cortat, Matile & Kopp, 1973; Cabib, 1975). Chitin synthase is more active during the development of the carpophore of *Coprinus cinereus* (Schaeff. ex Fr.) S. F. Gray (Gooday, 1973) and of the sporangiophore of *Phycomyces* (Jan, 1974).

Glucan synthases are mainly bound to the cell walls of *Phytophthora* (Wang & Bartnicki-Garcia, 1966) and *Neurospora crassa* Shear & Dodge (Mishra & Tatum, 1972). But, they are also found in cytoplasmic structures of yeasts (Sentandreu, Victoria-Elorza & Villanueva, 1975) and those of *Phytophthora* (Meyer, Parish & Hohl, 1976).

Properties and enzyme activity

Action of activators. Enzyme properties were determined using the particulate fraction obtained by centrifugation at $48\,000\,g$ for 30 min. According to substrate concentrations (10 μM, 1.2 mM), 80–92 % of the intracellular activity is bound to cytoplasmic structures (Fèvre, 1976).

As in green plants, enzyme activity was stimulated by the presence of activators (Fig. 11.14). Addition to the assay mixture of cellobiose (which stimulates β(1–3)glucan synthase in the *Avena* coleoptile: Tsai & Hassid, 1973), caused an increase in production of glucans with low (10 μM) or high (1.2 mM) UDP-glucose concentrations. MgCl$_2$ at low concentrations, which promotes β(1–4)-linked glucans in *Avena* (Tsai & Hassid, 1973) stimulated the glucan synthase activities but inhibition occurred when higher concentrations were used (Fèvre & Dumas, 1977; Fig. 11.14).

Identification of assay products. ^{14}C-labelled glucans and lipid fractions were separated according to their solubilities (Fig. 11.15). At a low concentration of UDP-glucose (10 μM) without MgCl$_2$, half of the synthesized products were soluble in methanol/chloroform. Addition of MgCl$_2$ (1.2 mM) reduced the formation of glycolipid and increased the hot alkali-insoluble fraction. With higher substrate concentrations (1.2 mM) the main product of synthesis was alkali-insoluble glucan which represented over 61 % of the enzyme activity (Fig. 11.15) (Fèvre & Dumas, 1977).

The hot alkali-insoluble glucan produced in the presence of MgCl$_2$ (1.2 mM) and UDP-glucose (1.2 mM) was treated with a commercial

Fig. 11.14. Effects of cellobiose and MgCl$_2$ on the glucan synthase activity of the particulate fraction (pellet, 48 000 *g*, 30 min) from *S. monoica*. Assay mixtures contained: ▲——▲, cellobiose and 1.2 mM UDP-glucose; O---O, cellobiose and 10 μM UDP-glucose; ■—·—■, MgCl$_2$ and 1.2 mM UDP-glucose.

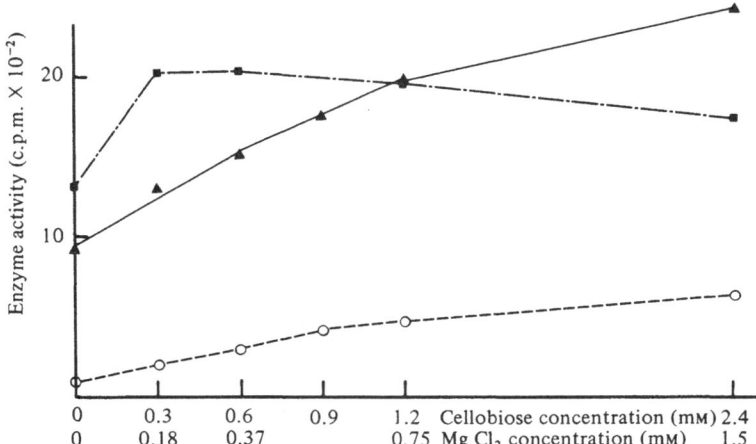

purified cellulase at pH 5 and 40 °C for 24 h. Paper chromatography revealed that this fraction contained ^{14}C-labelled compound with an $R_{glucose}$ corresponding to that of cellobiose (Fig. 11.16). Hence the enzyme preparation contained a $\beta(1-4)$glucan synthase. However, two other radioactive spots were also detected that could correspond to laminaribiose and laminaritriose, which would also indicate the concomitant action of a $\beta(1-3)$glucan synthase.

Presence of a thermolabile glucan synthase inactivator in the soluble fraction. The supernatant obtained after centrifugation at 48 000 g showed very weak specific activity. When it was added to the standard assay mixture containing the particulate fraction enzyme, the synthetic

Fig. 11.15. Solubility of radioactive products of the particulate fraction enzyme preparation obtained from *S. monoica* with UDP-glucose as substrate.
Labelled products were separated as four fractions: glycolipids, (methanol–chloroform soluble) hot water soluble, hot alkali soluble, and hot alkali insoluble. Activities were determined at two different substrate concentrations with or without $MgCl_2$.
Assay mixtures contained: (*a*) 1.2 mM UDP-glucose, 1.2 mM Mg Cl_2; (*b*) 1.0 μM UDP-glucose, 1.2 mM Mg Cl_2; (*c*) 1.0 μM UDP-glucose.

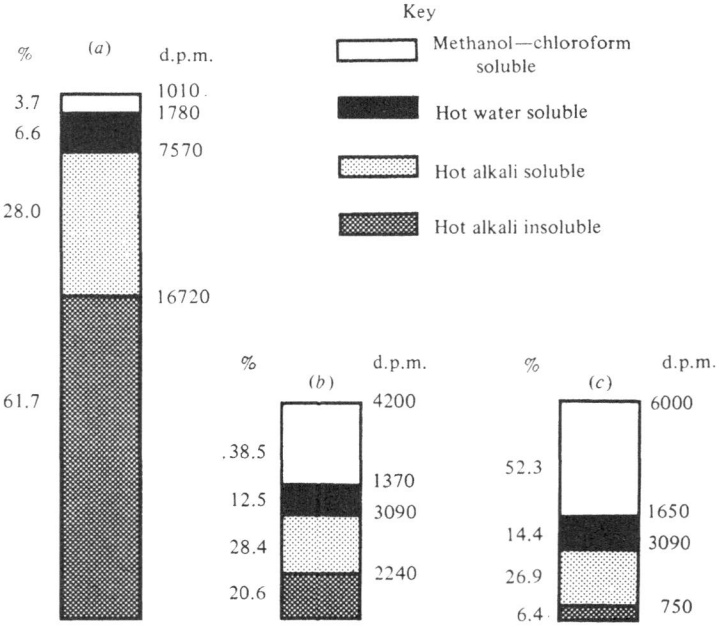

activity was reduced. If the supernatant was boiled before addition to the standard assay mixture, the loss of enzyme activity was lower and corresponded simply to the dilution of the assay mixture. When the standard assay mixture plus supernatant was incubated at 0 °C for 90 min, the inactivation of glucan synthase activity was more marked, and the inhibitory effect was even greater if the mixture was incubated at 30 °C for 90 min. The inactivating capacity of the supernatant was lost during cold storage: supernatant kept at −20 °C for two days had a lower inactivating effect on a fresh particulate fraction enzyme (Table 11.7).

Cellular localization of glucan synthases

Differential centrifugation. Wall pellet and particulate fraction were obtained by centrifugation of the mycelial homogenate at 1000 g and 48 000 g. Cell walls and particles contained most of the glucan synthase activity (Table 11.8). Wall-bound enzymes represented half the total

Fig. 11.16. Paper chromatography of the products of hydrolysis of the particulate fraction insoluble in hot alkali. Hydrolysis by purified cellulase (Worthington CSEII). Chromatography on Whatman No. 1 paper; propanol/ethyl acetate/water (7 : 1 : 2) as solvent.

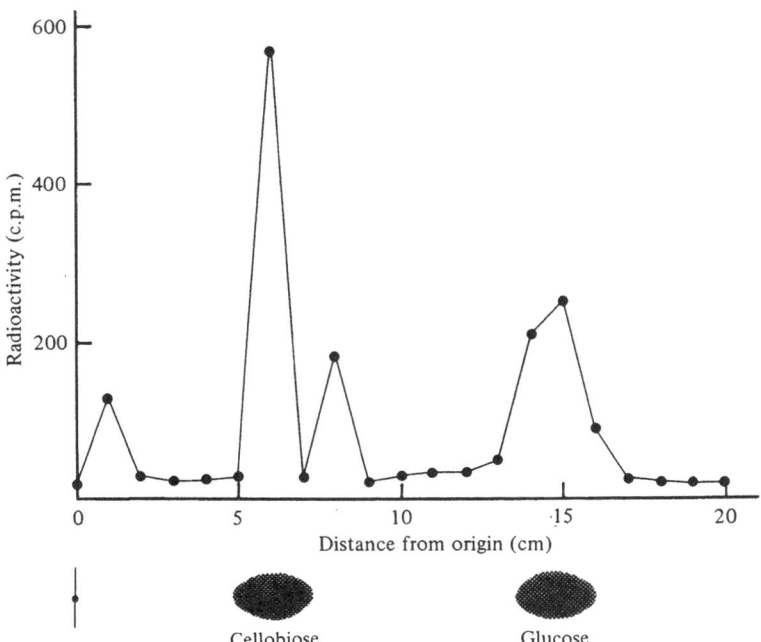

Table 11.7. *Presence in the soluble fraction of a thermolabile inactivator of particulate glucan synthases from* S. monoica

Conditions of assay	Enzyme activity	
	c.p.m.	% of control
Experiment I		
Pellet 100 μl + supernatant 100 μl (265 μg protein) (292 μg protein)	2817	49
Pellet 100 μl + supernatant 100 μl (heated for 10 min at 100 °C)	4575	79
Pellet 100 μl + supernatant 100 μl (mixture incubated for 1.5 h at 0 °C)	2082	36
Pellet 100 μl + supernatant 100 μl (mixture incubated for 1.5 h at 30 °C)	1125	19
Pellet 100 μl	5752	100
Experiment II		
Pellet 100 μl + buffer 100 μl (440 μg protein)	6961	88
Pellet 100 μl + supernatant 100 μl	5647	69
Pellet 100 ml + supernatant 100 μl (heated for 10 min at 100°C)	6733	86
Pellet 100 μl	7834	100

Particulate and supernatant fractions were prepared by centrifugation of mycelial homogenates at 48 000 g for 30 min. The supernatant of experiment I was stored for 2 days at -20 °C followed by 2 days at 4 °C and was then used in the assays of experiment II.

Table 11.8. *Comparison of the cellular localization of glucan synthase of branched and unbranched mycelia of* S. monoica

Fraction	Branched mycelium		Unbranched mycelium	
	% of total activity	Specific activity	% of total activity	Specific activity
Cell wall fraction (pellet 100 g for 10 min)	53	39945	57	14607
Particulate fraction (pellet 48 000 g for 30 min)	35	18960	11	3790
Soluble fraction (supernatant 48 000 g)	12	1056	32	1921

Unbranched mycelia were obtained by growth in the presence of DL–p-fluoro-phenylalanine (100 mg 1^{-1}). UDP-glucose was used at 1.2 mM in the assay. Specific activities are expressed as c.p.m. h^{-1} (mg protein)$^{-1}$

activity. The distribution and activity of glucan synthases were compared in different kinds of mycelial morphogenesis (branched and unbranched). In both types, most of the glucan synthase activity was associated with the cell walls, but the specific activity in this fraction was lower in unbranched mycelium. The particulate fraction from mycelium which grew more slowly without initiating branches represented a low part of the total activity and had a low specific activity.

Supernatants from both branched and unbranched mycelia showed low glucan synthase activity. However, the activity associated with the soluble fraction of the inhibited mycelium represented a significant part of the total activity.

Discontinuous density gradient centrifugations. Particulate fractions were layered on sucrose density gradients. The particles which sedimented at medium densities represented the greatest part of the glucan synthase activity found in the gradient (Fig. 11.17). The Golgi enzyme markers are also found at these densities. Enzyme activity was tested at two different substrate concentrations.

Fig. 11.17. Distribution of glucan synthase among membrane fractions of *S. monoica* in a discontinuous density gradient. Cytoplasmic particles (pellet, 48 000 *g* for 30 min) were layered on a discontinuous sucrose density gradient. Assays were done at two different substrate concentrations: (*a*) 0.3 nM UDP-glucose; (*b*) 1.2 mM UDP-glucose.

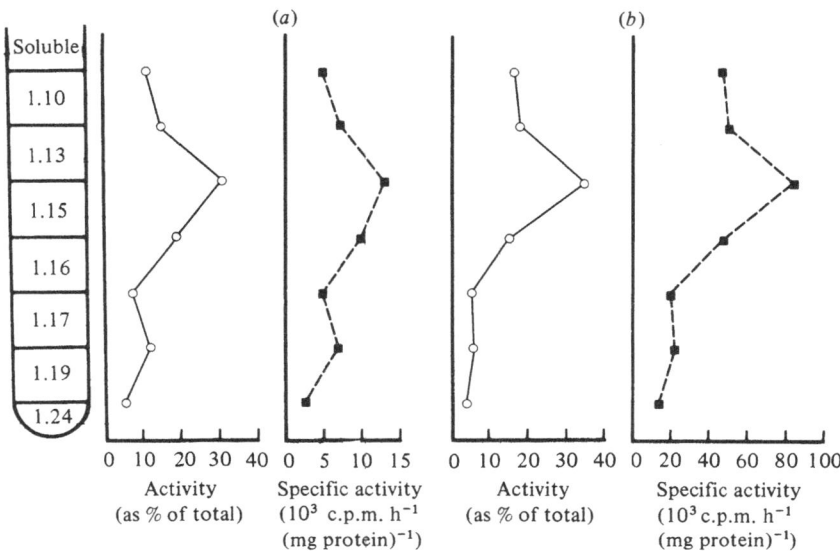

With 0.3 nM UDP-glucose, which according to Ordin & Hall (1968), Tsai & Hassid (1973) and van der Woude *et al.* (1974) promotes the synthesis of $\beta(1-4)$-linked glucans, higher specific activities were found at the same densities as the plasma membrane and above all in the Golgi apparatus.

With 1.2 mM UDP-glucose which, according to the same authors, stimulates $\beta(1-3)$glucan synthesis, light membrane (endoplasmic reticulum) and Golgi apparatus exhibit the majority of the activity.

Electron-microscopic examination of the particulate fractions

Fig. 11.18. Electron microscopy of dictyosomes and cytoplasmic vesicles obtained by discontinuous sucrose gradient, and equilibrated at density 1.15 g cm^{-3} Golgi-derived vesicles are indicated by arrows. (*a*) (*b*) lead citrate stain; (*c*) (*d*) polysaccharide staining according to the Thiéry procedure; (*e*) control for polysaccharide staining.

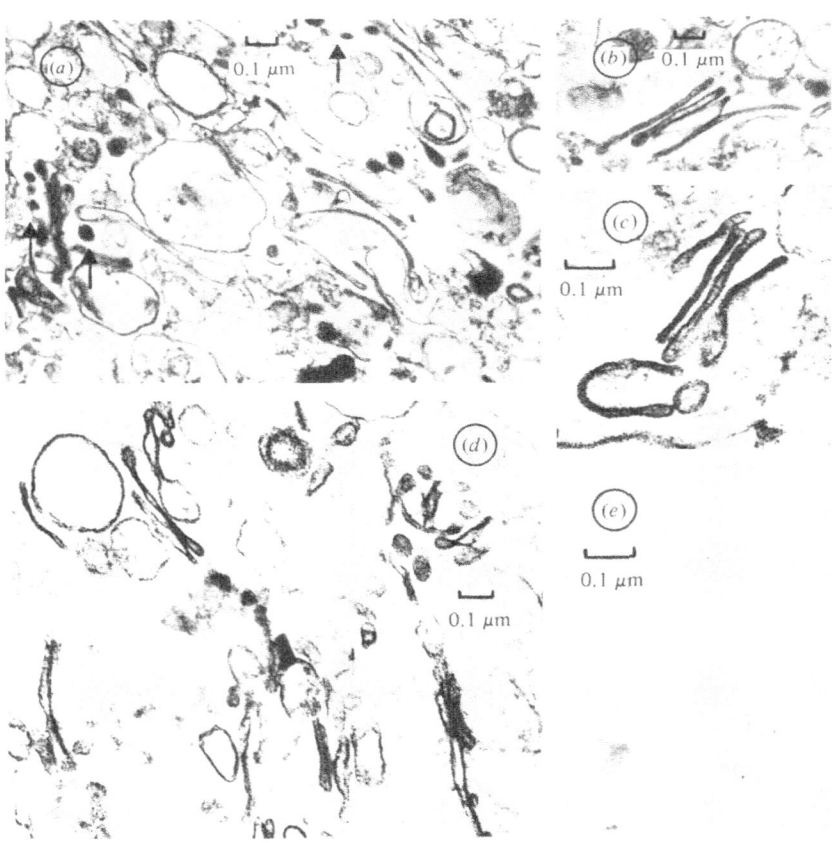

obtained in a sucrose density gradient confirmed my previous observations (Fèvre, 1977) since the fraction of medium density, 1.15 g cm^{-3}, was rich in dictyosomes and Golgi-derived vesicles (Fig. 11.18). Fractions of higher density (1.17 g cm^{-3}) contained vesicles (plasmalemma) and some isolated cisternae. Particles of lower density (1.10 g cm^{-3}) present in the endoplasmic reticulum-rich regions were vesicles of a different size (Fèvre & Dumas, 1977).

The 1.15 g cm^{-3} fraction was stained for polysaccharides. Silver grains were deposited on dictyosomes and on small vesicles. The large vesicles observed were stained and could have originated from expansion of cisternae during isolation procedures (Fig. 11.18). Thus, dictyosomes rich in synthase enzymes also contained polysaccharides, as demonstrated by the Thiéry procedure (1967).

Conclusions

Saprolegnia monoica contains enzymes which transfer glucose from labelled UDP-glucose to an unknown acceptor and lead to the production of radioactive glucans. The addition of MgCl$_2$ and cellobiose stimulated enzyme activity. High substrate concentrations increased synthesis of alkali-insoluble glucans which, as revealed by chromatography, contain β(1–4)polysaccharides.

A thermolabile component of the soluble fraction inactivated the particulate enzyme. Similar components which modify chitin synthase activity have been isolated from yeasts (Hasilik, 1974; Cabib, 1975) and demonstrated in filamentous fungi (McMurrough & Bartnicki-Garcia, 1973; Peberdy & Moore, 1975; Ruiz-Herrera & Bartnicki-Garcia, 1976).

Subcellular localization studies show that the enzymes were mainly bound to the cell wall. Intracellular enzymes are, above all, found in dictyosome fractions. Plasma membrane and endoplasmic reticulum exhibited differences in activity according to the substrate concentration, which may represent differences in the nature of the enzyme involved (van der Woude *et al.*, 1974; Shore & Maclachlan, 1975).

Biochemical and functional organization of the apex
Cytological organization

It is well known that the fungal hyphal tip is characterized by an accumulation of cytoplasmic vesicles.

Electron-microscopic studies (Fig. 11.19) of the apical tips of *Saprolegnia* have enabled us to recognize the secretory processes described

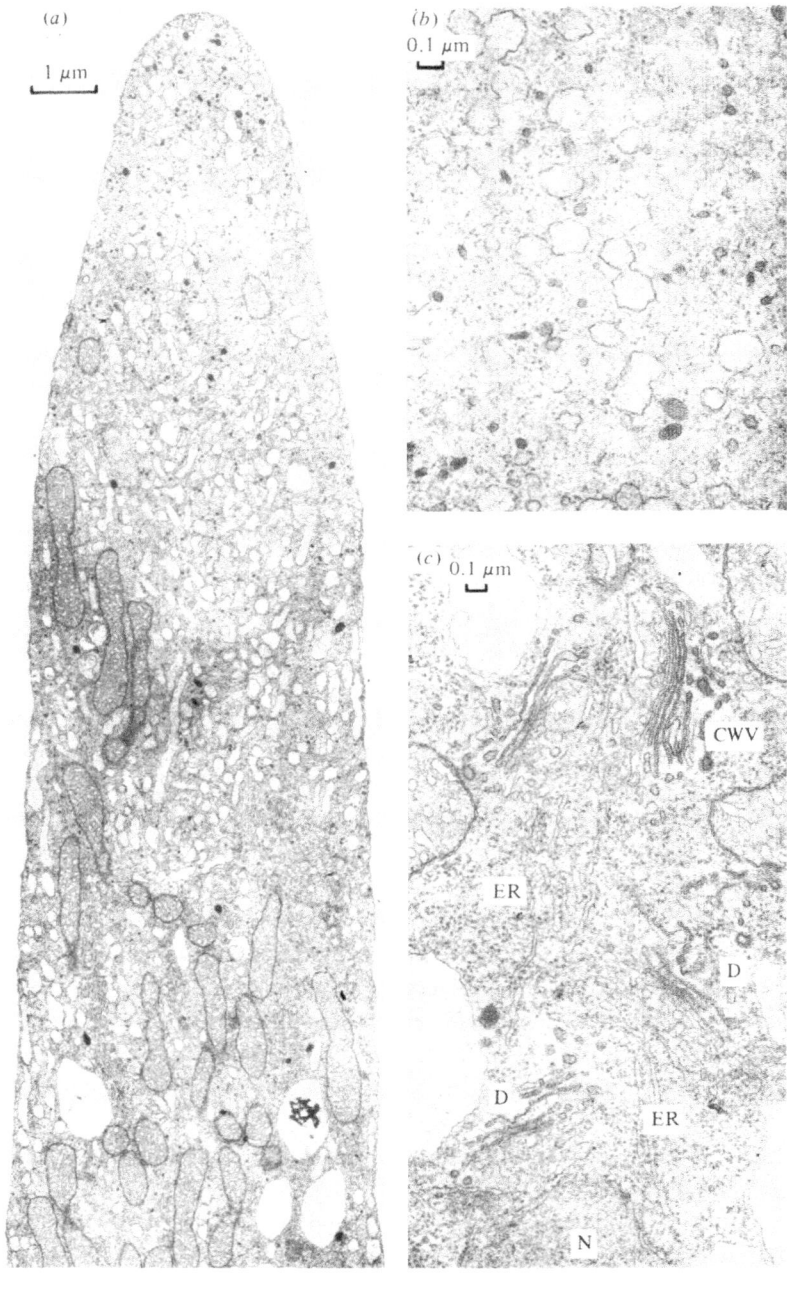

in Oomycetes (Grove, 1971; Heath *et al.*, 1971; Grove *et al.*, 1970).

Golgi membranes appear to originate from the endoplasmic reticulum and nuclei. The role of the Golgi complex in membrane differentiation is to transform the original endoplasmic reticulum membrane into the plasmalemma-compatible membrane of Golgi vesicles.

Two types of apical vesicles can be recognized. Small vesicles, which are sometimes continuous with dictyosome cisternae, have very dense contents. Large vesicles with a fibrous content could originate from a complete vesiculization of a mature cisterna. Both vesicle types are found free in the apical cytoplasm migrating towards the apex then fusing with the plasma membrane (Fèvre & Rougier, 1978).

Transport of the morphogen enzymes (Fig. 11.20)

Density gradient centrifugations have permitted the localization of cellulase, $\beta(1-3)$glucanase and glucan synthase in the Golgi-rich fraction. These enzymes were also detected in the particles associated with the endoplasmic reticulum and plasmalemma enzyme markers.

Microsomal enzymes are probably enzymes which are transported to the walls, whereas wall-bound enzymes act chiefly at the hyphal tips and branching sites.

From this cytological description of the fungal tip, it seems reasonable to postulate that morphogen enzymes synthesized in the endoplasmic reticulum are transferred to the Golgi apparatus. Dictyosomes rich in morphogen enzymes and $\beta(1-4)$-linked polysaccharides produce apical vesicles which secrete their contents into the walls by fusing with the plasma membrane. This membrane flow has also been found in budding yeast and in elongating green plant cells. In yeasts, the secretory budding vesicles contain glucanase, mannan and mannan synthase (Cortat, 1971; Cortat *et al.*, 1973). In green plants, endocellulase and β-glucan transferases are generated and transported in the same type of

---

Fig. 11.19. Cytology of the hyphal apex of *Saprolegnia* (Fèvre & Rougier, 1978). (*a*) Near median longitudinal section of a hyphal tip. The accumulation of vesicles is a characteristic of the apex. (*b*) Vesicles of the hyphal tip in a subapical zone. Small vesicles have a very dense content. Large vesicles have a granular content. (*c*) Dictyosomes in the subapical part of a hyphae. They are characterized by three or four cisternae and they produce small vesicles with dense content at their distal face.

Hyphae were fixed in glutaraldehyde then osmium, stained with uranyl acetate and lead citrate. N, nucleus; D, dictyosome; ER, endoplasmic reticulum; CWV, cell wall vesicles.

organelles (Bal *et al.*, 1976; Maclachlan, 1976) and both enzymes are present in the plasma membrane (van der Woude *et al.*, 1974; Koehler *et al.*, 1976).

Presumably, hydrolases and synthases of the organelles are active *in vitro* only because all the various components necessary for activity are provided in the assay mixture.

However, recent experiments with green plant cells have shown that $\beta(1-4)$glucans (probably hemicellulose but not nascent cellulose) have been found in dictyosome fractions of the *Pisum* stem (Ray, Eisinger & Robinson, 1976) and microsomal fraction of *Phaseolus* seedlings (Satoh, Matsuda & Tamaki, 1976).

Enzymes in the cell walls

Autoradiographs reveal localized wall deposition at hyphal tips and branching sites which are the zone of active synthases (Bartnicki-Garcia & Lippman, 1969; Gooday, 1971; Katz & Rosenberger, 1971). Low subapical incorporation of radioactive precursors indicates inefficient synthesis in this region, which could be explained by a lack of substrate. However, regulation of synthase activity could also be evoked.

Osmotic shock or cycloheximide treatment induce *N*-acetylglucosamine incorporation along the length of *Aspergillus* hypha. This must be due to the activation of the pre-existing enzymes (Katz & Rosenberger, 1971). Moreover, soluble extracts of *Saprolegnia* are capable of inhibiting glucan synthase activity *in vitro*. A sophisticated regulatory mechanism of chitin synthase activity during septum formation in yeasts has also been described by Cabib, Ulane & Bowers (1974).

Hydrolases are present in the cell walls. But in *Aspergillus nidulans* (Eidam) Winter (Polacheck & Rosenberger, 1975), as in *Saprolegnia monoica*, only the newly synthesized wall or apical parts of hyphae are subject to autolysis (Fèvre, 1977). $\beta(1-3)$glucanase and protease have been localized by immunofluorescence at the sites of branching and septum formation in *Neurospora* hyphae (Mehta & Mahadevan, 1975; Sukumaran & Mahadevan, 1975).

Enzyme trapped in the cell walls could be the consequence of wall rigidification in the growing hyphae. Only small molecules could then be released through the complex molecular sieve of the cell wall, as has been demonstrated for the invertase of *Neurospora* (Chang & Trevithick, 1974).

Cell wall growth. Hence, the original idea that expansion of cell wall is caused by local lysis and concomitant incorporation of wall polysaccharides appears to be true. Deviation of this equilibrium in one way or the other modifies hyphal morphogenesis (Robertson, 1965). Alteration of hyphal growth can also be obtained by using inhibitors of synthases and hydrolases.

The competitive inhibitor of chitin synthase, Polyoxin D, when added to the culture medium, affects the apical growth of a number of mycelial fungi (Bartnicki-Garcia & Lippman, 1972; Benitez, Villa & Garcia Acha, 1976), stops elongation of excised *Coprinus* stipes (Gooday, de Rousset-Hall & Hunsley, 1976) and prevents chitin-ring formation in the bud-scar of yeast (Bowers, Levin & Cabib, 1974). In the presence of this antibiotic, growing hyphal apices swell and burst, mycelial walls become irregular and lose their rigidity.

Fig. 11.20. Biochemical and functional organization of the apex of *Saprolegnia*. Morphogen enzymes synthesized in the endoplasmic reticulum are transferred to the Golgi apparatus. Dictyosomes produce apical vesicles which secrete their content into the wall by fusing with the plasma membrane. Do vesicles with different cytological characteristics have the same enzymatic equipment?

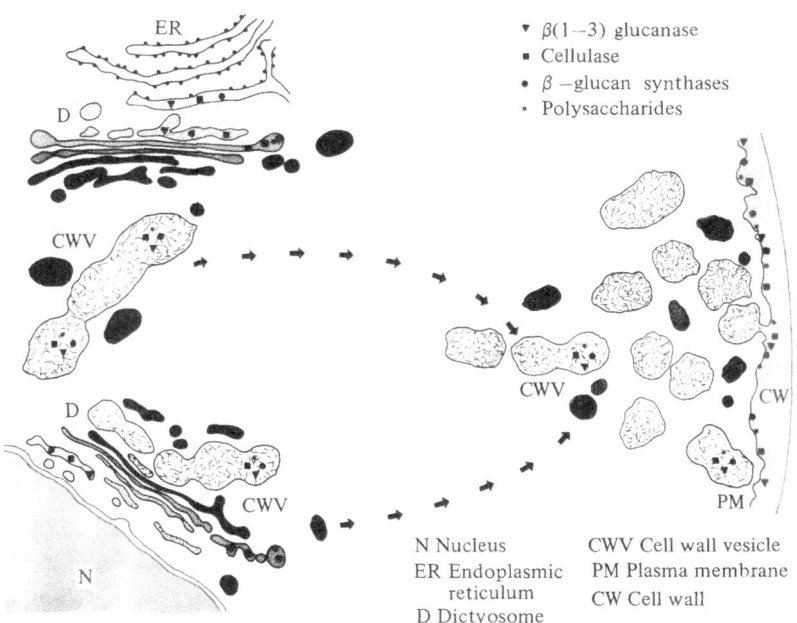

ER Endoplasmic reticulum
D Dictyosome

▾ $\beta(1-3)$ glucanase
■ Cellulase
● β —glucan synthases
· Polysaccharides

N Nucleus
ER Endoplasmic reticulum
D Dictyosome

CWV Cell wall vesicle
PM Plasma membrane
CW Cell wall

Glucono-δ-lactone, while inhibiting glucanase, leads to a loss of branching capacity of *Saprolegnia* and to an increased subapical incorporation of wall precursors. It also induces abnormal swelling of cells of the yeast *Pichia* (Villa *et al.*, 1976).

A deviation of the 'delicate balance between lysis and synthesis' could explain the morphogenetic effects of these drugs. In the presence of polyoxin D synthesis is inhibited and lysis is favoured leading to very fragile hyphae. Glucono-δ-lactone of the other hand, inhibits lysis without affecting synthesis. Inactivated glucanases cannot act on reinforced cell walls whose hyphae grow slowly without initiating branches.

In summary, the extension processes seem to involve the activity of endohydrolases very closely coordinated with, but obviously preceding, synthase activity. The hyphal tip receives a continual flow of vesicles which supplies the enzymes and precursors.

However, we do not know if:
Vesicles with differing cytological characteristics have the same content.
Synthases are active during their transport, permitting intussusception of polymers in the wall.
A regulatory mechanism exists to permit enzyme stimulation during branch formation.

References

Abd El-Al, A. T. H. & Phaff, H. J. (1968). Exo-β-glucanases in Yeast. *The Biochemical Journal*, **109**, 347–60.

Bal, A. K., Verma, D. P. S., Byrne, H. & Maclachlan, G. A. (1976). Subcellular localization of cellulases in auxin-treated pea. *Journal of Cell Biology*, **96**, 97–105.

Barras, D. R. (1972). A β-glucan endo-hydrolase from *Schizosaccharomyces pombe* and its role in cell wall growth. *Antonie van Leeuwenhoek. Journal of Microbiology and Serology*, **38**, 65–80.

Bartnicki-Garcia, S. (1973). Fundamental aspects of hyphal morphogenesis. In *Microbial Differentiation* (23rd Symposium of Society for General Microbiology) eds. J. M. Ashworth & J. E. Smith, pp. 245–67. Cambridge University Press.

Bartnicki-Garcia, S. (1973). Fundamental aspects of hyphal morphogenesis. In *Microbial* in *Mucor rouxii. Science*, **165**, 302–3.

Bartnicki-Garcia, S. & Lippman, E. (1972). Inhibition of *Mucor rouxii* by Polyoxin D: effects on chitin synthetase and morphological development. *Journal of General Microbiology*, **71**, 301–9.

Benitez, T., Villa, T. G., Garcia Acha, I. (1976). Effects of polyoxin D on germination, morphological development and biosynthesis of the cell wall of *Trichoderma viride*. *Archives of Microbiology*, **108**, 183–8.

Bowers, B., Levin, G. & Cabib, E. (1974). Effect of Polyoxin D on chitin synthesis and septum formation in *Saccharomyces cerevisiae*. *Journal of Bacteriology*, **119**, 564–75.

Brock, T. D. (1964). Enzyme synthesis during conjugation in the yeast *Hansenula wingei*.

Journal of Cell Biology, **23**, 15A.

Brown, C. M. & Hough, J. S. (1966). Protein disulphide reductase in yeast. *Nature (London)*, **211**, 201.

Bush, D. A. (1974). Autolysis of *Coprinus comatus* sporophores. *Experientia*, **30**, 984.

Cabib, E. (1975). Molecular aspects of yeast morphogenesis. *Annual Review of Microbiology*, **29**, 191–214.

Cabib, E., Ulane, R. & Bowers, B. (1974). A molecular model for morphogenesis. The primary septum of yeast. *Current Topics in Cellular Regulation*, **8**, 1–32.

Chang, P. L. Y. & Trevithick, J. R. (1972). Release of wall bound invertase and trehalase in *Neurospora crassa* by hydrolytic enzymes. *Journal of General Microbiology*, **70**, 13–22.

Chang, P. L. Y. & Trevithick, J. R. (1974). How important is secretion of exoenzymes through apical cell walls of fungi? *Archives of Microbiology*, **101**, 281–93.

Cohen, R. J. (1975). Some properties of chitinase from *Phycomyces blakesleeanus*. *Life Science*, **15**, 289–300.

Cortat, M. (1971). Localisation intracellulaire et fonction des β1–3 glucanases lors du bourgeonnement chez *Saccharomyces cerevisiae*. Thèse, Zurich.

Cortat, M., Matile, Ph. & Kopp, F. (1973). Intracellular localization of mannan synthetase activity in budding baker's yeast. *Biochemical and Biophysical Research Communications*, **53**, 482–9.

Dargent, R. (1975). Sur l'ultrastructure des hyphes en croissance de l'*Achlya bisexualis* Coker. Mise en évidence d'une sécrétion polysaccharidique et d'une activité phosphatasique alcaline dans l'appareil de golgi et au niveau des vésicules cytoplasmiques apicales. *Comptes Rendus Hebdomadaires des Séances de l'Académie des Sciences, Paris*, **280**, 1445–8.

Dow, J. H. & Rubery, P. H. (1975). Hyphal tip bursting in *Mucor rouxii*: antagonistic effects of calcium ions and acid. *Journal of General Microbiology*, **91**, 425–8.

Fèvre, M. (1968). Action de la D actinomycine et de la puromycine sur la croissance et la ramification du mycélium jeune de *Saprolegnia monoica* Pringsheim. *Comptes Rendus Hebdomadaires des Séances de L'Académie des Sciences, Paris*, **267**, 293–5.

Fèvre, M. (1969). Cellulase(s) et ramification chez *Saprolegnia monoica* Pringsheim *Comptes Rendus Hebdomadaires des Séances de l'Académie des Sciences, Paris*, **269**, 2347–50.

Fèvre, M. (1970). Contribution à l'étude du déterminisme de la ramification du mycélium de deux champignons *Saprolegnia monoica* Pringsheim et *Rhizoctonia solani* Kuhn. Thèse Doctorat de Spécialité, Université de Lyon.

Fèvre, M. (1972). Contribution to the study of the determination of mycelium branching of *Saprolegnia monoica* Pringsheim. *Zeitschrift für Pflanzenphysiologie*, **68**, 1–10.

Fèvre, M. (1976). Recherches sur le déterminisme de la morphogenèse hyphale. Aspects enzymatiques de la croissance et de la ramification des hyphes de *Saprolegnia monoica* Pringsheim. Thèse Doctorat d'Etat, Université de Lyon.

Fèvre, M. (1977). Subcellular localization of glucanase and cellulase in *Saprolegnia monoica* Pringsheim. *Journal of General Microbiology*, **103**, 287–95.

Fèvre, M. & Dumas, Chr. (1977). β-glucan synthases from *Saprolegnia monoica*. *Journal of General Microbiology*, **103**, 297–306.

Fèvre, M. & Rougier, M. (1978). Synthèse de la paroi apicale des hyphes de *Saprolegnia monoica*: données cytologiques et radioautographiques. *Colloque de la Société Française de Microscopie Electronique*, Nancy 1978.

Fèvre, M., Turian, G. & Larpent, J. P. (1974). Bourgeonnements et croissance hyphale fongiques. Homologies structurales et fonctionnelles, modèle *Neurospora et*

Saprolegnia. Physiologie Végétale, **13**, 23–38.

Fleet, G. H. & Phaff, H. J. (1974). Glucanases in *Schizosaccharomyces*. Isolation and properties of the cell wall associated β 1–3 glucanase. *Journal of Biological Chemistry*, **249**, 1717–28.

Fuhrman, G. F., Nehrli, E. & Boehm, C. (1975). Preparation and identification of yeast plasma membrane vesicles. *Biochimica and Biophysica Acta*, **363**, 295–310.

Gardiner, M. & Chrispeels, M. J. (1975). Involvement of golgi apparatus in the synthesis and secretion of hydroxyproline rich cell wall glycoproteins. *Plant Physiology*, **55**, 536–41.

Gooday, G. W. (1971). An autoradiographic study of hyphal growth of some fungi. *Journal of General Microbiology*, **67**, 125–33.

Gooday, G. W. (1973). Activity of chitin synthetase during the development of fruit bodies of the toadstool, *Coprinus cinereus. Biochemical Society Transactions*, **I**, 1105–07.

Gooday, G. W., de Rousset-Hall, A. & Hunsley, D. (1976). Effect of Polyoxin D on chitin synthesis in *Coprinus cinereus. Transactions of British Mycological Society*, **67**, 193–200.

Grove, S. N. (1971). Protoplasmic correlates of hyphal tip initiation and development in fungi. Ph. D. Thesis. Purdue University, USA.

Grove, S. N. & Bracker, C. E. (1970). Protoplasmic organization of hyphal tips among fungi: vesicles and spitzenkörper. *Journal of Bacteriology*, **104**, 2, 989–1009.

Grove, S. N., Bracker, C. E. & Morré, D. J. (1970). An ultrastructural basis for hyphal tip growth in *Pythium ultimum. American Journal of Botany*, **57**, 245–66.

Hasilik, A. (1974). Inactivation of chitin synthase in *Saccharomyces cerevisiae. Archives of Microbiology*, **101**, 295–301.

Heath, J. B., Gay, J. L. & Greenwood, A. D. (1971). Cell wall formation in the *Saprolegniales:* ctyoplasmic vesicles underlying developing walls. *Journal of General Microbiology*, **65**, 225–32.

Holtz, B. R., Stewart, P. S., Patton, S. & Schisler, L. C. (1972). Isolation and characterization of membranes from the cultivated mushroom. *Plant Physiology*, **50**, 541–546.

Iten, N. & Matile, Ph. (1970). Role of chitinase and other lysosomal enzymes of *Coprinus lagopus* in the autolysis of fruiting bodies. *Journal of General Microbiology*, **61**, 301–9.

Jan, Y. N. (1974). Properties and cellular localization of chitin synthase in *Phycomyces blakesleeanus. Journal of Biological Chemistry*, **249**, 1973–9.

Kanetsuna, F., Carbonell, L. M., Azumai, I. & Yamamura, Y. (1972). Biochemical studies on the thermal dimorphism of *Paracoccidioides brasiliensis. Journal of Bacteriology*, **110**, 208–18.

Katz, D. & Rosenberger, R. F. (1971). Hyphal wall synthesis in *Aspergillus nidulans:* effect of protein synthesis inhibition and osmotic shock on chitin insertion and morphogenesis. *Journal of Bacteriology*, **108**, 184–90.

Koehler, D. C., Leonard, R. T., van der Woude, W. J., Linkins, A. E. & Levis, L. N. (1976). Association of latent cellulase activity with plasma membranes from Kidney Bean abscission zones. *Plant Physiology*, **58**, 324–30.

Kröning, A. & Egel, (1974). Autolytic activities associated with conjugation and sporulation in fission yeast. *Archiv für Mikrobiologie* **99**, 241–9.

Kulaev, I. S. (1973). The enzyme of polyphosphate metabolism in protoplast and some sub-cellular structures of *Neurospora crassa*. In *Yeast, Mould and Plant Protoplasts*, eds. J. R. Villanueva, I. Garcia-Achar, S. Gascon & F. Uruburu, pp. 259–73. New York, London: Academic Press.

Larpent, J. P. (1966). Caractères et déterminisme des corrélations d'inhibition dans le mycélium jeune de quelques champignons. Thèse Université Clermont-Ferrand. *Annales des Sciences Naturelles. Botanique,* **VII**, 1–130.

Leonard, R. T. & van der Woude, W. J. (1976). Isolation of plasma membranes from corn roots by sucrose density gradient centrifugation. An anomalous effects of ficoll. *Plant Physiology,* **57**, 105–14.

Maclachlan, G. A. (1976). A potential role for endocellulase in cellulose biosynthesis. *Applied Polymer Symposium,* **28**, 645–58.

Maddox, I. S. & Hough, J. S. (1971). Yeast glucanase and mannanase. *Journal of the Institute of Brewing,* **77**, 44–7.

Mahadevan, P. R. & Mahadkar, U. R. (1970). Role of enzymes in growth and morphology of *Neurospora crassa*: cell wall bound enzymes and their possible role in branching. *Journal of Bacteriology,* **101**, 941–7.

Marriott, M. S. (1975). Enzymatic activity of purified plasma membranes from the yeast and mycelial forms of *Candida albicans*. *Journal of General Microbiology,* **89**, 345–52.

Mehta, N. M. & Mahadevan, P. R. (1975). Proteases of *Neurospora crassa*: their role in morphology. *Indian Journal of Experimental Biology,* **13**, 131–4.

Meyer, R., Parish, R. N. & Hohl, H. R. (1976). Hyphal tip growth in *Phytophthora*: gradient distribution and ultrahistochemistry of enzymes. *Archives of Microbiology,* **110**, 215–24.

Mishra, N. C., Tatum, E. L. (1972). Effect of L-sorbose on polysaccharide synthases of *Neurospora crassa*. *Proceedings of the National Academy of Sciences, USA,* **69**, 313–17.

Morré, D. J. (1971). Isolation of Golgi apparatus. *Methods in Enzymology,* **22**, 130–48.

Mullins, J. T. & Ellis, A. E. (1974). Sexual morphogenesis in *Achlya*: ultrastructural basis for the hormonal induction of antheridial hyphae. *Proceedings of the National Academy of Sciences, USA,* **71**, 1347–50.

Nickerson, W. J. & Falcone, G. (1956a). Enzymatic reduction of disulfide bonds in cell wall protein of baker's yeast. *Science,* **124**, 318–19.

Nickerson, W. J. & Falcone, G. (1956b). Identification of protein disulfide reductase as a cellular division enzyme in yeasts. *Science,* **124**, 722–3.

Nolan, R. A. & Bal, A. K. (1974). Cellulase localization in hyphae of *Achlya ambisexualis*. *Journal of Bacteriology,* **117**, 840–3.

Nombela, C., Uruburu, F. & Villanueva, J. R. (1974). Studies on membranes isolated from extracts of *Fusarium culmorum*. *Journal of General Microbiology,* **81**, 247–54.

Nurminen, T., Oura, E. & Suomalainen, H. (1969). The enzymatic composition of the isolated cell wall and plasma membrane of baker's yeast. *Biochemical Journal,* **116**, 61–9.

Ordin, L. & Hall, M. A. (1968). Cellulose synthesis in higher plants from UDP-glucose. *Plant Physiology,* **43**, 473–6.

Page, W. J. & Stock, J. J. (1972). Isolation and characterization of *Microsporum gypseum* lysosomes: role of lysosomes in macroconidia germination. *Journal of Bacteriology,* **110**, 354–62.

Page, W. J. & Stock, J. J. (1974). Sequential action of cell wall hydrolases in the germination and outgrowth of *Microsporum gypseum* macroconidia. *Canadian Journal of Microbiology,* **20**, 483–9.

Patni, N. J., Billmire, E. & Aaronson, S. (1974). Isolation of the *Ochromonas danica* plasma membrane and identification of several membrane enzymes. *Biochimica et Biophysica Acta,* **373**, 347–55.

Peberdy, J. F. & Moore, P. M. (1975). Chitin synthase in *Mortierella vinacea*: properties, cellular location and synthesis in growing cultures. *Journal of General*

Microbiology, **90**, 228–36.

Polacheck, Y. & Rosenberger, R. F. (1975). Autolytic enzymes in hyphae of *Aspergillus nidulans*: their action on old and newly formed walls. *Journal of Bacteriology*, **121**, 332–37.

Powell, J. T. & Brew, K. (1974). Glycosyltransferases in the Golgi membranes of *Onion* stem. *Biochemical Journal*, **142**, 203–9.

Ray, P. M., Shininger, T. L. & Ray, M. M. (1969). Isolation of β-glucan synthase particles from plant cells and identification with golgi membranes. *Biochemistry*, **64**, 605–12.

Ray, P. M., Eisinger, W. R. & Robinson, D. G. (1976). Organelles involved in cell wall polysaccharide formation and transport in Pea cells. *Berichte der Deutschen botanischen Gesellschaft*, **89**, 121–46.

Robertson, N. F. (1965). The mechanism of cellular extension and branching. In *The Fungi* vol. 1, eds. G. C. Ainsworth & A. S. Sussman, pp. 613–24. New York, London: Academic Press.

Rolland, J. C. & Pilet, P. E. (1974). Implications du plasmalemme et de la paroi dans la croissance des cellules végétales. *Experientia* **30**, 441–451.

Ruiz-Herrera, J. & Bartnicki-Garcia, S. (1976). Proteolytic activation and inactivation of chitin synthetase from *Mucor rouxii*. *Journal of General Microbiology*, **97**, 241–9.

Satoh, S., Matsuda, K. & Tamaki, K. (1976). β 1–4 glucan occuring in homogenate of *Phaseolus aureus* seedlings. Possible nascent stage of cellulose biosynthesis in vivo. *Plant & Cell Physiology*, **17**, 1243–1254.

Scarborough, G. A. (1975). Isolation and characterization of *Neurospora crassa* plasma membranes. *Journal of Biological Chemistry*, **250**, 1106–1111.

Selitrennikoff, C. P., Nelson, R. E. & Stegel, R. N. (1974). Phase specific genes for macroconidiation in *Neurospora crassa*. *Genetics* **78**, 679–690.

Sentandreu, R., Victoria-Elorza, M. & Villanueva, J. R. (1975). Synthesis of yeast wall glucan. *Journal of General Microbiology* **90**, 13–20.

Shore, G. & Maclachlan, G. A. (1975). The site of cellulose synthesis – Hormone treatment alters the intracellular location of Alkali-Insoluble β 1–4 glucan (cellulose) synthase activities. *Journal of Cell Biology*, **64**, 557–571.

Sietsma, J. H. & Haskins, R. H. (1968). The incorporation of cholesterol by *Pythium* sp. PRL 2142 and some of its effects on cell metabolism. *Canadian Journal of Biochemistry*, **46**, 813–818.

Sukumaran, C. P. & Mahadevan, P. R. (1975). Localization of enzymes in the cell walls of *Neurospora crassa*. *Indian Journal of Experimental Biology* **13**, 127–130.

Suomalainen, H. & Nurminen, T. (1973). Structure and function of the yeast cell envelope. In *Yeast, Mould and Plant Protoplasts*, eds. J. R. Villanueva, I. Garcia-Achar, S. Gascón & F. Uruburu, pp. 167–86. New York, London: Academic Press.

Thiéry, J. P. (1967). Mise en évidence des polysaccharides sur coupes fines en microscopie électronique. *Journal de Microscopie*, **6**, 987–1017.

Thomas, D. des S. & Mullins, J. T. (1967). Role of enzymatic wall softening in plant morphogenesis. Hormonal induction in *Achlya*. *Science*, **156**, 84–5.

Thomas, D. des S. & Mullins, J. T. (1969). Cellulase induction and wall extension in the water mold *Achlya ambisexualis*. *Physiologia Plantarum*, **22**, 347–53.

Tolbert, N. E. (1974). Isolation of subcellular organelles of metabolism on isopycnic sucrose gradients. *Methods in Enzymology*, **31**, 734–46.

Tsai, C. M. & Hassid, W. Z. (1973). Substrate activation of β 1–3 glucan synthetase and its effect on the structure of β-glucan obtained from UDP-D-glucose and particulate enzyme of oat coleoptiles. *Plant Physiology*, **51**, 998–1001.

van der Woude, W. J., Lembi, C. A., Morré, D. J., Kindinger, J. I. & Ordin, L. (1974). β-glucan synthase of plasma membrane and Golgi apparatus from *Onion* stem. *Plant Physiology*, **54**, 333–40.

Villa, T. G., Notario, V., Benitez, T. & Villanueva, J. R. (1976). On the effect of glucono-δ-lactone on the yeast *Pichia polymorpha*. *Archives of Microbiology*, **109**, 157–61.

Wang, M. C. & Bartnicki-Garcia, S. (1966). Biosynthesis of β 1–3 and β 1–6 linked glucan by *Phytophthora cinnamomi* hyphal walls. *Biochemical and Biophysical Research Communications*, **24**, 832–7.

Wessels, J. G. H. (1965). Morphogenesis and biochemical processes in *Schizophyllum commune*. *Wentia*, **13**, 1–113.

Wessels, J. G. H. & Koltin, Y. (1972). R-glucanase activity and susceptibility of hyphal walls to degradation in mutants of *Schizophyllum* with disrupted nuclear migration. *Journal of General Microbiology*, **71**, 471–5.

12
Endogenous lytic enzymes and wall metabolism

R.F.ROSENBERGER
National Institute of Medical Research, Division of Genetics, Mill Hill, London NW7 1AA, UK

Introduction

Almost all fungi synthesize mechanically strong, rigid walls which protect the fragile protoplast from osmotic and other damage. The walls also maintain the characteristic shapes of mycelia, fruiting bodies and spores; shapes which have evolved to help an organism exploit its ecological niche. However, being surrounded by such a rigid 'exoskeleton', fungal hyphae can only grow and differentiate by cutting and modifying the polymers making up their wall. Such wall modifications appear to be the function of endogenous lytic enzymes.

Enzymes capable of hydrolysing wall polymers have invariably been found in fungi when a sufficiently intensive search has been made. They include hydrolases acting on chitin (Mahadevan & Mahadkar, 1970; de Vries & Wessels, 1973; Polacheck & Rosenberger, 1975; see also Chapters 8 and 10), $\beta(1-3)$glucan (Mahadevan & Mahadkar, 1970; Wessels & Koltin, 1972; Fleet & Phaff, 1974; Polacheck & Rosenberger, 1975), $\beta(1-4)$glucan (Thomas, Lutzak & Manavatha, 1974; Fèvre, 1976), $a(1-3)$glucan (Zonneveld, 1972) and proteins (Cabib & Farkas, 1971; Cohen, 1973; Page & Stock, 1974); for some polymers both endo- and exo-acting enzymes have been described (Barras, 1972; Fleet & Phaff, 1974). Some of the fungal hydrolases have also been extensively purified and their properties described in detail (Faith, Neubeck & Reese, 1971; Fleet & Phaff, 1974, 1975; Enari & Markhanen, 1977). Information on the nature of potentially lytic enzymes is thus available. However, very much less is known about their mode of action and function in wall metabolism and about the mechanisms which control their activity on the hyphal envelope.

The main reason for our limited knowledge of the way lytic enzymes act in wall extension is the chemical and physical complexity of fungal walls. The information available on wall structure and synthesis is not yet sufficient to predict which chemical bonds need to be broken to allow the insertion of new envelope. Fungi also synthesize catabolic enzymes which hydrolyse exogenous polysaccharides and allow the monomers to be used as carbon sources (Faith *et al.*, 1971). Thus, not every glucanase or chitinase found in mycelia need be concerned with wall metabolism. To clarify the basis of lytic enzyme activity, various workers have studied the location of lysins (enzymes which hydrolyse wall polymers) in hyphae, their interaction with the wall and the way growth conditions affect their synthesis. This chapter will be concerned with such studies and many of the data and much of the discussion will be based on work performed by Dr. Y. Polachek on *Aspergillus nidulans* (Eidam) Winter in the author's laboratory at the Department of Microbiological Chemistry, Hebrew University-Hadassah Medical School, Jerusalem.

The intracellular distribution of lytic enzymes in fungi
Lysins and cytoplasmic vesicles
Morphological studies have shown that small vesicles, which are scattered throughout the cytoplasm of hyphae, aggregate at points of wall synthesis (Grove & Bracker, 1970; Collinge & Trinci, 1974). The vesicles fuse with the plasma membrane at these points, presumably liberating their contents. Further, cytochemical investigations have indicated that at least some of the vesicles contain lytic enzymes (Nolan & Bal, 1974; Page & Stock, 1974). Taken together, these findings suggest a mechanism by which lysins can be safely transported to specific sites on the hyphal wall.

Direct proof that vesicles contain lytic enzymes requires the isolation of intact vesicles in amounts sufficient for biochemical studies. Such isolations present technical difficulties, since treatments which break hyphae usually disrupt internal vacuoles. Purification of vesicles has thus been limited to organisms from which protoplasts can be readily prepared or which can be broken by gentle mechanical means. Successful isolations have been reported from yeasts (Cabib & Farkas, 1971; Cortat, Matile & Wiemken, 1972), *Saprolegnia monoica* Pringsheim (Fèvre, 1976) and a mutant of *Aspergillus nidulans* (Polacheck & Rosenberger, 1978) which will readily form protoplasts. In each case the vesicles contain 'cryptic' lysins which hydrolysed their substrate after the membrane had been ruptured or dissolved. The activities found included β(1–3)glucanase (yeasts, *S. monoica* and *A. nidulans*), proteases (yeasts and *A. nidulans*), cellulase (*S. monoica*) and chitinase (*A. nidulans*).

Only a small number of organisms have so far been investigated and generalizations may thus prove to be premature. With this reservation in mind, the above results appear to outline the fate of endogenous lytic enzymes from the time of their synthesis to their arrival at points of wall growth. There are clearly many gaps in this outline. Little is known about the mechanisms fungi use to introduce lytic enzymes into vesicles. While yeasts can apparently direct internal vesicles to specific points on the wall (Cabib & Farkas, 1971), it is not known how such movement is achieved. However, the existence of vesicles containing lysins does provide a basis for understanding the controls over these enzymes up to the point when they reach their target and interact with the envelope.

Wall-bound lytic enzymes

When hyphae are disrupted mechanically, the supernatants obtained by differential centrifugation contain lytic enzymes. These have presumably originated from ruptured vesicles. In addition, however, lysins are present in the particulate fraction and remain attached to wall fragments during purification and repeated washing with concentrated salt solutions. This partition of lytic activity into free (also vesicular, when breakage has been sufficiently gentle) and bound enzymes has been found in yeasts (Fleet & Phaff, 1974), *Neurospora crassa* Shear & Dodge (Mahadevan & Mahadkar, 1970), *Saprolegnia monoica* (Fèvre, 1976) and *Aspergillus nidulans* (Polacheck & Rosenberger, 1975). Since envelope-associated enzymes may have particular significance in wall metabolism, we have used purified walls of *A. nidulans* in attempts to investigate them.

When such walls are incubated, the bound enzymes detach by digesting part of the supporting polymers and their activities can then be assayed (Polacheck & Rosenberger, 1978). The mixture of detached enzymes degraded chitin, β(1–3)glucan, protein, aryl-β- and aryl-α-glucosides and aryl-β-N-acetylglucosamine. Electrophoresis of such an enzymes degraded chitin, β(1–3)glucan, protein, aryl-β- and aryl-α-chitinase and one band with laminarinase activity, two bands hydrolysing aryl-β-glucoside, and two bands splitting aryl-β-N-acetylglucosamine (see Fig. 12.1 and also Polacheck & Rosenberger, 1978). Similar activities were present in the cytoplasmic fraction, and Tables 12.1 and 12.2 compare the amounts and specific activity of the lysins in the soluble, membrane and wall fractions obtained from the same mycelium. A considerable part of each enzyme activity was bound to the walls, and the specific activity of the attached enzymes was markedly higher than that in other fractions (note that only activities on identical substrates can be compared; the enzyme units given in Tables 12.1 and 12.2 differ from substrate to substrate). The low activity in the membrane fraction shows that the enzymes in the wall preparation did not originate from contaminating membranes; such walls, in fact, contain little membrane (Polacheck & Rosenberger, 1975).

Fèvre (1976) and Mahadevan & Mahadkar (1970) have shown that

Fig. 12.1. Polyacrylamide gel electrophoresis of lysins bound to walls of *Aspergillus nidulans*. The proteins detached from purified walls were electrophoresed as described by Polachek & Rosenberger (1978*a*); gels were cut into 2 mm slices and the enzyme activities in slices determined as described by Polacheck & Rosenberger (1978*a*). The whole gel is drawn at the top; each numbered line represents a protein band staining with Coomassie blue. Enzyme activities shown are for the following substrates: ●——●, aryl-β-glucoside; △·—·△, aryl-β-N-acetylglucosamine; ○---○, colloidal chitin.

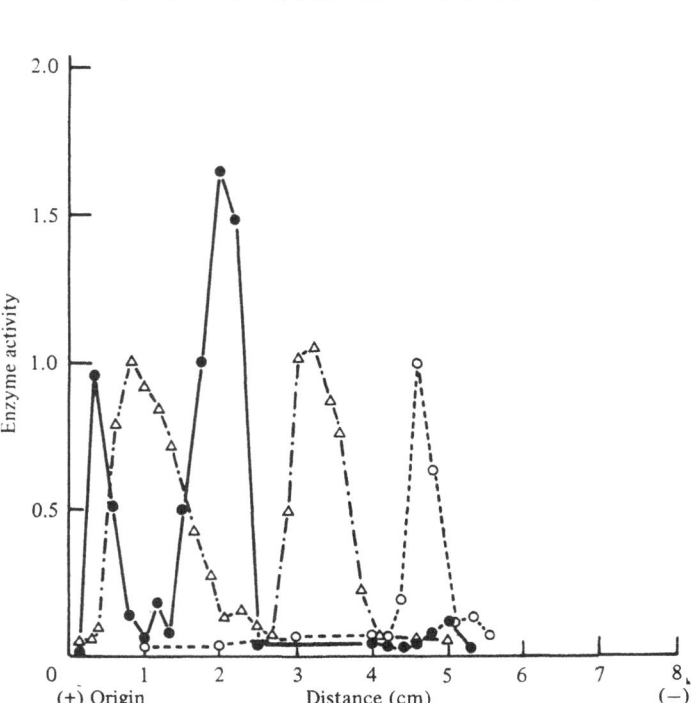

the amounts of wall-bound lysins in *Saprolegnia monoica* and *Neurospora crassa* can vary markedly and appear to be positively correlated with the degree of hyphal branching. In exponentially growing *Aspergillus nidulans*, the quantity of lysins bound per unit of wall appeared reasonably constant. However, the amounts of both soluble and wall-bound lysins increased sharply when the carbon source was exhausted (Table 12.3). Cohen (1973) demonstrated that resuspension of exponentially growing *A. nidulans* mycelium in medium lacking a carbon source led to a marked derepression of protease synthesis; this is also true of other lytic enzymes. The synthesis of both free and bound lysins thus appears to be controlled by catabolite repression (Faith *et al.*, 1971) and the significance of this regarding the involvement of lysins in wall metabolism and branching will be discussed later.

Table 12.1. *Lytic enzyme activity in fractions obtained from broken hyphae of* Aspergillus nidulans *by differential centrifugation*

Fraction	Total enzyme units[a]	Specific activity[b]
Wall	2900	1050
Soluble	2761	27
Membrane	22	4
Culture medium	826	45

[a] Enzyme activity measured with whole wall as substrate. One enzyme unit liberates soluble components from the wall at the rate of 1 μg h^{-1} at 37 °C.
[b] Enzyme units/total protein.

Table 12.2. *Distribution of specific lysins in fractions of broken hyphae of* Aspergillus nidulans

Enzyme substrate	Wall fraction Total enzyme units[a]	Specific activity[b]	Soluble fraction Total enzyme units[a]	Specific activity[b]
Chitin	22	10.5	125	1.4
Laminarin	945	446	6800	55
Protein	8.6	3	14	0.1

[a] Enzyme units are as defined in Polacheck & Rosenberger (1978).
[b] Enzyme units/total protein.

Nature of the lytic enzyme–hyphal wall association

Both free and bound lysins are found in the same preparation of broken hyphae; this distribution could have several explanations. Hyphae may contain distinct populations of lytic enzymes which have quite different affinities for the envelope. Alternatively, all the enzymes may have a high affinity for wall polymers but the exposure of a limited number of binding sites allows only a proportion of these to attach. Finally, attachment may not be due to enzyme–substrate interactions occurring when internal structure is destroyed by mechanical breakage, but to biochemical processes which take place in the growing hypha.

To clarify the nature of the enzyme-wall association, we looked for conditions which would lead to a detachment of lytic enzymes. Purified walls from *Aspergillus nidulans* were treated with varying concentrations of LiCl (1–8 M), sodium dodecylsulphate (SDS) (0.1–1 %), Triton X100 (0.01–1 %), dithiothreitol or 0.1 % cetyltrimethyl ammonium bromide (CETAB) (Table 12.4). The lytic enzymes proved remarkably resistant to inactivation by these agents; lysins could still be recovered from treated walls by autodigestion and only 4 % SDS partially inactivated soluble enzyme (Table 12.4). Among the reagents tested, only the cationic detergent CETAB proved effective and solubilized virtually all of the bound lytic activity (Table 12.4).

The detachment of bound enzymes by 0.1 % CETAB is strong evidence that these are not covalently linked to the wall and that the association is mediated by a hydrophobic component. Since an anionic

Table 12.3 *Comparison of lysin activities in carbon-starved*[a] *and exponentially growing cultures of* Aspergillus nidulans

Enzyme substrate	Wall fraction $\left(\dfrac{\text{starved}}{\text{exponential}}\right)$	Soluble fraction $\left(\dfrac{\text{starved}}{\text{exponential}}\right)$
Chitin	32	140
Laminarin	3.3	4.6
Protein	9.4	51
β-Glucoside	2.8	8.1
Whole walls	3.8	14.3

[a] Half of a culture growing on mineral–glucose medium was washed and resuspended in medium lacking a carbon source. After a further 5 h incubation at 37°C, both cultures were harvested and the amounts of lytic enzymes per hyphal dry weight estimated. Figures are the ratio of activities in the two cultures.

and a neutral detergent (SDS and Triton X-100) had little effect, the hydrophobic component is likely to be a negatively charged lipid, presumably a phospholipid(s). The lysin–wall association in *Aspergillus nidulans* thus differs markedly from that found in Gram-positive bacteria (Shockman, Daneo-Moore & Higgins, 1974). In the latter organisms, lysins bind tightly to homologous walls because of a high affinity for their substrate. Detergents are thus not effective in removing them, while treatment with 8 M LiCl, which neutralizes the strong electrostatic attraction (and does not solubilize *A. nidulans* lysins), will detach them from the wall.

The behaviour of soluble lysins provides additional evidence that enzyme–substrate interactions are not the mechanisms responsible for the lysin–wall association. When free lysins are mixed with purified walls, some enzyme absorption takes place. However, enzymes absorbed *in vitro* like this can be detached by 0.3 M LiCl, while the association *in vivo* is stable in 8 M LiCl. We have also followed the kinetics of enzyme detachment during autodigestion. Unlike bacterial walls, total solubilization of *Aspergillus nidulans* walls is not required for enzyme detachment; free lysins accumulate when only a small fraction of the wall has been hydrolysed (Table 12.5). It is tempting to speculate that the fraction hydrolysed is the same lipids which are dissolved by CETAB.

To determine if enzyme–lipid–wall complexes can arise as artefacts of hyphal breakage, we disrupted intact hyphae in the presence of purified walls freed from enzymes and radioactively labelled with [14C]glucose (Polacheck & Rosenberger, 1978). After breakage, the total wall fraction was purified and allowed to autolyse. The exogenously added

Table 12.4. *Solubilization of bound enzymes by LiCl and detergents.*

Reagent	Enzymes[a] detached	Enzymes[a] remaining bound	Extraction (%)	Inhibition of soluble enzyme (%)
None	0	282	0	0
8 M LiCl	4	248	2	7
1 % Triton X-100	14	261	5	2
4 % SDS	17	258	6	40
0.1 % CETAB	254	28	90	9

[a] Figures are units of lysin determined with whole wall as substrate.

walls had bound very little enzyme during breakage, since practically no ^{14}C-labelled compounds were released during the first stages of autolysis. Thus, the lipid-mediated association of lysins and walls appears to be due to biochemical processes in the growing hypha; processes which affect only a portion of the total lysins present.

Sites of lytic enzyme attachment in relation to the hyphal apex

Our approach to determining the sites of lytic enzyme attachment has been based on the following rationale. Growing hyphae insert new wall only at their apices and the apical wall can therefore be radioactively labelled to distinguish it from older, lateral wall. Suppose such hyphae are disrupted, and the wall fraction purified and allowed to autolyse. It seems reasonable to assume that the bound enzymes will first of all hydrolyse those sections of the wall to which they are attached. The isotope content of the soluble fragments liberated during the first stages of autolysis should thus indicate how the attached enzymes are distributed between apical and older walls.

Tables 12.6 and 12.7 describe the results of such an experiment. Hyphae were first grown with [3H]glucose and then with [14C]glucose as sole carbon source. Hydrolysis of the purified walls with strong mineral acid showed the true proportions of [3]H (lateral wall) and [14]C (apical wall) (Polacheck & Rosenberger, 1975; 1978). Exogenously added lysins preferentially hydrolysed the apical wall (Table 12.6); we have previously shown that apical walls of *Aspergillus nidulans* are more susceptible to lytic enzymes (Polacheck & Rosenberger, 1975). However, the attached lysins behaved quite differently (Table 12.7). During

Table 12.5 *Release of bound lysins by autodigestion of supporting wall*

Time of incubation (min)	Bound enzyme[a]	Detached enzyme[a]	Proportion of wall solubilized (%)
0	720	0	–
15	595	120	–
30	550	160	–
60	435	276	0.8
120	350	320	1.5
240	85	534	2.2

[a] Figures are units of lysin determined with whole wall as substrate.

the first minutes of autolysis, the $[^{14}C]/[^{3}H]$ ratio in the soluble fragments was much lower than that produced by soluble lysins and lower than that from acid hydrolysis. As autolysis proceeded and bound lysins detached (Table 12.5), the $[^{14}C]/[^{3}H]$ ratio rose, as would be expected from the greater activity of soluble enzymes on the apical wall. Thus, lysins appear to be found to the older, lateral rather than the new, apical wall.

Table 12.6. *Hydrolysis of differentially labelled purified* Aspergillus nidulans *walls by exogenously added, soluble lysins, and HCl*

Autolysis time (min)	^{3}H (lateral wall)	^{14}C (apical wall)	$^{14}C/^{3}H$
15	7 160	28 500	3.98
30	11 830	54 470	4.6
60	26 970	91 235	3.38
HCl	253 860	266 550	1.05

Cultures were first grown with $[^{3}H]$ glucose and then $[^{14}C]$glucose. After purification, part of the walls were heated to inactivate attached lysins (this table) and part allowed to autolyse (Table 12.7). Soluble autolysin was added to the heated walls. Figures represent c.p.m. (i.e. soluble components) after removal of insoluble wall.

Table 12.7. *Hydrolysis of differentially labelled, purified* Aspergillus nidulans *walls by bound lysins and HCl*

Autolysis time (min)	^{3}H (lateral wall)	^{14}C (apical wall)	$^{14}C/^{3}H$
15	31 782	20 990	0.66
30	39 673	30 863	0.77
60	47 043	44 392	0.94
120	55 329	66 916	1.21
240	68 768	101 656	1.48
HCl	219 174	230 133	1.05

Walls prepared as in Table 12.6. Figures represent c.p.m. after removal of insoluble wall.

Discussion

The evidence showing that part of the endogenous lytic enzymes are firmly bound to the wall has been summarized in previous sections. In this discussion an attempt will be made to evaluate the significance of the lysin–wall association and to fit it into an overall picture of lysin activity and control. It may be assumed that the lytic enzymes synthesized in the cytoplasm are packaged into membrane-bound organelles. Such vesicles constitute a mechanism for controlling lysin activity in the cytoplasm and for directing the enzymes to their target. It is natural to think that the enzyme–wall association is the result of events occurring after the vesicles have reached the envelope. This view is supported by the demonstration that the binding of lysins is not an artefact of hyphal breakage. Mahadevan & Mahadkar (1970) and Fèvre (1976) have suggested that the bound enzymes are those lysins which have been liberated at points of wall synthesis and are acting on wall polymers. They base this suggestion on the correlation between the degree of branching and the amounts of bound lysins in *Neurospora crassa* and *Saprolegnia monoica*. In these organisms, the amounts of attached enzymes appeared to increase with the number of branches, that is with the number of wall-synthesizing sites. In *Aspergillus nidulans*, however, we have found firstly, that the lysin–wall association cannot be explained by enzyme–substrate interactions and secondly, that the lysins are mainly attached to old lateral wall. This leads us to consider a quite different interpretation of the wall–enzyme association.

The bound lytic enzymes can be dissociated from the wall by a cationic detergent and once detached are not able to reform the original association. The hydrophobic binding is therefore not mediated by integral parts of the enzyme molecules or the wall; it appears to be due to a lipid which can be removed by treatments which do not break chemical bonds. A simple explanation for this would be that whole vesicles containing lysins are trapped in the wall. This leads us to propose a model where vesicles reaching the envelope follow one of two pathways. Those arriving at the actual sites of wall synthesis would release their contents (by mechanisms as yet unknown) and the free lysins would participate in wall metabolism. Others, possibly those reaching the wall at sites just behind the apex where wall-thickening is occurring (Trinci, 1978), would be trapped intact. These latter vesicles will become associated with old wall as the hypha extends and they are postulated to contain the bound enzymes. It appears relevant that at least some morphological studies have demonstrated vesicle-like structures in walls (Page & Stock, 1972, 1974; Nolan & Bal, 1974).

Any model dealing with wall-bound lysins needs to explain why intact hyphae show very little or no wall turnover while purified walls autolyse (Polacheck & Rosenberger, 1977). On the assumption that bound enzymes are those lysins active during apical wall extension, it can be argued that cell rupture has destroyed a delicate balance between wall synthesis and hydrolysis. With the demonstration that bound enzymes are attached to old wall and that the binding is not an artefact of cell breakage, a different explanation is required. The activity of attached lysins must be inhibited in the growing hypha and this inhibition released by cell rupture. In the model proposed above, the bound lysins are contained in vesicles and would be inactive in the growing hypha. Cell breakage could release the enzymes by damage to the vesicle membrane. Such damage might be mechanical or could be due to the release of phospholipase from broken internal organelles. It is tempting to be highly speculative and to consider that such a hypothetical phospholipase may have an important function in wall metabolism. By analogy with the model for septum synthesis in yeast (Cabib & Farkas, 1971), vesicles containing such a phospholipase could be directed to a particular point on the lateral wall. Liberation of the phospholipase and hydrolysis of the membranes of vesicles trapped in the wall may then be sufficient to initiate branching. The proteases liberated from the vesicles could activate polymer synthases contained in the plasma membrane as inactive zymogens and the liberated chitinase and glucanases could prepare the wall for insertion of new material.

The correlation between the amounts of bound enzymes and the number of wall-synthesizing sites does not, at first sight, appear to fit the above model. However, in *Aspergillus nidulans* at least, changes in nutritional conditions can produce striking changes in the amounts of lysins. The synthesis of these enzymes appears to be controlled by catabolite repression, and carbon starvation leads to a 30-fold increase in the amount of bound chitinase (Table 12.3). It is possible that growth conditions which either alter the degree of branching or change the growth form may affect the severity of catabolite repression. For example, highly branched mycelium can be very compact and the diffusion of nutrients to hyphae in the centre of a colony may be slow. In such a case, the increased amounts of lysins need not be a direct consequence of branching and would be compatible with the proposed model.

References

Barras, D. R. (1972). A β-glucan endo-hydrolase from *Schizosaccharomyces pombe* and its role in wall growth. *Antonie Van Leeuwenhoek, Journal of Microbiology and Serology*, **38**, 65–80.

Cabib, E. & Farkas, V. (1971). The control of morphogenesis: an enzymatic mechanism for the initiation of septum formation in yeast. *Proceedings of the National Academy of Sciences, USA*, **68**, 2052–6.

Cohen, B. L. (1973). Regulation of intracellular and extracellular neutral and alkaline proteases in *Aspergillus nidulans*. *Journal of General Microbiology*, **79**, 311–20.

Collinge, A. J. & Trinci, A. P. J. (1974). Hyphal tips of spreading colonial mutants and wild-type of *Neurospora crassa*. *Archiv für Mikrobiologie*, **99**, 353–68.

Cortat, M., Matile, P. & Wiemken, A. (1972). Isolation of glucanase-containing vesicles from budding yeast. *Archiv für Mikrobiologie* **82**, 189–205.

de Vries, O. M. H. & Wessels, J. G. H. (1973). Release of protoplasts from *Schizophyllum commune* by combined action of purified $a(1–3)$ glucanase and chitinase derived from *Trichoderma viride*. *Journal of General Microbiology* **76**, 319–30.

Enari, T.-M. & Markhanen, P. (1977). Production of cellulolytic enzymes by fungi. *Advances in Biochemical Engineering*, **5**, 1–24.

Faith, W. T., Neubeck, C. E. & Reese, E. T. (1971). Production and applications of enzymes. *Advances in Biochemical Engineering*, **1**, 79–110.

Fèvre, M. (1976). Recherches sur le déterminisme de la morphogenèse hyphale. D.Sc. Thesis, Universite Claude Bernard, Lyon, France.

Fleet, G. H. & Phaff, H. A. (1974). Glucanases in *Schizosaccharomyces*. Isolation of the cell wall association $\beta(1–3)$glucanases. *Journal of Biological Chemistry*, **249**, 1717–28.

Fleet, G. H. & Phaff, H. J. (1975). Glucanases in *Schizosaccharomyces*. Isolation and properties of an exo β-glucanase from the cell extracts and culture fluid of *S. japonicus*, var. *versatilis*. *Biochimica et Biophysica Acta*, **401**, 318–32.

Grove, S. N. & Bracker, C. E. (1970). Protoplasmic organisation of hyphal tips among fungi: vesicles and spitzenkorper. *Journal of Bacteriology*, **104**, 989–1009.

Mahadevan, P. R. & Mahadkar, U. R. (1970). Role of enzymes in growth and morphology of *Neurospora crassa*: cell wall bound enzymes and their possible role in branching. *Journal of Bacteriology*, **101**, 941–7.

Nolan, R. A. & Bal, A. K. (1974). Cellulase localisation in hyphae of *Achlya ambisexualis*. *Journal of Bacteriology*, **117**, 840–3.

Page, W. J. & Stock, J. J. (1972). Isolation and characterisation of *Microsporum gypseum* lysosomes: role of lysosomes in macroconidia germination. *Journal of Bacteriology*, **110**, 354–62.

Page, W. J. & Stock, J. J. (1974). Sequential action of cell wall hydrolases in the germination and outgrowth of *Microsporum gypseum* macroconidia. *Canadian Journal of Microbiology*, **20**, 483–9.

Polacheck, Y. & Rosenberger, R. F. (1975). Autolytic enzymes in hyphae of *Aspergillus nidulans*: their action on old and newly-formed walls. *Journal of Bacteriology*, **121**, 332–7.

Polacheck, Y. & Rosenberger, R. F. (1977). *Aspergillus nidulans* mutant lacking $a(1–3)$-glucan, melanin and cleistothecia. *Journal of Bacteriology*, **132**, 650–6.

Polacheck, Y. & Rosenberger, R. F. (1978). The distribution of autolysins in hyphae of *Aspergillus nidulans*: existence of a lipid mediated attachment to hyphal walls. *Journal of Bacteriology*, **135**, 741–54.

Shockman, G. D., Daneo-Moore, L. & Higgins, M. L. (1974). Problems of cell wall and membrane growth, enlargement and division. *Annals of the New York Academy of Sciences,* **235,** 161–97.

Thomas, D. des S., Lutzak, M. & Manavatha, E. (1974). Cytochalasin selectively inhibits synthesis of a secretory protein, cellulase, in *Achlya. Nature,* **249,** 140–1.

Trinci, A. P. J. (1978). Wall and hyphal growth. *Science Progress,* **65,** 75–99.

Wessels, J. G. H. & Koltin, Y. (1972). R-glucanase activity and susceptibility of hyphal walls to degradation in mutants of *Schizophyllum* with disrupted nuclear migration. *Journal of General Microbiology,* **71,** 471–5.

Zonneveld, B. J. M. (1972). A new type of enzyme, an exo-splitting $a(1-3)$ glucanase from non-induced cultures of *Aspergillus nidulans. Biochimica et Biophysica Acta,* **258,** 541–7.

13
Membrane transport and hyphal growth

D. H. JENNINGS

Department of Botany, The University, PO Box 147, Liverpool L69 3BX, UK

Introduction

It is a self-evident but important truth that if a fungus is to grow it must absorb nutrients from the external medium. These nutrients, which can be inorganic ions or organic molecules, must traverse the outer membrane, and with very few exceptions they do so via proteins or carriers. This chapter is an attempt to show that carriers may do more than just supply the metabolic machinery of the mycelium; they may play an integral role in controlling the rate of hyphal growth.

It should not surprise us that carriers should possess regulatory properties. We should not forget that the presence of a differentially permeable membrane is itself an aspect of regulation. It is well known that a lipid membrane presents a large activation energy barrier for a polar molecule. Raven (1977) has pointed out that this barrier compares with the activation energy which must be lowered before 'homogeneous-phase' reactions of cell metabolism can take place. Movement across membranes is catalysed by carriers; homogeneous-phase reactions by enzymes.

Further, a fungal hypha is in most instances in contact with the external environment. The carriers in the outer membrane are the means by which the chemical constituents of the environment make contact with the internal milieu of a hypha. We know, for instance, that carriers have differing affinities for compounds of similar chemical composition. Such a situation can be important in relation to the adaptation of a fungus to its environment. Thus the possession of a carrier having differing affinities for sugars allows a fungus to utilize them in the most efficient manner, namely sequentially. The first sugar to be utilized will be that with the highest affinity for the carrier, the order of utilization of the remainder being in the order of decreasing affinity.

I have also argued previously (Jennings, 1974) that for orderly functioning under natural conditions, the transport systems of a fungal hypha need to be under internal control. Without such control too much solute might enter the hypha, leading to excess synthesis of certain soluble metabolites or the production of too high an internal osmotic pressure, with consequent bursting of the hyphal tips (Robertson, 1959). Thus we would anticipate control of transport by metabolism. Indeed, an interesting picture is building up of the interaction between fermentation and hexose transport in *Saccharomyces*, in which glucose-6-phosphate has an important regulatory role (Azam & Kotyk, 1969; Kotyk & Kleinzeller, 1967; Becker & Betz, 1972; Serrano & De la Fuente, 1974).

Thus, on almost *a priori* grounds one might expect a complex interaction between transport processes and growth. I shall now present the facts that are relevant to the case that I want to make. In order not to confuse them with the hypotheses that I present later, I am purposely giving the factual information a separate section.

I have been highly selective in the facts which I am presenting. In no way am I giving a review of all that is known about solute transport in filamentous fungi. The reader is referred to the reviews by Armstrong (1972), Oxender (1972), Jennings (1973, 1974, 1976*a*, *b*), Burnett (1976), Pateman & Kinghorn (1976) and Whitaker (1976) for other information. Studies on solute transport in yeasts should not be ignored; they throw valuable light on what might be happening in filamentous fungi (Suomalainen & Oura, 1971; Barnett, 1976).

Experimental facts
Distribution of potassium and sodium in a hypha
Five years ago (Jennings, 1973) I pointed out that a hypha does not appear to behave uniformly along its length with respect to potassium and sodium. Germinating spores of *Neurospora crassa* Shear & Dodge (Slayman & Tatum, 1964) and regenerating hyphal fragments of *Dendryphiella salina* (Sunderland) Pugh & Nicot, Jennings & Aynsley, 1971) have a considerably lower potassium : sodium ratio initially than later on during the growth (Table 13.1).

One is, of course, inferring spatial distribution of ions from temporal changes. The advent of the techniques of X-ray microanalysis allows data for spatial distribution of ions in a hypha to be determined directly.

However, for the data to be acceptable, certain stringent criteria must be met (Saubermann & Echlin, 1975). Recently I and my colleagues (Galpin *et al.*, 1978) have described a procedure involving freeze-quenching which meets the necessary criteria. Certainly we believe that we have developed a method in which there is an extremely rapid spread of freezing in the hyphae. There is a very limited amount of handling of the specimen with, we believe, very little ice crystal formation, thawing or condensation.

As a result of this technique we were able to show the following (Table 13.2):

(i) The apex of the hypha of *D. salina* is much less selective for potassium against sodium than those regions which are behind the apex.

(ii) The region 1–50 μm behind the apex may be the most selective for potassium against sodium.

(iii) The total content of the two monovalent cations appears to remain constant along the hypha.

Galpin *et al.* (1978) argued that these data relate to the ion content of the cytoplasm. Other studies (Jones & Jennings, 1965; Slayman & Slayman, 1970) have shown that only a small proportion of the

Table 13.1. *Potassium : sodium ratios in mycelium of* Neurospora crassa *and* Dendryphiella salina *at different times during batch growth*

N. crassa[a]		*D. salina*[b]	
Time (h)	K/Na	Time (h)	K/Na
0	8.5	10	0.0049
2	6.5	24	0.11
4	4.5	30	0.42
6	15.0	48	0.25
8	12.0	54	0.085[c]
10	12.5		
12	14.0		
14	16.5		

Data for *N. crassa* calculated from Fig. 1 of Slayman & Tatum (1964); *N. crassa* grown in medium containing 36.8 mol l⁻¹ potassium and 8.4 mol l⁻¹ sodium. Data for *D. salina* from Jennings & Aynsley (1971); *D. salina* grown in medium containing 0.11 mol l⁻¹ potassium and 7.6 mol l⁻¹ sodium.
[a] End of lag period at 3.5 h.
[b] End of lag period at 7.5 h.
[c] All the potassium in the medium depleted by this time.

potassium in a hypha is associated with the wall. In any case, prior to freeze-quenching the hyphae were washed for 5 min at 4 °C with 10 mM calcium chloride; it is difficult to envisage many monovalent cations remaining associated with the wall under these conditions.

It is interesting to see that changes in potassium and sodium content and in the potassium : sodium ratio with time follow a similar sequence as occurs along hyphae (Table 13.3).

Table 13.2. *Counts for potassium and sodium, and potassium :sodium ratios at stated regions of hyphae of* Dendryphiella salina

Location	K	Counts (200 s^{-1}) Na	Total	K/Na
Apex	966 ± 76	1208 ± 217	2174 ± 274	0.91 ± 0.14
1–50 μm	1534 ± 91	518 ± 87	2058 ± 167	3.30 ± 0.49
200 μm	1128 ± 66	53 ± 155	1836 ± 137	2.00 ± 0.54

The above figures are presented to illustrate in a numerical way the conclusions of Galpin *et al.* (1978) discussed in the text. The relatively high variance is a result of pooling data from different hyphae. The reader is referred to the above paper about how the original data were analysed statistically.

Figures are mean plus or minus standard error.
D. salina grown on one carbon stub in 10 % natural seawater.

Table 13.3. *Concentrations of potassium and sodium and their ratio in the mycelium of* Dendryphiella salina

Age (h)	K (mmol per 100 g dry wt)	Na (mmol per 100 g dry wt)	K/Na
0	0.017 ± 0.002	0.014 ± 0.002	1.2
4	0.036 ± 0.001	0.013 ± 0.004	2.8
6	0.030 ± 0.001	0.007 ± 0.004	4.3
12	0.030 ± 0.005	0.007 ± 0.004	4.3
24	0.040 ± 0	0.013 ± 0.004	31

D. salina grown in shake culture. Data represent the means of three cultures (± standard deviation). The vertical lines join values which are not significantly different at 0.05 level of significance.

Potassium-42 flux studies

Jennings & Aynsley (1971) showed that loss of radioactivity from the mycelium of *D. salina* preloaded with [42]K, into non-radioactive medium under virtually steady-state conditions (no growth; no net movement of potassium) was not a simple exponential function of time. Radioactivity was lost more rapidly in the first 6–10 h than subsequently, when it occurred at a slower constant rate (Fig. 13.1).

The mycelium thus appears to behave as a system in which [42]K is located in two or more compartments. Over the first 6–10 h, there is doubt about the exact characteristics of potassium exchange between the mycelium and the medium. Thereafter, however, there is a log-linear loss of radioactivity from the mycelium. On the assumption that there is a uniform concentration of potassium within the mycelium, an efflux rate was calculated. The value of 0.1–0.17 pmol cm^{-2} s^{-1} is comparable to the value of 0.4 pmol cm^{-2} s^{-1} for the passive leak of potassium from mycelium of *N. crassa* determined by Slayman & Slayman (1968) using three indirect approaches.

The characteristics of exchange of [42]K just described cannot be explained by two or more mycelial compartments containing different amounts of potassium. For this to happen the concentration in the larger

Fig. 13.1. Radioactivity in the mycelium of *Dendryphiella salina* as a percentage of that initially present after transference from a medium containing 2.5 mM potassium chloride labelled with [42]K to an inactive medium of similar chemical composition. The dotted line is the extrapolation of the linear portion of the curve. (From Jennings & Aynsley, 1971).

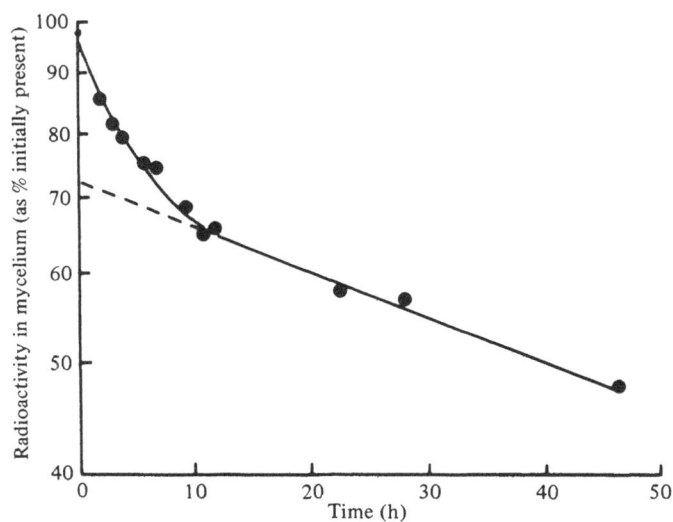

compartment would have to be at least twenty times greater than in the smaller compartment(s). The X-ray microanalytical data and gross chemical analyses show that this is not possible. Another explanation must be sought. The most likely is that some part of the mycelium has a higher passive permeability to potassium than the remainder of the mycelium.

Some clues come from a comparison of data obtained by Slayman and Tatum (1965) for *Neurospora crassa*. They showed that the steady-state potassium exchange across the plasmalemma at pH 5.8 shows only first-order kinetics. There is no evidence at all of more than one compartment. I suggested (Jennings, 1976a) that the difference may be due to the bursting of the hyphal tips of *N. crassa* (Robertson, 1959) during washing of the mycelium with distilled water prior to the measurement of potassium tracer influx.

Plasmalemma electrical potential measurements

Only Slayman and his colleagues have made any measurements of the electrical potential across the plasmalemma of hyphae of filamentous fungi, but there is no doubt of the significance of the findings. They include the first unequivocal demonstration of an electrogenic pump in plant cells (Slayman, 1965a, b). The following observations are of special relevance to this discussion; they also give part of the biophysical and biochemical basis to the known facts of the exchange of potassium for hydrogen or sodium ions by mycelium of *N. crassa* (Slayman & Slayman, 1968).

(i) Measurements of the potential along a hypha (Slayman & Slayman, 1962) show that as one approaches the tip there is a progressive drop from around -120 mV at about 8 mm to -25 mV at the tip itself (Fig. 13.2).

(ii) The potential is brought about by an electrogenic pump driven by ATP (Slayman, Long & Lu, 1973), the current being carried by protons which are extruded into the external medium in exchange for potassium (Slayman, 1970).

(iii) When glucose or 3-*O*-methyl glucose are added to the medium, the membrane becomes depolarized (Slayman & Slayman, 1974). This is interpreted as being the consequence of the co-movement of the sugar and protons across the membrane such that the latter can be regarded as short-circuiting the proton extrusion pump. Fig. 13.3 shows the situation diagrammatically. There is indirect evidence that amino acids and phosphate are also handled by the fungus in this manner (Slayman & Gradmann, 1975).

(iv) In *poky* mutants there are mitochondria which are defective in cytochromes b a and a₃ but which compensate by the presence of an alternative cyanide-insensitive oxidase (Slayman *et al.*, 1975). When cyanide is presented to wild-type *N. crassa* the membrane potential declines with ATP levels. On the other hand, in *poky* it oscillates near the resting level (Gradmann & Slayman, 1975). The oscillation cannot be explained in terms of oscillation of

Fig. 13.2. The plasmalemma potential difference at various points from the growing tip along a hypha of *Neurospora crassa*. (From Slayman & Slayman, 1962, copyright 1962 by the American Association for the Advancement of Science.)

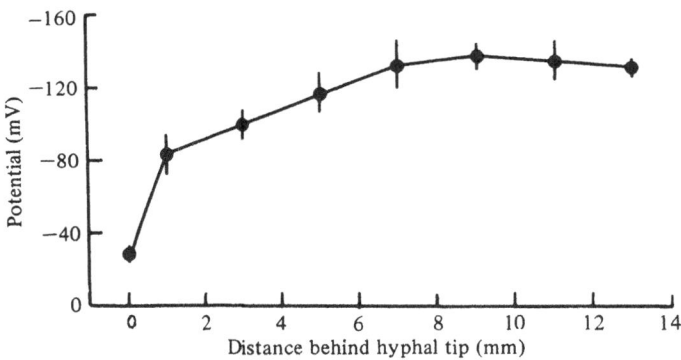

Fig. 13.3. A model for transport across the plasma membrane of *Neurospora crassa*. (From Slayman & Gradmann, 1975).

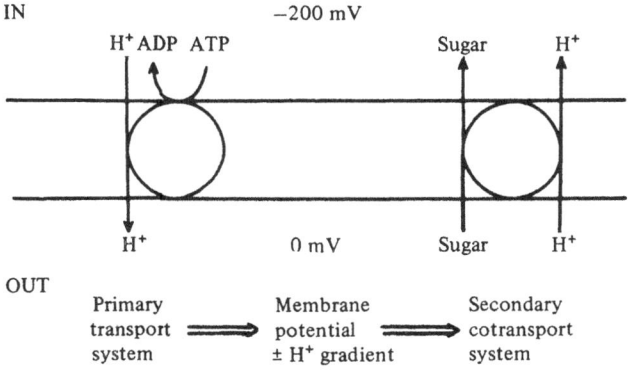

ATP concentration within the mycelium. Several lines of indirect reasoning led to the hypothesis that cyclic (3',5'-) adenosine monophosphate (cAMP) might be involved. Among a number of drugs known to alter the cAMP level in *N. crassa*, serotonin, quinidine and histamine lower it, whilst caffeine raises it. The first three compounds suppress the oscillations, whilst caffeine does the reverse, doubling the amplitude and number of observable cycles and increasing the frequency by 25 %.

cAMP and transport processes

Recently, Pall (1977) has shown that the level of cAMP in *Neurospora crassa* can be increased by the following treatments: large influxes of metabolizable compounds or non-metabolizable analogues, e.g. glucose or 3-*O*-methyl glucose; rapid temperature drops; and addition of agents which uncouple oxidative phosphorylation (Fig. 13.4). All these treatments have been shown to depolarize the membrane. Pall feels that the simplest interpretation is that membrane depolarization, by stimulating membrane-bound adenyl cyclase (Flawia & Torres, 1972; Scott, 1976), produces the increased cAMP levels. Control of the adenylate cyclase activity by the membrane potential would occur if the enzyme was regulated by small charged molecules, the concentration of which in the membrane was under the influence of the membrane potential.

Fig. 13.4. The concentration of cAMP in mycelium of *Neurospora crassa* at varying times after addition of 1 mmol l^{-1} glucose to the medium. (From Pall, 1977.)

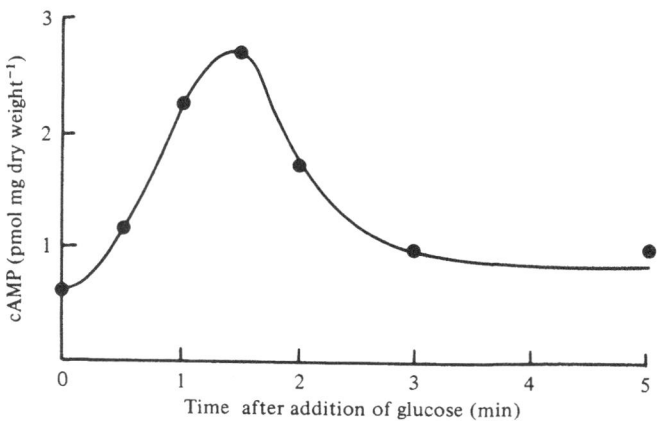

Light-microscopical histochemical study of the distribution of adenosine triphosphatase

Galpin & Jennings (1975*a*) made a study on the hyphae of *Dendryphiella salina* growing either in media in which potassium chloride (2.0 mM) was the major salt present, or in seawater. Histochemical assays showed that adenosine triphosphatase had enhanced activity in seawater-grown cultures. The addition of potassium or sodium ions to the assays of material from the potassium chloride cultures also increased this activity. There was very much less activity in the young parts of the hyphae compared with the older areas of the mycelium. Galpin & Jennings (1975*a*) believed that their observations were compatible with the concept that ATPases are involved in cation transport. This view was supported by similar studies on sporangiophores of *Phycomyces blakesleeanus*; these studies have shown that ATPase distribution is correlated with internal potassium content (Galpin & Jennings, 1975*b*). The data are certainly consistent with the information for *Neurospora crassa* presented above that ATP provides the energy for the exchange of potassium with hydrogen (or sodium) ions across the plasma membrane.

cAMP and branching

There is now a considerable amount of evidence that certain sugars, particularly L-sorbose, have a paramorphogenic effect on *Neurospora crassa*. L-sorbose restricts radial growth in solid cultures and increases hyphal branching but does not reduce dry weight production in liquid cultures (Tatum, Barratt & Cutter, 1949). Known biochemical changes have been summarized by Scott, Mishra & Tatum (1973). L-sorbose and 3-*O*-methyl glucose have similar effect on *Dendryphiella salina* (Galpin, Jennings & Thornton, 1977). Theophylline and phaseollin act in a similar manner, and further experiments with theophylline showed that it acted synergistically with the two sugars. The data were interpreted in terms of the paramorphogens acting via cAMP.

Hypotheses (with a few more facts)

I start by examining the permeability of the membrane at the hyphal tip. The X-ray microanalytical data for *Dendryphiella salina* would suggest that it is more permeable than the remainder of the hypha with respect to sodium and potassium, which accounts for the much lower potassium : sodium ratio at the tip. This offers an explanation of the fact that the mycelium does not behave as a single compartment system with respect to ^{42}K – radioactivity being released more readily from the apex. There is a flow of potassium along the hypha to the apex as a result of this higher permeability and the presence of a potassium–hydrogen/sodium exchange pump located further back. Potassium is pumped into the hyphae in exchange for hydrogen or sodium ions.

The net consequence of this is a current of potassium along a hypha to the tip. As will be discussed below, the current is generated by the ion pumps located in the plasmalemma pumping potassium into the hypha. The flow of potassium may reflect an electric current moving through the hypha. But caution is needed here. Conceivably an anion might move at the same rate as the potassium, so that there was no net movement of charge. Equally the movement of sodium into the hypha could have the same result. On the other hand, if the movement of potassium were from one fixed negative charge (on proteins and lipids) to another, then an electric current could flow. Essentially the current would be generated by the potassium–hydrogen/sodium exchange pumps, the energy ultimately coming from ATP.

Evidence is now accumulating that developing systems generally drive steady ion currents through themselves and thus produce substantial electrical fields within themselves (Jaffe & Nuccitelli, 1977). It has already been suggested that such currents are present in fungi and may be responsible for moving vesicles to the tip (Bartnicki-Garcia, 1973; Harold, 1977). However, care is needed in extrapolating from systems for which there are measurements of the strength and direction of the electric current flow to growing hyphal apices, for which there is virtually no information of this kind. All that we know is that in the mature vegetative hypha of *Neurospora crassa* there is a voltage-dependent current of around $15\,\mu A$ running out through the potassium–hydrogen/sodium ion exchange pump (which has a relatively high electrical resistance) due to the extrusion of protons (Slayman & Gradmann, 1975). Nevertheless, given the low passive permeability of the plasmalemma to potassium away from the tip and a high passive permeability at the tip itself, it is not unreasonable to suppose that a current could be flowing along the hypha to the tip. The current would be brought about by potassium diffusing down its electrochemical potential gradient.

Though I have given a value for the current carried by protons through the pump in mature parts of a hypha of *Neurospora crassa*, even that information may not be relevant to the apex. Thus the potential difference across the plasmalemma drops as one moves to the tip. The very low potential difference across the tip itself may be due to greater puncture injury here as the electrode is inserted (Jaffe & Nuccitelli, 1977). Alternatively, the low potential difference may not be an artefact and may indeed be a consequence of a much higher permeability to ions across the plasmalemma in this region.

But why does the potential difference fall as one moves to the tip? At this stage one can only speculate. Some possibilities spring to mind. In the following list, the word pump refers to the potassium–hydrogen/sodium exchange pump which is electrogenic and makes a substantial contribution to the plasmalemma potential difference:

(i) There are fewer pump molecules per unit area of membrane as one moves towards the tip.

(ii) There are the same number of pump molecules per unit area along the whole hyphal length, but as one moves to the tip their activity is reduced. This could occur for the following reasons: there might be a higher concentration of cAMP, or a lower concentration of ATP, within the cytoplasm of the younger regions, the former compound acting solely in a regulatory manner, the latter as the energy supply for the pump; alternatively there might be a lower production of hydrogen ions in the same region.

(iii) It is also possible that there are the same number of pump molecules per unit area and that these all have the same activity. However, as one moves to the tip the membrane is becoming increasingly depolarized by an increasingly larger flux of protons which accompany the increasingly larger flux of glucose necessary to meet the metabolic needs of the cytoplasm in the hyphal apex.

Can any of these suggestions be excluded? If we accept the histochemical study of Galpin & Jennings (1975a) we can exclude (iii). That study showed that there was very much less ATPase in younger parts of a hypha of *Dendryphiella salina*. If (iii) were true there would be uniform staining for ATPase along the hypha. The decreased amount of stain in the younger parts is compatible with either (i) or (ii), except that since the concentration of ATP used in the histo-chemical assay is above that which would cause substrate limitation, we can discount the possibility that there is such limitation in the growing hypha.

Even at this stage, although we cannot distinguish between possibilities (i) and (ii) the consequences could be the same in one respect, namely that there would be a reduced flux of protons (and therefore electric current) out of the younger parts of a hypha. This could mean that its cytoplasm is more electrically insulated from the external medium and, therefore, there is a better chance of any electrical current flow to the tip.

Whenever there is a flow of solute there will be a flow of water. So if there is a flow of ions to the tip there will be a flow of water. Such coupling of flows can be described by irreversible thermodynamics. The flow of liquid could bring about movement of vesicles to the tip. This suggestion has been made previously by me (Jennings, 1973; Jennings *et al.*, 1974). The possibility was raised that within the hypha there is a standing-gradient osmotic flow (Diamond & Bossert, 1967) brought about by the greater activity of the pumps away from the hyphal tip. Though, as I have just indicated, the suggestion about the distribution of pump activity along a hypha remains valid, no further evidence has been brought forward in support of the presence of a standing-gradient osmotic flow in a hypha. Indeed, the constancy of total monovalent cation concentration at any one point along a hypha would seem to me to argue against the hypothesis.

However, there is no doubt that water flow can occur along a hypha. There is emphatic evidence from studies on drop formation at hyphal tips of *Serpula lacrimans* (Jennings *et al.*, 1974; Jennings, 1976*a*). These drops are formed when hyphae grow on a non-absorptive surface; they seem to be produced as a result of the generation of a hydrostatic pressure within the mycelium caused by absorption of sugars at the food base leading to the production of a high osmotic pressure in the hyphae. I do not want to suggest that any water movement along the hyphae to the tip in filamentous fungi is necessarily brought about by these means, but the observations do indicate that a flow of water can be directed preferentially to the tip.

Not only is one now inclined to doubt the presence of a standing-gradient osmotic flow within a hypha, but it is possible that the theoretical basis behind the concept is incorrect. Hill (1975*a, b*) has argued that the known observations for which the concept has been developed are better explained by the presence of electro-osmotic coupling at the boundary membranes. There is good evidence that electro-osmosis occurs in giant algal cells (Barry & Hope, 1969*a, b, c*). Therefore, it is not inconceivable that it can occur in a filamentous fungus – the case can certainly be argued on the basis of the similarity between the electrical potential across the plasma membrane and the passive permeability of that membrane to the major ions traversing it. If there were an electro-osmotic flow of water it would be brought about by the activity of the pump, water movement accompanying the movement of potassium into the hypha. Therefore, if one regulates the activity of the pump, one will regulate the electro-osmotic component of water flow.

We need to keep in mind the possibility that the motive force for the movement of vesicles to the tip may reside in some other process than electric current or fluid flow. A contractile system, e.g. microfilaments, is a likely candidate; such a possibility certainly cannot be ignored. Were it to exist, one would conceive that electric current or fluid flow would then act as a mechanism for orientation of the movement of the vesicles parallel to the rigid walls of the hyphae.

Whatever the motive force for movement of vesicles to the tip, there must be some mechanism by which the growth at the tip regulates either the flow of electric current, so there is not an untoward increase in concentration of ions at the tip, or the flow of fluid and an equivalent increase in solute concentration. A signal must move down the hypha from the tip to the pump molecules to regulate their activity commensurate with the rate of growth. There is evidence that the tip has biochemical features distinct from that part of the hypha behind (Turian, 1976) and so it can be conceived of as being a distinct control point within a hypha. What might be the signal? I would like to suggest that cAMP is a possible candidate. We know it can modulate pump activity. There is evidence that when the concentration in the mycelium rises the rate of hyphal extension is reduced and there is an increased frequency of branching (Galpin *et al.*, 1977). It could be postulated that the activity of the pump is reduced. In consequence of this, the directional effects on vesicle movement brought about by pump activity will no longer be present. The concentration of vesicles could build up away from the tip and, according to the model presented in Chapter 16 by Prosser, hyphal branches will be more readily produced.

A summary of my views is given in Fig. 13.5. They centre around a

Fig. 13.5. Proposed major ion and water flows across the plasma membrane and along a growing hypha, indicating their interaction with cAMP.

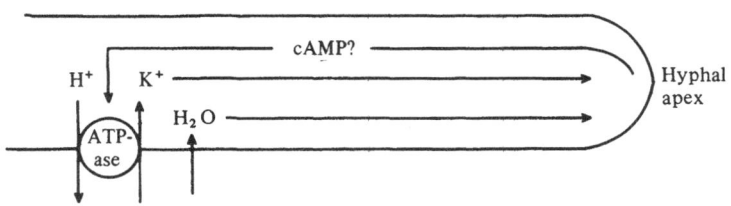

flow of potassium to the hyphal tip from pumps some way back. It is possible, however, that in *Dendryphiella salina* it is the flow of sodium, and in a terrestrial fungus such as *Neurospora crassa* it is the flow of protons in the reverse direction, which should be considered. But that is further speculation which it would be imprudent to make without further experimental facts. Perhaps I have speculated too much already; however, I believe it is necessary in order to indicate how growth and solute transport might be interrelated. At any rate, I hope what I have written will encourage further work on the problem and remind those who work in the field of fungal membrane transport that it is now time to move away from studies on the gross transport properties of mycelia to investigations which focus on the first 50μm or so of a growing hypha.

References

Armstrong, W. McD. (1972). Ion transport and related phenomena in yeast and other micro-organisms. *Transport and Accumulation in Biological Systems*, ed. E. J. Harris, pp. 407–45. London: Butterworths.

Azam, F. & Kotyk, A. (1969). Glucose-6-phosphate as regulator of monosaccharide transport in baker's yeast. *F.E.B.S. Letters*, **2**, 333–5.

Barnett, J. A. (1976). The utilization of sugars by yeasts. *Advances in Carbohydrate Chemistry and Biochemistry*, **32**, 126–234.

Barry, P. H. & Hope, A. B. (1969a). Electro-osmosis in *Chara* and *Nitella* cells. *Biochimica et Biophysica Acta*, **193**, 124–8.

Barry, P. H. & Hope, A. B. (1969b). Electro-osmosis in membranes: effects of unstirred layers and transport numbers. I. Theory. *Biophysical Journal*, **9**, 700–28.

Barry, P. H. & Hope, A. B. (1969c). Electro-osmosis in membranes: effects of unstirred layers and transport numbers. II. Experimental. *Biophysical Journal*, **9**, 729–57.

Bartnicki-Garcia, S. (1973). Fundamental aspects of hyphal morphogenesis. In *Microbial Differentiation* (23rd Symposium of the Society for General Microbiology), eds. J. M. Ashworth & J. E. Smith, pp. 245–67. Cambridge University Press.

Becker, J-U. & Betz, A. (1972). Membrane transport as controlling pacemaker of glycolysis in *Saccharomyces carlsbergensis*. *Biochimica et Biophysica Acta*, **274**, 584–97.

Burnett, J. H. (1976). *Fundamentals of Mycology*, 2nd edition. London: Edward Arnold.

Diamond, J. M. & Bossert, W. H. (1967). Standing gradient osmotic flow. A mechanism for coupling of water and solute transport in epithelia. *Journal of General Physiology*, **50**, 2061–83.

Flawia, M. M. & Torres, H. N. (1972). Adenylate cyclase activity in *Neurospora crassa* I. General properties. *Journal of Biological Chemistry*, **247**, 6873–9.

Galpin, M. F. J. & Jennings, D. H. (1975a). Histochemical study of the hyphae and the distribution of adenosine triphosphatase in *Dendryphiella salina*. *Transactions of the British Mycological Society*, **65**, 477–83.

Galpin, M. F. J. & Jennings, D. H. (1975b). Potassium and ATPase in *Phycomyces*. *Transactions of the British Mycological Society*, **66**, 151–3.

Galpin, M. F. J., Jennings, D. H. & Thornton, J. D. (1977). Hyphal branching in *Dendryphiella salina*: effect of various compounds and the further elucidation of the effect of sorbose and the role of cyclic AMP. *Transactions of the British Mycological Society,* **69,** 175–82.

Galpin, M. F. J., Jennings, D. H., Oates, K. & Hobot, J. (1978). Localization by X-ray microanalysis of soluble ions, particularly potassium and sodium, in fungal hyphae. *Experimental Mycology* **2,** 258–69.

Gradmann, D. & Slayman, C. L. (1975). Oscillation of an electrogenic pump in the plasma membrane of *Neurospora. Journal of Membrane Biology,* **23,** 181–212.

Harold, F. M. (1977). Ion currents and physiological functions in microorganisms. *Annual Review of Microbiology,* **31,** 181–203.

Hill, A. E. (1975*a*). Solute–solvent coupling in epithelia: a critical examination of the standing gradient osmotic flow theory. *Proceedings of the Royal Society, London,* **B190,** 99–114.

Hill, A. E. (1975*b*). Solute–solvent coupling in epithelia: an electro-osmotic theory of fluid transfer. *Proceedings of the Royal Society, London,* **B190,** 115–34.

Jaffe, L. F. & Nuccitelli, R. (1977). Electrical controls of development. *Annual Review of Biophysics and Bioengineering,* **6,** 445–76.

Jennings, D. H. (1973). Cations and filamentous fungi: invasion of the sea and hyphal functioning. In *Ion Transport in Plants,* ed. W. P. Anderson, pp. 323–35. New York, London: Academic Press.

Jennings, D. H. (1974). Sugar transport into fungi: an essay. *Transactions of the British Mycological Society,* **62,** 1–24.

Jennings, D. H. (1976*a*). Transport and translocation in filamentous fungi. In *The Filamentous Fungi,* vol. 2 *Biosynthesis and Metabolism,* eds. J. E. Smith & D. R. Berry, pp. 32–64. London: Edward Arnold.

Jennings, D. H. (1976*b*). Transport in fungal cells. In *Encyclopedia of Plant Physiology,* new series vol. 2 *Transport in Plants, A Cells,* eds. U. Lüttge & M. G. Pitman, pp. 189–228. Berlin: Springer-Verlag.

Jennings, D. H. & Aynsley, J. S. (1971). Compartmentation and low temperature fluxes of potassium in mycelium of *Dendryphiella salina. New Phytologist,* **70,** 713–23.

Jennings, D. H., Thornton, J. D., Galpin, M. F. J. & Coggins, C. R. (1974). Translocation in fungi. In *Transport at the Cellular Level* (28th Symposium of the Society for Experimental Biology) eds. M. A. Sleigh & D. H. Jennings, pp. 139–56. Cambridge University Press.

Jones, E. B. G. & Jennings, D. H. (1965). The effect of cations on the growth of fungi. *New Phytologist,* **64,** 86–100.

Kotyk, A. & Kleinzeller, A. (1967). Affinity of the yeast membrane carrier for glucose and its role in the Pasteur effect. *Biochimica et Biophysica Acta,* **135,** 106–11.

Oxender, D. L. (1972). Amino acid transport in micro-organisms. In *Metabolic Pathways,* vol. 4 ed. L. E. Hokin, pp. 133–85. New York, London: Academic Press.

Pall, M. L. (1977). Cyclic AMP and the plasma membrane potential in *Neurospora crassa. Journal of Biological Chemistry,* **252,** 7146–50.

Pateman, J. A. & Kinghorn, J. R. (1976). Nitrogen metabolism. In *The Filamentous Fungi,* vol. 2 *Biosynthesis and Metabolism* (eds. J. E. Smith & D. R. Berry), pp. 159–237. London: Edward Arnold.

Raven, J. A. (1977). Regulation of solute transport at the cell level. In *Integration of activity in the higher plant* (31st Symposium of the Society for Experimental Biology), ed. D. H. Jennings, pp. 73–99. Cambridge University Press.

Robertson, N. F. (1959). Experimental control of hyphal branching forms in hyphomycetous fungi. *Journal of the Linnean Society, London,* **56,** 207–11.

Saubermann, A. J. & Echlin, P. (1975). The preparation, examination and analysis of frozen hydrated tissue sections by scanning transmission electron microscopy and X-ray microanalysis. *Journal of Microscopy,* **105,** 155–91.

Scott, W. A. (1976). Adenosine 3'-5'-cyclic monophosphate deficiency in *Neurospora crassa. Proceedings of the National Academy of Sciences, USA,* **73,** 2995–9.

Scott, W. A., Mishra, N. C. & Tatum, E. L. (1973). Biochemical genetics of morphogenesis in *Neurospora. Brookhaven Symposia in Biology,* **25,** 1–18.

Serrano, R. & De la Fuente, G. (1974). Regulatory properties of the constitutive hexose transport in *Saccharomyces cerevisiae. Molecular and Cellular Biochemistry,* **5,** 161–71.

Slayman, C. L. (1965*a*). Electrical properties of *Neurospora crassa.* Effects of external cations on the intracellular potential. *Journal of General Physiology,* **49,** 69–92.

Slayman, C. L. (1965*b*). Electrical properties of *Neurospora crassa.* Respiration and the intracellular potential. *Journal of General Physiology,* **49,** 93–116.

Slayman, C. L. (1970). Movement of ions and electrogenesis in micro-organisms. *American Zoologist,* **10,** 377–92.

Slayman, C. L. & Gradmann, D. (1975). Electrogenic proton transport in the plasma membrane of *Neurospora. Biophysical Journal,* **15,** 968–71.

Slayman, C. L. & Slayman, C. W. (1962). Measurement of membrane potentials in *Neurospora. Science,* **136,** 876–7.

Slayman, C. L. & Slayman, C. W. (1968). Net uptake of potassium in *Neurospora.* Exchange for sodium and hydrogen ions. *Journal of General Physiology,* **52,** 424–43.

Slayman, C. L. & Slayman, C. W. (1974). Depolarisation of the plasma membrane of *Neurospora* during active transport of glucose: evidence for a proton dependent cotransport system. *Proceedings of the National Academy of Sciences, USA* **71,** 1935–9.

Slayman, C. L., Long, W. S. & Lu, C. Y-H. (1973). The relationship between ATP and an electrogenic pump in the plasma membrane of *Neurospora crassa. Journal of Membrane Biology,* **14,** 305–38.

Slayman, C. W., Rees, D. C., Orchard, P. P. & Slayman, C. W. (1975). Generation of adenosine triphosphate in cytochrome-deficient mutants of *Neurospora. Journal of Biological Chemistry,* **250,** 396–408.

Slayman, C. W. & Slayman, C. L. (1970). Potassium transport in *Neurospora.* Evidence for a multisite carrier at high pH. *Journal of General Physiology,* **55,** 758–86.

Slayman, C. W. & Tatum, E. L. (1964). Potassium transport in *Neurospora.* I. Intracellular sodium and potassium concentration and cation requirements for growth. *Biochimica et Biophysica Acta,* **88,** 578–82.

Slayman, C. W. & Tatum, E. L. (1965). Potassium transport in *Neurospora.* II. Measurement of steady-state potassium fluxes. *Biochimica et Biophysica Acta,* **102,** 149–60.

Suomalainen, H. & Oura, E. (1971). Yeast nutrition and solute uptake. In *The Yeasts,* vol. **2,** eds. A. H. Rose & J. S. Harrison, pp. 3–74. New York, London: Academic Press.

Tatum, E. L., Barratt, R. W. & Cutter, V. M. (1949). Chemical induction of colonial paramorphs in *Neurospora* and *Syncephalastrum. Science,* **109,** 509–11.

Turian, G. (1976). Reducing power of hyphal tips and vegetative apical dominance in fungi. *Experientia,* **32,** 989–91.

Whitaker, A. (1976). Amino acid transport in fungi: an essay. *Transactions of the British Mycological Society,* **67,** 365–76.

14
Regulation of macromolecular composition during growth of *Neurospora crassa*

L. ALBERGHINA,* E. STURANI,†
M. G. COSTANTINI,† E. MARTEGANI†
AND R. ZIPPEL*

*Cattedra di Biochimica comparata, Facoltà di Scienze, Università di Mialno, Italy
†Centro del C.N.R. di Biologia Cellulare e Molecolare delle Piante, Università di Milano, Milano Italy.

Introduction

The biochemistry of bacterial growth has been studied extensively in recent years, and much information is now available about the macromolecular composition of prokaryotic cells grown under different conditions (Maaløe, 1969; Skjold, Juares & Hedgcoth, 1973; Kieldgaard & Gausing, 1974). Mechanisms to regulate macromolecular synthesis in bacteria have been suggested by Cooper & Helmstetter (1968), Donachie (1968) and Dennis & Bremer (1974). Since few comparable studies have been made with eukaryotic cells, we decided to analyse the biochemistry of the growth of *Neurospora crassa* Shear & Dodge; this organism was chosen for its well-known genetic and biochemical versatility. In this chapter, growth and macromolecular synthesis in *N. crassa* are described and a model is advanced which relates and gives structure to these observations and suggests possible regulatory mechanisms.

Levels and rates of synthesis of DNA, RNA and protein

The growth rate‡ of *Neurospora crassa* (wild type strain 74A St Lawrence) in submerged, shake-flask culture depends at any given temperature upon the composition of the medium (Alberghina, 1973). By choosing appropriate carbon and nitrogen sources it is possible to elicit a wide range of exponential growth rates so that at 30°C the doubling time of the organism can be varied from about 1 to 8 h (Alberghina, Sturani & Gohlke, 1975). Fig. 14.1 shows that the

‡ Exponential rates of growth are expressed as number of doublings in biomass during one hour of culture (μ) or as the rate constant of exponential growth (λ, min^{-1}), being $\lambda = [(\ln 2)/(60)]\mu$.

macromolecular composition of *N. crassa* is influenced by growth rate; there is an appreciable increase in ribosome level (ribosomes per unit of protein) with increase in growth rate, but the levels of protein (protein per genome equivalent of DNA) and of transfer RNA (tRNA) (tRNA per genome equivalent of DNA) increase only slightly with growth rate. The tRNA : ribosome ratio, which is low (about 12 to 15) at growth rates greater than 0.5 doublings h^{-1}, increases to about 35 when the growth rate is reduced (Alberghina *et al.*, 1975).

We have determined protein stability in *Neurospora crassa* cultivated at different growth rates. Cultures were pulse-labelled with L[*carboxy*-^{14}C]leucine, chased with a high concentration of [^{12}C]leucine, and the radioactivity retained in protein (i.e. into material precipitable by hot trichloroacetic acid) was determined at intervals (Alberghina & Martegani, 1977*a*). Fig. 14.2 shows that proteins are fairly stable in mycelia grown upon media containing glucose or acetate (doubling times of 2 to 3 h), whereas they are less stable at lower growth rates (cultures grown on glycerol or ethanol). The half-lives calculated from the slopes of the

Fig. 14.1. Macromolecular composition of *Neurospora crassa* cultivated at different growth rates. The concentration of tRNA is given per genome equivalent (●) and per ribosome (▲). The protein level (□) is expressed as the number of amino acids bound into protein per genome equivalent. The number of ribosomes (○) is given per amino acid bound into protein. R-protein (■) indicates ribosomal protein, as the percentage of total protein. The rate of exponential growth is given as μ (number of doublings in biomass per h).

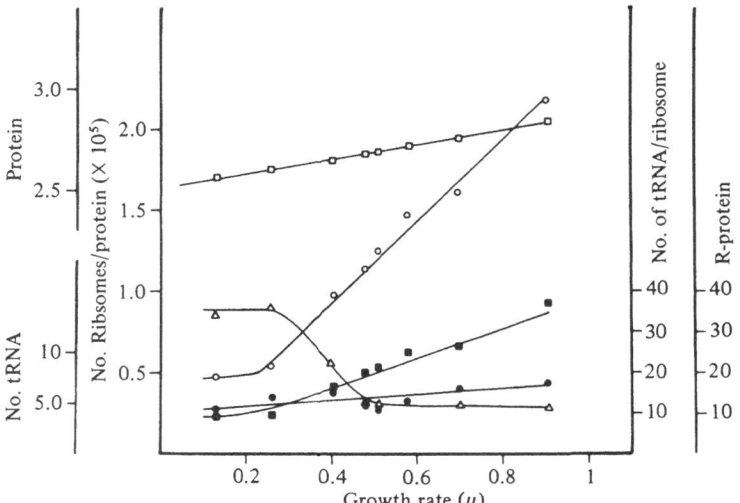

lines shown in Fig. 14.2 range from 57 h for cultures grown on glucose to 24 h for cultures grown on ethanol. A fast-decaying and a slow-decaying protein component can be distinguished in cultures grown on glycerol; the half-life of the fast-decaying component (about 4 % of the total protein) is 0.75 h, whilst that of the slow decaying component is 60 h (Alberghina & Martegani, 1977*a*). The rates of protein degradation in mycelia grown at different growth rates were calculated (see footnote to Table 14.1 for details) from the half-lives and are compared in Table 14.1 with the rates of protein accumulation. Table 14.1 shows that the rate of protein degradation is only 3 to 5 % of the rate of protein accumulation in cultures grown on glucose or acetate but increases to about 30 % in cultures grown on ethanol.

These data were also used to determine ribosomal activity at various growth rates. The total rate of protein synthesis can be calculated from

Fig. 14.2. Protein stability for mycelia grown at different rates. (*a*) Glucose ($\mu = 0.51$); (*b*) acetate ($\mu = 0.41$); (*c*) glycerol ($\mu = 0.26$); (*d*) ethanol ($\mu = 0.13$). The cultures were pulse-labelled with L-[1-^{14}C]leucine (5×10^{-7} M, specific activity 10 Ci mol^{-1} for cultures in glucose and acetate; 6×10^{-8} M, specific activity 80 Ci mol^{-1} for cultures growing in glycerol and ethanol). After 30 to 45 min, when almost all the radioactivity had been incorporated into protein, a chase with 10^{-4} M [^{12}C]leucine was performed, and the radioactivity retained into the hot TCA-precipitable material was determined. Time zero is the moment of the chase.

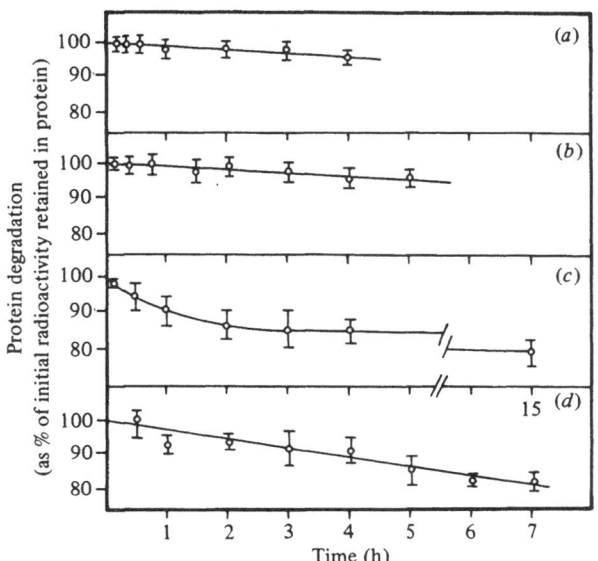

Table 14.1. *Rate of protein synthesis and average ribosomal activity in* Neurospora crassa *grown on various media*

Medium	Rate of growth[a], μ (doublings in biomass per hour)	Rate of protein accumulation[b] (amino acids per min)	Rate of protein degradation[c] (per genome)	Average ribosomal activity (amino acids, polymerized per min per ribosome)
Nutrient broth	0.91	3.00×10^8	ND	480
Glucose + casamino acids	0.65	2.06×10^8	ND	480
Acetate + casamino acids	0.58	1.82×10^8	ND	460
Glucose	0.51	1.57×10^8	5.5×10^6	480
Glycerol + casamino acids	0.48	1.47×10^8	ND	475
Acetate	0.41	1.24×10^8	5.4×10^6	500
Ethanol	0.13	0.38×10^8	1.23×10^7	420

ND, not determined. [a]As determined by Alberghina *et al.* (1975). [b]Calculated from the protein levels reported in Fig. 14.1 multiplied by $[(1n2)/60] \times \mu$. [c]Calculated, assuming that protein degradation follows a negative exponential equation (Goldberg & Dice, 1974), from the protein levels in Fig. 14.1 multiplied by $1/T_2$ estimated from the slope of the straight lines of Fig. 14.2, considering one class of decaying labelled protein.

the rate of protein accumulation and the rate of protein degradation. This value divided by the ribosome content gives the average ribosomal activity (Table 14.1); ribosomal activity does not vary with growth rate. That the efficiency of protein synthesis is the same at all growth rates is further indicated by the observations that (i) the percentage of ribosomes active in protein synthesis (i.e. numbers of ribosomes in polysomes expressed as a percentage of the total number of ribosomes) is the same at all growth rates (Alberghina et al., 1975); (ii) the relative content of polyadenylate-containing RNA (poly(A+)RNA) is fairly constant (about 4.7 % of total RNA) at different growth rates (Table 14.2; Costantini et al., 1978); since poly (A+) RNA is a relevant fraction of messenger RNA in Neurospora crassa (Mirkes & McCalley, 1976) and other eukaryotic cells (Carlin, 1978).

From the observation that ribosome efficiency does not vary with growth rate, it follows that in Neurospora crassa there must be a direct relationship between growth rate and ribosomal level; Fig. 14.3 shows the relationship between ribosome level (ρ, the number of ribosomes per unit of protein) and growth rate (λ, min^{-1}) in a number of organisms. In N. crassa and Escherichia coli (Migula) Castellani & Chalmers there is direct proportionality between growth rate and

Table 14.2. *Polyadenylate–containing RNA in* Neurospora crassa *cultivated at three different growth rates* (see Table 14.1)

Carbon source in medium	Poly(A+) RNA (as % of total RNA)
Glucose	4.75 ± 0.26
Acetate	4.45 ± 0.34
Glycerol	4.86 ± 0.33

Mycelia growing exponentially in Vogel's mineral medium contain glucose, acetate or glycerol as carbon source, labelled from the moment of inoculation with $[5 - {}^3H]$ uridine (3×10^{-4}M, 4 Ci mol^{-1}), or alternatively with $[{}^{32}P]$ orthophosphate (1 Ci mol^{-1}), were collected, and the total RNA was extracted according to Singer & Penman (1973). The total RNA was then fractionated on oligodeoxythymidilate cellulose into poly(A+) and poly (A$^-$) RNA. The data are expressed as the cold trichloroacetic acid (TCA) precipitable radioactivity present in the poly (A+) fraction as percentage of the TCA precipitable radioactivity of total RNA recovered from the column.

ribosomal RNA (rRNA) level, i.e. a doubling in growth rate results in a doubling in rRNA level. As observed by Maaløe (1969) it follows that in these organisms the average ribosomal activity does not vary with growth rate (Table 14.1). Instead, in yeast a doubling in growth rate results in only a relatively small increase in ribosome level, suggesting that at low growth rates rRNA is present in excess. It has been shown that in yeast the percentage of ribosomes bound in polysomes is less at low than at high growth rates, whilst the rate of protein elongation per active ribosome is fairly constant at all growth rates (Waldron, Jund & Lacroute, 1974, 1977).

Regulation of the rate of ribosomal RNA synthesis

The level of ribosomes in mycelia growing exponentially is primarily determined by the rate at which their component parts are synthesized. Fig. 14.4 shows that the rates of rRNA synthesis and rRNA

Fig. 14.3. Ribosomal level as a function of the rate of growth in different organisms. The ribosomal level, ρ, is expressed as the number of ribosomes per amino acid bound into protein. λ (min^{-1}) is the growth rate. The data of Kieldgaard & Gausing (1974) were used to estimate the value of ρ in *E. coli* B/r cells. The data of Skjold, Juares & Hedgcoth (1973) were used to estimate the value of ρ in *E. coli* 15 T$^-$ cells. The data of Waldron & Lacroute (1975) ($\bullet$) and of Boehlke & Friesen (1975) ($\mathbf{O}$) were used to estimate the value of ρ in yeast cells. The data of Mauck & Green (1973) were used to estimate the value of ρ in mouse fibroblasts.

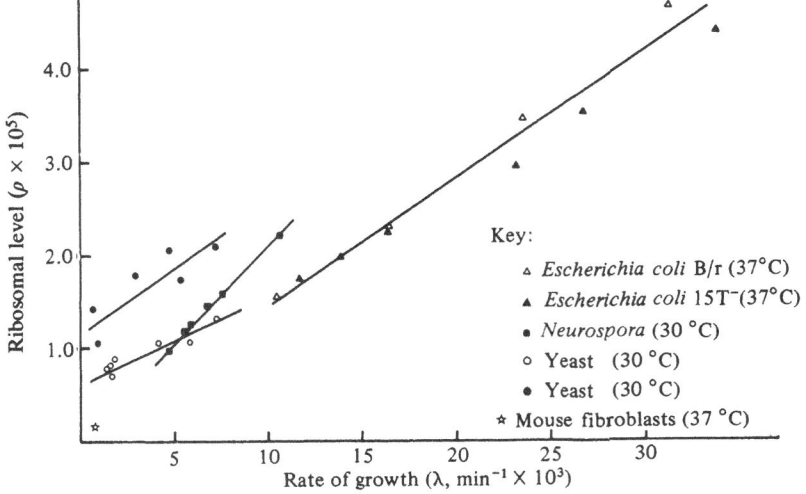

Key:

$\triangle$ *Escherichia coli* B/r (37°C)

$\blacktriangle$ *Escherichia coli* 15T$^-$(37°C)

$\bullet$ *Neurospora* (30 °C)

o Yeast   (30 °C)

• Yeast   (30 °C)

☆ Mouse fibroblasts (37 °C)

methylation in *Neurospora crassa* are functions of the square of the final ribosome level (ρ). Thus a doubling of ρ (and hence of growth rate) means that there has to be a four-fold increase in the rate of rRNA synthesis.

That control of rRNA synthesis is a key point in the regulation of growth rate of *Neurospora crassa* is further indicated by the results obtained from growth transition experiments. During a shift-down in growth rate (from growth on glucose to growth on glycerol) the following responses are observed (Sturani, Magnani & Alberghina, 1973); (i) net synthesis of RNA almost stops for about 2 h after glucose exhaustion, and then resumes at the same rate as that observed during growth on glycerol alone; (ii) the rate of methylation of rRNA is reduced after glucose is exhausted but that of soluble RNA (sRNA) (largely tRNA) is only partially inhibited; (iii) the rates at which DNA and protein accumulate in the culture are only slightly affected during the transition period. During a shift-up in growth rate (from growth on acetate to growth on glucose) the following responses are observed (Sturani *et al.*, 1976): (i) net synthesis of RNA increases

Fig. 14.4. Rate of synthesis (O) and of methylation (●) of ribosomal RNA in cells having different ribosomal levels. The rate of synthesis of rRNA is expressed as nucleotides polymerized per min per 1 μg of protein. The data are calculated from those of Alberghina *et al.* (1975). The rate of methylation of rRNA is expressed as number of methyl groups incorporated per min per 1 μg of protein. These data are calculated from those of Sturani *et al.* (1976). The steady state ribosomal level, ρ, is expressed as a number of ribosomes per amino acid bound into protein.

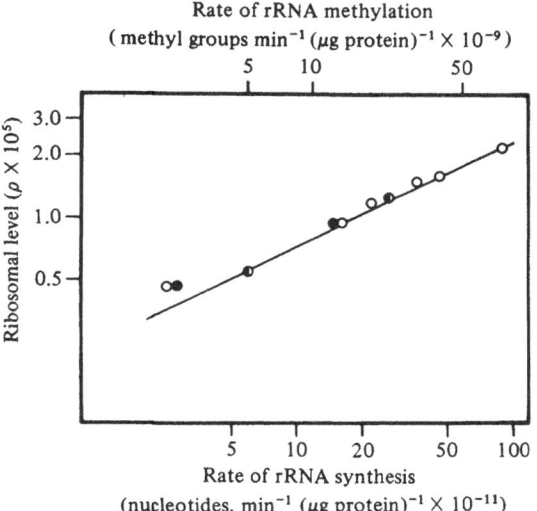

markedly 30 min after glucose addition, initially at a rate significantly greater than that measured during growth on glucose alone; (ii) after a lag of 30 min there is a three-fold increase in the rate of rRNA methylation which eventually attains a rate higher than that observed during growth on glucose alone; (iii) the rates of DNA and protein accumulation continue at the preshift rates for about 2 h but then increase with increase in growth rate.

During both types of growth transition described above, there is a rapid change in the rate of rRNA synthesis so that after a relatively short period the culture has attained the macromolecular composition characteristic of the new medium. Furthermore, for both types of growth transition the rate of rRNA methylation typical of the new medium is established when the ribosome level has adjusted to that characteristic of the new growth condition. This suggests that a compensatory mechanism regulates the rate of rRNA synthesis according to the ribosome level present.

A model for cellular growth

The available data, which cover many aspects of the biochemistry of the growth of *Neurospora crassa*, have not yet been incorporated into a general hypothesis. However, since mathematical models are the specific methodological tools employed to study the dynamics of complex systems, we have used them to study growth regulation in *N. crassa*.

A model for the regulation of growth of mycelia in *Neurospora crassa* was first proposed by Alberghina (1975). This model was subsequently developed to include ribosome and protein synthesis, DNA replication and nuclear division (Alberghina & Martegani, 1977b; Alberghina & Mariani, 1978). The model is illustrated in Fig. 14.5.

In the model, net increase in protein (P) depends upon the balance between the rates of protein synthesis and degradation. The net rate of protein synthesis (expressed as amino acids polymerized per min) is given by:

$$\frac{dP}{dt} = K_2 R - P/T_2 \tag{1}$$

where R is the number of ribosomes, K_2 is their average efficiency (expressed as amino acids polymerized per min per ribosome) and T_2 is the time constant (min) of the rate of protein degradation (given by a negative exponential equation, see footnote to Table 14.1).

The net rate of ribosome synthesis is given by:

$$\frac{dR}{dt} = K_1(\rho P - R) - R/T_1 \tag{2}$$

Ribosome degradation (like protein degradation) is expressed by a negative exponential equation with time constant T_1 (min). The rate of ribosome synthesis is given by a more complex term that incorporates the aspects of rRNA regulation discussed above.

Neurospora crassa cultures have characteristic time-invariant values of ρ (number of ribosomes per unit of protein (Fig. 14.3)) that correspond to specific rates of rRNA synthesis (Fig. 14.4). During balanced exponential growth, the rate of ribosome synthesis has to maintain the level of ribosomes in the mycelia close to the characteristic ρ value, although at the same time the level of protein rises continuously due to the activity of the ribosomes. It follows that the rate of ribosome synthesis (per unit volume of culture) has to increase in proportion to the increase in protein, and this fact is expressed in equation (2) by the terms of ρP. On the other hand, the dynamics of shift-down and shift-up growth transitions suggest that a compensatory mechanism (i.e. a negative feedback) may control ribosome synthesis. This hypothesis is expressed in equation (2) as $K_1(\rho P - R)$. The relations between the kinetics of net protein synthesis, DNA replication and nuclear division in *Neurospora crassa* are assumed to be similar to those described for

Fig. 14.5. The model for cellular growth in *Neurospora crassa*. (See text for explanation).

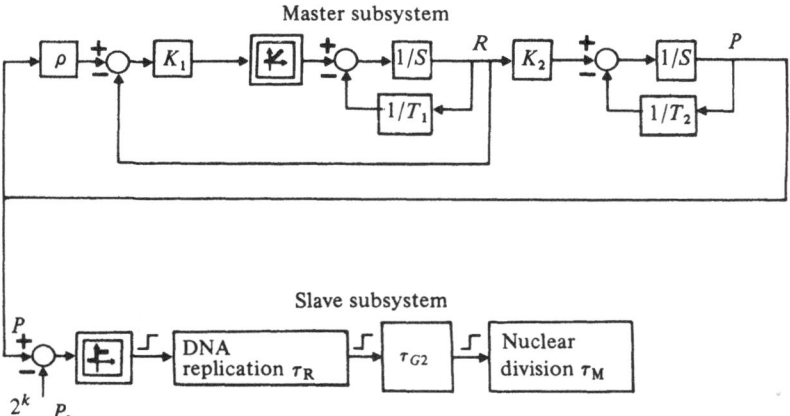

other eukaryotic cells (Mitchison, 1971; Hartwell, 1974; Hartwell *et al.*, 1974; Fantes & Nurse, 1977; Johnston, Pringle & Hartwell, 1977). They are discussed in more detail elsewhere (Alberghina & Mariani, 1978). Thus, replication of chromosomal DNA is initiated when the concentration of protein reaches a threshold value, P_s. When

$$P/P_s = 2^k \tag{3}$$

with P_s = constant, k = 0, 1, 2 . . . (2^k being the number of origins of genome replication and thus k = 0 in haploid *Neurospora crassa*); DNA replication is initiated and proceeds at a constant rate, K_3. After a period of time

$$\tau_R = \frac{1}{K_3} = \text{constant} \tag{4}$$

DNA replication stops. After a further delay

$$\tau_{G2} = \text{constant} \tag{5}$$

nuclear division takes place.

The model is presented in Fig. 14.5. The upper part of the figure represents the relations described by equations (1) and (2). The lower part of the figure shows the events of DNA replication and nuclear division as described in equations (3), (4) and (5). The model is divided into two subsystems: the master, whose state variables are R and P, and the slave, which is activated only when the protein reaches the threshold level $2^k P_s$; the slave subsystem describes DNA replication, the G_2 period and nuclear division.

It has been shown that the dynamics of the entire system are determined by the master subsystem, and depend upon the parameters K_2, ρ and T_2 (Alberghina & Mariani, 1978). The growth rate constant (λ, min^{-1}) is given by the relation:

$$\lambda = K_2\rho - \frac{1}{T_2} \tag{6}$$

We have verified equation (6) experimentally (Martegani & Alberghina, 1977; Table 14.3).

The model also describes adequately the kinetics of ribosome and protein synthesis during a shift-down in growth rate (see above). The experimentally determined values of K_2, T_2 and ρ, at different growth rates, are shown in Fig. 14.6. The dynamics of ribosome and protein synthesis have been simulated using the model and, as shown in Fig.

14.7, the simulations are in good agreement with the experimental data (Alberghina & Martegani, 1976).

Thus, the model provides an adequate description of the growth kinetics of *Neurospora crassa*. It suggests that growth rate is regulated by the kinetics of ribosome and protein synthesis. Johnston *et al.* (1977), using a different approach, have come to a similar conclusion. Thus, the studies on the factors that regulate the dynamics of mycelial growth should be restricted to an analysis of the first subsystem of the model.

A negative feedback for ribosome synthesis?

One interesting element in the first subsystem of the model is the suggestion that a negative feedback regulates ribosome formation. The presence of a negative feedback mechanism is, of course, consistent (as shown in the simulation of a shift-down in growth rate in Fig. 14.7) with the observed kinetics of ribosome accumulation during nutritional-induced shifts in growth rate, but it is not yet supported by more direct experimental evidence. The following experiments were performed to test the hypothesis. If a negative feedback is present a derangement of its function might be brought about by an inhibition of ribosome activity. The inhibitor chosen was cycloheximide, which acts at the level of the 60 S subunit of ribosomes (Siegel & Sisler, 1965; Rao & Grollman, 1967).

Low concentrations of cycloheximide severely reduce the growth (measured from absorbance of the culture at 450 nm) of cultures of

Table 14.3. *Observed and predicted growth rates of* Neurospora crassa mycelia *growing at two temperatures on minimal medium with glucose as carbon source*

Temperature (°C)	K_2 (amino acids polymerized per min per ribosome)	ρ (number of ribosomes per amino acid bound into protein)	T_2	Predicted growth rate	Observed growth rate
25	316	1.48×10^{-5}	6000	4.56×10^{-3}	4.88×10^{-3}
8	70	0.92×10^{-5}	∞	6.44×10^{-4}	5.75×10^{-4}

The values of K_2, ρ and T_2 were determined by independent measurements (Martegani & Alberghina, 1977) and were then used to predict the organism's growth rate (λ) using equation (6). The observed growth rate was determined from the increase in absorbance at 450 nm of the culture.

Neurospora crassa for 1 to 2 h. Subsequently, growth proceeds exponentially at a rate which is slower than the original growth rate (Fig. 14.8). Cycloheximide $(0.02 \, \mu g \, ml^{-1})$ decreases the rate of protein accumulation, but for about 1 h RNA accumulation is unaffected (Fig. 14.9); there is also an increase in the RNA:protein ratio (Figs. 14.9, 14.10). Thus cycloheximide induces a shift-down in growth rate, but there is a change in macromolecular composition which is similar to that observed during a shift-up in growth rate. The 'extra' RNA that accumulates in the presence of cycloheximide has the same composition as that synthesized during growth in its absence; rRNA constitutes

Fig. 14.6. Changes in the values of the parameters K_2, T_2, and ρ during a shift-down in growth rate (from growth on glucose to growth on glycerol). A culture of *Neurospora crassa* growing in medium containing 0.01% glucose and 2% (w/v) glycerol utilizes the glucose until it is exhausted (time zero); a diauxic lag occurs during which the macromolecular syntheses are differently affected (see text and Sturani *et al.*, 1973). K_2, average ribosomal activity, is expressed as amino acids polymerized per min per ribosome and has been determined as previously reported (Alberghina & Martegani, 1977a). T_2, the time constant of protein degradation, expressed in min, was determined as reported in Fig. 14.2. ρ is the number of ribosomes per amino acid bound into protein. (n) is newly-made protein, (o) is old protein.

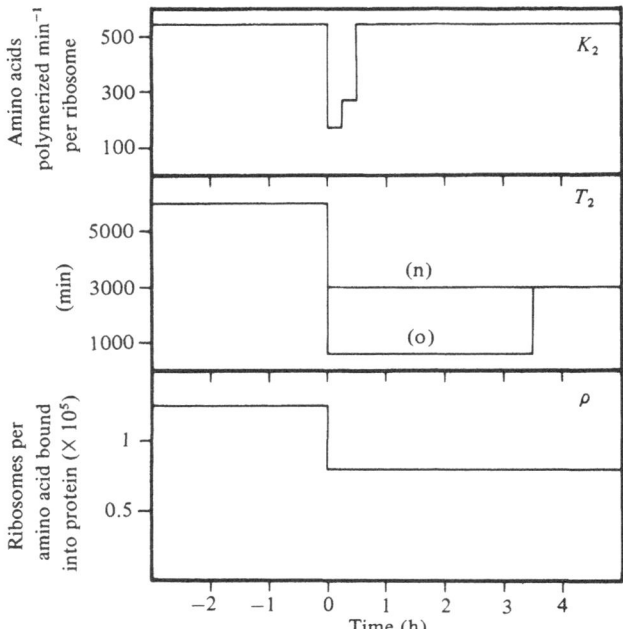

about 80 % of the total RNA both in control and in cyclohexmide-treated cultures (Table 14.4). Moreover, the relative rates of methylation of both rRNA and sRNA (expressed as pmoles of methyl groups incorporated during a 10 min pulse with [^{3}H-*methyl*]methionine in rRNA and in sRNA for 1 A$_{260nm}$ unit of unfractionated RNA) are unaffected by the treatment with cycloheximide at least for 90 min (Table 14.5).

We have carried out experiments to determine if the observed stimulation of rRNA synthesis by cycloheximide is paralleled by a stimulation of ribosomal protein synthesis to yield enhanced production of complete ribosomes. At intervals after the addition of cycloheximide, 30 min pulses of L[1−^{14}C]leucine were given to control and cycloheximide-treated cultures; the radioactivity that was incorporated into ribosomal

Fig. 14.7 Simulation of ribosome and protein accumulation during a shift-down in growth rate. Protein (▲) and RNA (●) accumulation during a shift-down transition from growth on glucose to growth on glycerol (see Fig. 14.6). In this experiment the exhaustion of glucose occurs at 3 h 10 min. The continuous lines are dynamics computed according to the equations of the model, and the circles and triangles indicate the experimental points, obtained as incorporation of [^{32}P]orthophosphate and L-[1-^{14}C]leucine, respectively, into RNA and protein (Alberghina & Martegani, 1976).

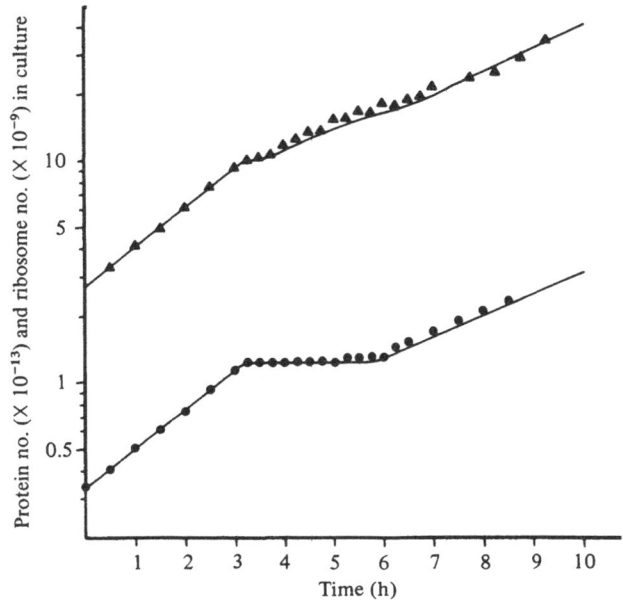

particles and the 15 000 g supernatant (S-15) was determined. Table 14.6 shows that the radioactivity incorporated into ribosomal protein (expressed as a percentage of the radioactivity incorporated into S-15 protein) increases from 28 % in the control to 35 % in cycloheximide-treated cultures. Thus, following cycloheximide treatment there is an increase in the relative rate of synthesis of ribosomal protein comparable to that observed for rRNA.

Table 14.4. *Effect of cycloheximide on the sRNA:rRNA ratio in* Neurospora crassa

Growth condition	sRNA/rRNA	rRNA (as % of total RNA)
Cycloheximide absent	0.24 ± 0.02	81
Cycloheximide present $(0.02 \, \mu g \, ml^{-1})$	0.28 ± 0.02	78

Mycelia were labelled from the moment of inoculation with [^{32}P]orthophosphate (0.27 Ci mol^{-1}). Control cultures growing on glucose and cultures treated for 2 h with cycloheximide (0.02 μg ml^{-1}) were collected; the total RNA was extracted and fractionated into soluble and ribosomal RNA on sucrose gradients as previously described (Alberghina *et al.*, 1975).

Table 14.5. *Effect of cycloheximide on the rates of methylation of soluble and ribosomal RNA in* Neurospora crassa

	pmol [^{3}H]methyl groups per $A_{260 \, nm}$ incorporated into	
	Soluble RNA	Ribosomal RNA
Cycloheximide absent	42	52
Cycloheximide present (15 min)	43	47
Cycloheximide present (30 min)	39	47
Cycloheximide present (90 min)	45	55

Mycelia exponentially growing in glucose or treated for 15, 30 or 90 min with cycloheximide (0.02 μg ml^{-1}) were labelled for 10 min with L[^{3}H-methyl]methionine (5 $\times$ 10^{-6} M, 100 Ci mol^{-1}). The total RNA was then extracted and separated into soluble and ribosomal RNA on sucrose gradients (Sturani *et al.*, 1973). The data are expressed as pmole [^{3}H]methyl groups incorporated during the 10 min pulse into sRNA and rRNA, per $A_{260 \, nm}$ unit of unfractionated RNA.

Fig. 14.8. Effect of the addition of different concentrations of cycloheximide (CHI) on the growth of *Neurospora* cultures. Symbols ●, control (glucose); □, 0.02 μg ml⁻¹ CHI; O, 0.05 μgml⁻¹ CHI; ▲, 0.1 μg ml⁻¹ CHI.

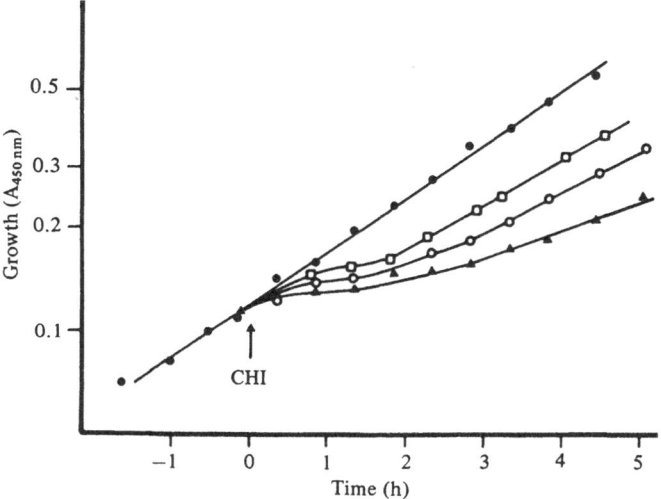

Fig. 14.9. Effect of cycloheximide on protein and RNA accumulation in *Neurospora crassa*. Cycloheximide (0.02 μg ml⁻¹) was added at zero time to cultures of *Neurospora* growing exponentially in glucose medium. One culture was supplemented from the moment of the inoculation with [2-¹⁴C]uridine (10⁻⁴ M, 0.1 Ci mol⁻¹) to monitor RNA accumulation (▲). Another culture was supplemented from the moment of the inoculation with L-[1-¹⁴C]leucine (5 × 10⁻⁴ M, 0.1 Ci mol⁻¹) in order to follow protein accumulation (O).

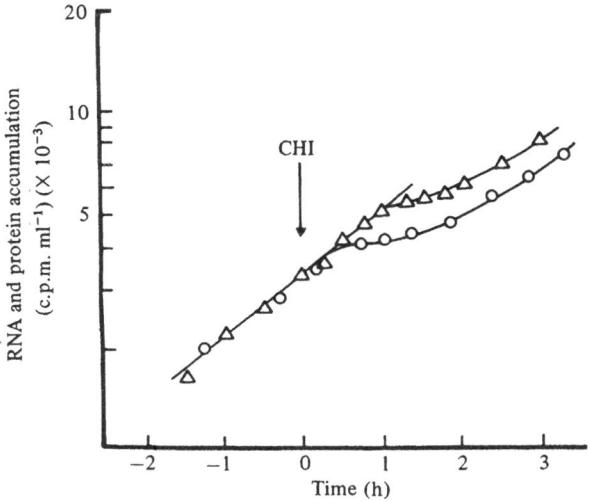

While cycloheximide stimulates the relative rate of ribosome synthesis (both rRNA and r-protein), it nevertheless inhibits the net rate of protein accumulation (Fig. 14.9). We therefore investigated the effect of cycloheximide on ribosome activity using a procedure described by Martegani & Alberghina (1977). First the rate of protein synthesis (per ml of culture) was determined as the rate of L[1-^{14}C]leucine incorporation into protein; the method employed is described in the legend to Fig. 14.11. The rate of L-leucine incorporation increases in the control culture with time in a way which parallels the increase in biomass. During the first 30 min after the addition of cycloheximide there is a strong inhibition of leucine incorporation which is subsequently partially released (Fig. 14.11). As previously shown (Martegani & Alberghina, 1977), it is possible to estimate the rate of protein synthesis from the rate of leucine incorporation, since it is known that leucine constitutes 8.6 % of the total amino acid residues in *Neurospora crassa* protein (Rho & De Bush, 1971). In addition, from the information contained in Table 14.4 and Figs. 14.9 and 14.10, it is possible to calculate the ribosome content (per ml of culture) at the various times after cycloheximide addition. Further, the average ribosomal efficiency can

Fig. 14.10. Effect of cycloheximide on RNA:protein ratio (symbols: O, control; ●, + cycloheximide) and on the rate of growth. The concentration of cycloheximide was 0.02 μg ml^{-1}. RNA and protein were determined by standard colorimetric techniques. The shaded area represents the observed variation in RNA: protein ratio at different growth rates.

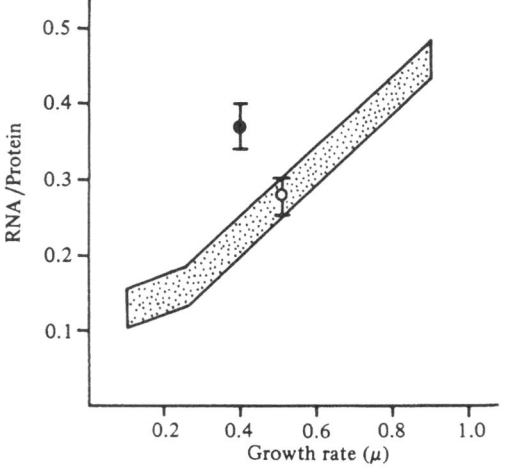

Table 14.6. *Effect of cycloheximide on the relative rate of synthesis of ribosomal proteins in* Neurospora crassa

Period of labelling (min)	Radioactivity from L-[1 −¹⁴C]leucine incorporated into ribonucleoprotein (as % total radioactivity incorporated into S–15 proteins)	
	Cycloheximide absent	Cycloheximide present
0–30	27.0	30.3
30–60	28.3	36.5
60–90	29.0	34.6

A culture growing exponentially in glucose was separated into six aliquots. Cycloheximide (0.02 μg ml⁻¹) was added to half the cultures at time 0. At various times (0, 30 and 60 min) the cultures were labelled with L-[1 −¹⁴C]leucine (10⁻⁵ M, 2 Ci mol⁻¹) for 30 min. At the end of the labelling period the cultures were collected, extracted (Sturani *et al.*, 1976), and the 15 000 *g* supernatant (S–15) was fractionated on sucrose gradient in order to separate the ribosomal particles from the soluble proteins. The radioactivity incorporated into ribonucleoproteins and into soluble proteins was determined.

Fig. 14.11. Effect of cycloheximide on the rate of protein synthesis. Pulses of 10⁻⁵ M L-[1−¹⁴C]leucine (2 Ci mol⁻¹) were given to cultures exponentially growing in glucose (white) and to cultures to which cycloheximide (0.02 μg ml⁻¹) had been added at time zero (shaded).

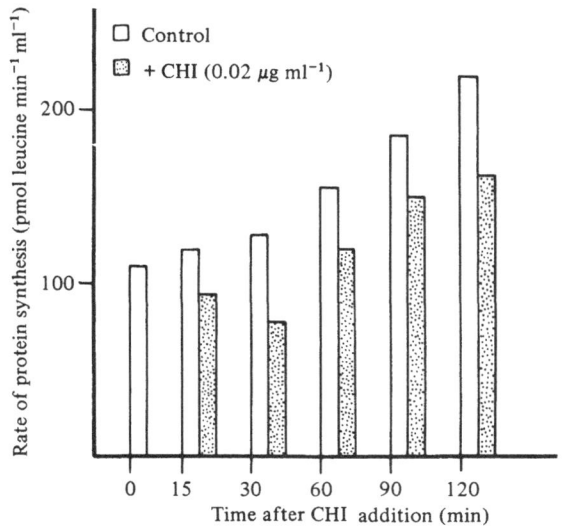

be calculated from the rate of protein synthesis (per ml of culture) and the ribosome content. Figure 14.12 shows how average ribosomal activity varies after cycloheximide addition; during the first 30 min it drops to about 60 % of the original value, then recovers and stabilizes during the subsequent period of exponential growth (Figs. 14.8 and 14.9) to about 75 % of the control value.

Inhibitors of protein synthesis (chloramphenicol, fusidic acid) also stimulate rRNA synthesis and cause an increase in ribosome content in *Escherichia coli* (Bennett & Maaløe, 1974; Shakulov & Klyachko, 1975). Such a response may have a selective advantage since it stimulates ribosome production under conditions where those already present are not working properly. It has been proposed that a decrease in the concentration of guanosine-5'-diphosphate,3'-diphosphate may be involved in this stimulation of ribosome synthesis. In addition a reduction in the rate of protein synthesis may result in an increase in the level of intracellular amino acids that would repress a number of biosynthetic pathways. Consequently a larger share of transcriptional activity may shift to make rRNA (Bennett & Maaløe, 1974). It is unlikely that the first mechanism operates in *Neurospora crassa*, since guanosine-5'-diphosphate,3'-diphosphate is not detectable in this fungus, nor in many other eukaryotic cells under conditions which provoke a large increase

Fig. 14.12. Effect of cycloheximide on the average ribosomal activity. The average ribosomal activity is expressed as amino acids polymerized per min per ribosome. At time zero, 0.02 μg ml^{-1} of cycloheximide were added to a culture growing exponentially in glucose.

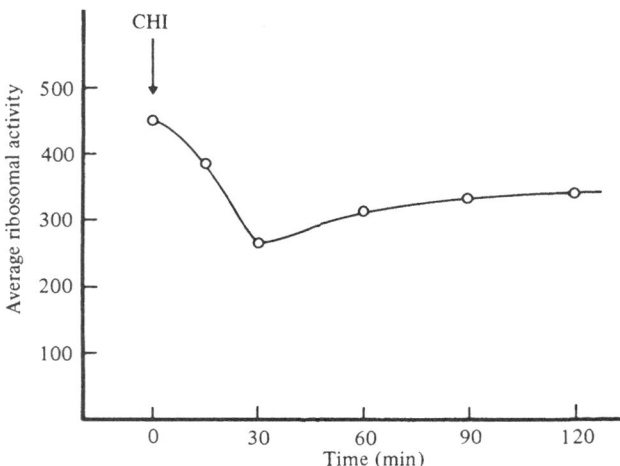

in bacteria (Mamont *et al.*, 1972; Alberghina *et al.*, 1973; Buckel & Böch, 1973; Kudrna & Edlin, 1975; Brandhorst & Fromson, 1976; Pirrone *et al.*, 1976). The second mechanism is based on the possibility that RNA polymerase in *Escherichia coli* can shift from mRNA to rRNA synthesis (Nierlich, 1972; Bremer, Berry & Dennis, 1973). This possibility, however, can also be ruled out in eukaryotes in which separate RNA polymerases are known to synthesize mRNA and rRNA (Buhler *et al.*, 1976). Thus there remains the possibility that the stimulation of ribosome synthesis by cycloheximide is due to its interference with a negative feedback mechanism controlling ribosome synthesis. According to this hypothesis, a stimulation of ribosome production would be expected if ribosomes inhibited by cycloheximide are no longer detected by the negative feedback mechanism. A similar hypothesis has been suggested to explain the stimulation of rRNA synthesis by chloramphenicol in *Escherichia coli* (Shakulov & Klyachko, 1975). Although there is as yet little experimental evidence of the molecular basis of this control mechanism, it should be noted that chloramphenicol prevents the binding of the protein elongation factor, Tu, to ribosomes (Pestka, 1971) and that the same Tu factor stimulates rRNA synthesis *in vitro* (Travers, 1976).

In conclusion, the agreement between the experimental and predicted kinetics for steady-state and transient-growth conditions, together with correlation between the inhibition of ribosome activity and the stimulation of ribosome synthesis indicate that the model has a reasonable degree of self-consistency and that it is able to relate in quantitative terms the synthesis of various macromolecules. However, this consistency does not conclusively prove the validity of all the assumptions proposed in the model. Nevertheless, there is sufficient evidence to accept the model for the moment as a reasonable hypothesis to explain the control of mycelial growth.

Nutrients and the regulation of growth

At any given temperature the composition of the medium will affect the rate of growth of mycelium of *Neurospora crassa*. The intermediary metabolism of mycelia grown on different media may differ. For instance, a culture grown on glucose has an almost identical growth rate (Table 14.1) and macromolecular composition (DNA, RNA and protein (Fig. 14.1)) to a culture grown on glycerol plus casamino acids. It is thought that regulatory molecules are able to integrate macromolecular synthesis with the substrates produced by intermediary metabolism (Nierlich, 1974). On the other hand, a whole set of metabolic functions (protein and ribosome synthesis and degradation, DNA replication) are modulated in their activity when there is a change in the concentration of these regulatory signals (Tomkins, 1975).

Thus, nutritional conditions should change the concentration of regulatory molecules, which then modulate the rates of the various processes of the subsystems shown in Fig. 14.5. We have shown that the dynamics of growth are regulated by the master subsystem, and that in *Neurospora crassa* changes in growth rate are primarily determined by the changes in ρ. In our attempt to determine the regulatory signals which control ρ (i.e. as shown in Fig. 14.4, the rate of rRNA synthesis), we have investigated whether or not the concentration of ribonucleoside triphosphates (substrates of the reaction that synthesizes rRNA) have a regulatory function in *N. crassa* (Constantini, Zippel & Sturani, 1977) similar to that proposed for Ehrlich cells (Grummt & Grummt, 1976). Fig. 14.13 shows the rates of rRNA methylation (Fig. 14.13*a* and *c*) and the levels of the four ribonucleoside triphosphates (Fig. 14.13*b* and *d*) during a shift-up in growth rate (Fig. 14.13*a* and *b*) and during a shift-down in growth rate (Fig. 14.13*c* and *d*). During a shift-up in growth rate, there are increases in nucleotide levels and the rate of rRNA methylation, both of which are proportional to the increases in the rate of rRNA synthesis. However, during a shift-down in growth rate the nucleotide levels oscillate independently; this observation suggests that the rate of ribosome synthesis is not modulated by nucleotide concentration.

Several unusual nucleotides have been implicated in the regulation of rRNA synthesis in *Escherichia coli*. Guanosine-5'-diphosphate,3'-diphosphate has been shown to be a negative regulator of the synthesis of rRNA (Van Ooyen, Gruber & Jørgensen, 1976), whilst the so-called 'phantom spot' detected by Gallant, Shell & Bittner (1976) might be a positive regulator. Both nucleotides are absent in *Neurospora crassa*, or at least they are below the level of detection of the analysis (Alberghina *et al.*, 1973; Martegani & Alberghina, manuscript in preparation). We conclude that growth regulation in eukaryotes and bacteria involve different regulatory molecules.

Acknowledgements. The authors wish to thank Mr A Grippo for his expert technical assistance and Ms B. Johnston for preparing the manuscript. This work has been supported in part by a grant from the Consiglio Nazionale delle Ricerche (to L.A.).

Fig. 14.13. Levels of the ribonucleoside triphosphates and rate of RNA methylation during a shift-up and a shift-down transition of growth. (*a*), (*b*) Shift-up from acetate to glucose. To a culture exponentially growing in acetate, 2 % glucose was added at time zero. (*a*) The rate of RNA methylation was determined giving 5 min pulses with L-[^{3}H-methyl] methionine (5×10^{-6} M, specific activity 50 Ci mol^{-1}) and measuring the cpm incorporated into 1 A$_{260nm}$ unit of purified RNA during the 5 min pulse (for details see Sturani *et al.*, 1976). (*b*) To determine the levels of the nucleoside triphosphates (NTP), the cultures were grown from the moment of the inoculation in the presence of [^{32}P]orthophosphate (specific activity 4 Ci mol^{-1}). At different times, aliquots of the cultures were collected and the nucleotides extracted and separated by chromatography on polyethylenimine cellulose pre-coated sheets. The spots of the four nucleotides were scraped off, and the radioactivity was counted (for details see Constantini *et al.*, 1977). (*c*), (*d*) Shift-down from glucose to glycerol. The shift-down was obtained as indicated in the legend of Fig. 14.6. Time zero is the moment in which the diauxic lag begins. (*c*) For the determination of the rate of RNA methylation, pulses of 5 min were given at different times with L-[^{14}C-methyl]methionine (5×10^{-6} M, 2.5 Ci mol^{-1}) (for details see Sturani *et al.*, 1973). (*d*) The levels of the nucleotides were determined as described above. Symbols: ▲, rate of methylation of RNA in the course of the shift-up (*a*) and during the shift-down (*c*). O, ATP; ●, UTP; △, GTP; □, CTP levels during the shift-up (*b*) and the shift-down (*d*).

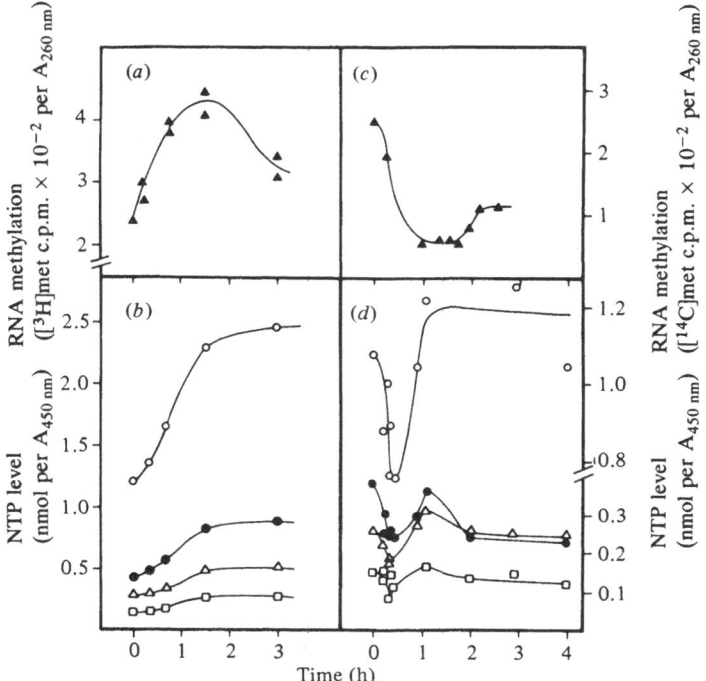

References

Alberghina, F. A. M. (1973). Growth regulation in *Neurospora crassa*. Effects of nutrients and temperature. *Archiv. für Mikrobiologie*, **89**, 83–94.

Alberghina, F. A. M. (1975). Dynamics of cellular growth. *Biosystems*, **7**, 183–8.

Alberghina, L. & Mariani, L. (1978). Control of cell growth and division. In *Biomathematics and Cell Kinetics*, eds. A. J. Valleron & P. D. M. MacDonald, pp. 133–46. Amsterdam: Elsevier–North-Holland.

Alberghina, F. A. M. & Martegani, E. (1976). Steady and transitory states in cellular growth. *Cybernetica*, **19**, 229–48.

Alberghina, F. A. M. & Martegani, E. (1977*a*). Protein degradation in *Neurospora crassa*. In *Intracellular Protein Catabolism*, vol. 2, eds. V. Turk & N. Marks, pp. 67–72. New York: Plenum Press.

Alberghina, L. & Martegani, E. (1977*b*). Modeling *Neurospora* growth. *Neurospora Newsletters*, **24**, 10–11.

Alberghina, F. A. M., Schiaffonati, L., Zardi, L. & Sturani, E. (1973). Lack of guanosine tetraphosphate accumulation during inhibition of RNA synthesis in *Neurospora crassa*. *Biochimica et Biophysica Acta*, **312**, 435–9.

Alberghina, F. A. M., Sturani, E. & Gohlke, J. R. (1975). Levels and rates of synthesis of ribosomal ribonucleic acid, transfer ribonucleic acid and protein in *Neurospora crassa* in different steady states of growth. *Journal of Biological Chemistry*, **250**, 4381–8.

Bennett, P. M. & Maaløe, O. (1974). The effect of fusidic acid on growth, ribosome synthesis and RNA metabolism in *E. coli*. *Journal of Molecular Biology*, **90**, 541–61.

Boehlke, K. W. & Friesen, J. D. (1975). Cellular content of ribonucleic acid and protein in *Saccharomyces cerevisiae* as a function of exponential growth rate: calculation of the apparent peptide chain elongation rate. *Journal of Bacteriology*, **121**, 429–33.

Brandhorst, B. & Fromson, D. (1976). Lack of accumulation of ppGpp in sea urchin embryos. *Developmental Biology*, **48**, 458–60.

Bremer, H., Berry, L. & Dennis, P. P. (1973). Regulation of ribonucleic acid synthesis in *E. coli* B/r: an analysis of a shift-up. II. Fraction of RNA polymerase engaged in the synthesis of stable RNA at different steady-state growth rates. *Journal of Molecular Biology*, **75**, 161–79.

Buckel, P. & Böch, A. (1973). Lack of accumulation of unusual guanosine nucleotides upon amino acid starvation of two eukaryotic organisms. *Biochimica et Biophysica Acta*, **324**, 184–7.

Buhler, J. M., Iborra, F., Sentenac, A. & Fromageot, P. (1976). Structural studies on yeast RNA polymerases. *Journal of Biological Chemistry*, **251**, 1712–17.

Carlin, R. K. (1978). The poly(A) segment of mRNA: I. Evolution and function and the evolution of viruses. *Journal of Theoretical Biology*, **71**, 323–38.

Costantini, M. G., Zippel, R. & Sturani, E. (1977). Levels of the ribonucleoside triphosphates and rate of RNA synthesis in *Neurospora crassa*. *Biochimica et Biophysica Acta*, **476**, 272–8.

Costantini, M. G., Zippel, R., Martegani, E. & Sturani, E. (1978). Polyadenylate-containing RNA in *Neurospora crassa* in different steady states of growth. Edinburgh Meeting of the Federation of European Societies of Plant Physiology (In press).

Cooper, S. & Helmstetter, C. E. (1968). Chromosome replication and the division cycle of *Escherichia coli* B/r. *Journal of Molecular Biology*, **31**, 519–40.

Dennis, P. P. & Bremer, H. (1974). Macromolecular composition during steady-state growth of *Escherichia coli* B/r. *Journal of Bacteriology*, **119**, 270–81.

Donachie, W. D. (1968). Relationship between cell size and time of initiation of DNA replication. *Nature (London),* **219,** 1077–9.

Fantes, P. & Nurse, P. (1977). Control of cell size at division in fission yeast by a growth-modulated size control over nuclear division. *Experimental Cell Research,* **107,** 377–86.

Gallant, J., Shell, L. & Bittner, R. (1976). A novel nucleotide implicated in the response of *E. coli* to energy source downshift. *Cell,* **7,** 75–84.

Goldberg, A. L. & Dice, J. F. (1974). Intracellular protein degradation in mammalian and bacterial cells. *Annual Review of Biochemistry,* **43,** 835–69.

Grummt, I. & Grummt, F. (1976). Control of nucleolar RNA synthesis by the intracellular pool size of ATP and GTP. *Cell,* **7,** 447–53.

Hartwell, L. H. (1974). *Saccharomyces cerevisiae* cell cycle. *Bacteriological Reviews,* **38,** 164–98.

Hartwell, L. H., Culotti, J., Pringle, J. R. & Reid, B. J. (1974). Genetic control of the cell division cycle in yeast. *Science,* **183,** 45–51.

Johnston, G. C., Pringle, J. R. & Hartwell, L. H. (1977). Coordination of growth with cell division in the yeast *Saccharomyces cerevisiae. Experimental Cell Research,* **105,** 79–98.

Kieldgaard, N. O. & Gausing, K. (1974). Regulation of biosynthesis of ribosomes. In *Ribosomes* eds. M. Nomura, A. Tissiére & P. Langyel, pp. 369–92. New York: Cold Spring Harbor Laboratory Press.

Kudrna, R. & Edlin, G. (1975). Nucleotide pools and regulation of ribonucleic acid synthesis in yeast. *Journal of Bacteriology,* **121,** 740–42.

Maaløe, O. (1969). An analysis of bacterial growth. *Developmental Biology,* Suppl. 3, 33–58.

Mamont, P., Hershko, A., Kram, R., Schachter, L., Lust, J. & Tomkins, G. (1972). The pleiotypic response in mammalian cells: search for an intracellular mediator. *Biochemical and Biophysical Research Communications,* **48,** 1378–84.

Martegani, E. & Alberghina, L. (1977). Low temperature restriction of the rate of protein synthesis in *Neurospora crassa. Experimental Mycology,* **1,** 339–51.

Mauck, J. C. & Green, H. (1973). Regulation of RNA synthesis in fibroblasts during transition from resting to growing state. *Proceedings of the National Academy of Sciences, USA,* **70,** 2819–22.

Mirkes, P. E. & McCalley, B. (1976). Synthesis of polyadenylic acid-containing ribonucleic acid during germination of *Neurospora crassa* conidia. *Journal of Bacteriology,* **125,** 174–80.

Mitchison, J. M. (1971). *The Biology of the Cell Cycle.* Cambridge University Press.

Nierlich, D. P. (1972). Regulation of ribonucleic acid synthesis in growing bacterial cells. I. Control over the total rate of RNA synthesis. *Journal of Molecular Biology,* **72,** 751–64.

Nierlich, D. P. (1974). Regulation of bacterial growth. *Science,* **184,** 1043–50.

Pestka, S. (1971). Inhibitors of ribosome functions. *Annual Review of Microbiology,* **25,** 487–562.

Pirrone, A. M., Roccheri, M. C., Bellanca, V., Aciemo, P. & Giudice, G. (1976). Studies on the regulation of ribosomal RNA synthesis in the sea urchin development. *Developmental Biology,* **49,** 311–20.

Rao, S. S. & Grollman, A. P. (1967). Cycloheximide resistance in yeast: a property of the 60 S ribosomal subunit. *Biochemical and Biophysical Research Communications,* **29,** 696–704.

Rho, H. M. & De Busk, A. G. (1971). NH$_2$ terminal residues of *Neurospora crassa* proteins. *Journal of Bacteriology,* **107,** 840–5.

Shakulov, R. S. & Klyachko, E. V. (1975). Stimulation of RNA synthesis by chloramphenicol. *Biochemistry,* **40,** 220–2. (Russian translation.)

Siegel, M. R. & Sisler, H. D. (1965). Site of action of cycloheximide in cells of *Saccharomyces pastorianus. Biochimica et Biophysica Acta,* **103,** 558–67.

Singer, R. H. & Penman, S. (1973). Messenger RNA in HeLa cells: kinetics of formation and decay. *Journal of Molecular Biology,* **78,** 321–34.

Skjold, A. C., Juares, H. & Hedgcoth, C. (1973). Relationship among DNA, RNA and specific transfer RNAs in *E. coli* 15T⁻ at various growth rates. *Journal of Bacteriology,* **115,** 177–87.

Sturani, E., Magnani, F. & Alberghina, F. A. M. (1973). Inhibition of ribosomal RNA synthesis during a shift-down transition of growth in *Neurospora crassa. Biochimica et Biophysica Acta,* **319,** 153–64.

Sturani, E., Costantini, M. G., Zippel, R. & Alberghina, F. A. M. (1976). Regulation of RNA synthesis in *Neurospora crassa.* An analysis of a shift-up. *Experimental Cell Research,* **99,** 245–52.

Tomkins, G. M. (1975). The metabolic code. *Science,* **189,** 760–3.

Travers, A. (1976). RNA polymerase specificity and the control of growth. *Nature (London),* **263,** 641–6.

Van Ooyen, A. J. J., Gruber, M. & Jørgensen, P. (1976). The mechanism of action of ppGpp on rRNA synthesis *in vitro. Cell,* **8,** 123–8.

Waldron, C., Jund, R. & Lacroute, F. (1974). The elongation rate of proteins of different molecular weight classes in yeast. *FEBS Letters,* **46,** 11–16.

Waldron, C., Jund, R. & Lacroute, F. (1977). Evidence for an high proportion of inactive ribosomes in slow-growing yeast cells. *Biochemical Journal,* **168,** 409–15.

Waldron, C. & Lacroute, F. (1975). Effect of growth rate on the amounts of ribosomal and transfer ribonucleic acid in yeast. *Journal of Bacteriology,* **122,** 855–65.

15
The duplication cycle and branching in fungi

A.P.J.TRINCI

Microbiology Department, Queen Elizabeth College, Campden Hill Road, London W8 7AH, UK

Introduction

The events which occur during the cell cycle of a uninucleate, eukaryotic cell are integrated with one another in a manner which ensures an ordered progression through the cycle. During the course of a single cycle, the structural and functional capacities of the cell are usually doubled. Thus the mean duration of a cycle is a direct measure of the doubling time of the population.

Growth of an exponential culture of a mould is 'balanced' (Trinci, 1978*a*); all extensive properties of 'balanced' cultures, such as macromolecules (e.g. DNA, RNA, protein, carbohydrate, lipid, etc), organelles (e.g. nuclei, mitochondria, ribosomes, etc.) and enzyme activities (e.g. respiratory enzymes), increase at the same rate (Campbell, 1957). The increase of such extensive properties in a mycelium, as in a single cell, must be integrated in a manner which brings about an ordered progression of events during growth. Thus, it is likely that there is a discrete cycle of cytological and biochemical events during mycelial growth which is analogous to the cell cycle of uninucleate cells. Since cell separation does not occur in filamentous fungi, this hypothetical cycle of events has been called a duplication cycle (Fiddy & Trinci, 1976*a*; Trinci, 1978*a*). The main events in the cell cycle of uninucleate, eukaryotic cells (micro-organisms and tissue cultures of mammalian cells) will be described, before considering whether or not a similar cycle can be recognized during the growth of moulds.

The cell cycle of uninucleate eukaryotic cells

Progress through the cell cycle of a eukaryotic cell is usually measured in relation to the processes of DNA replication and cell division. On the basis of these events Howard & Pelc (1953) divided the cell cycle into the phases of G1, S, G2 and D. The basic features of the eukaryotic cell cycle are illustrated in Fig. 15.1. They have been described in detail by Mitchison (1971) and Prescott (1976).

The G1 period

The G1 period spans the gap between the completion of cell division and the beginning of DNA replication; it is assumed to contain a succession of events which culminate in the initiation of DNA replication. Protein synthesis is essential for transit of the cell through the G1 period. Since populations of nutrient-starved or inhibited cells often accumulate in G1, it is commonly assumed that there is a checkpoint(s) in this period at which the cell decides whether or not to proceed with the cycle. Further, mating factors ('a' and 'a') have been isolated from *Saccharomyces cerevisiae* Hansen which block the cell cycle of the opposite mating type at or near the point in G1 at which nutrient-starved cells accumulate (Hartwell, 1974; Johnson, Pringle &

Fig. 15.1. Generalized cell-cycle of a unicellular eukaryotic cell; based upon Prescott (1976).

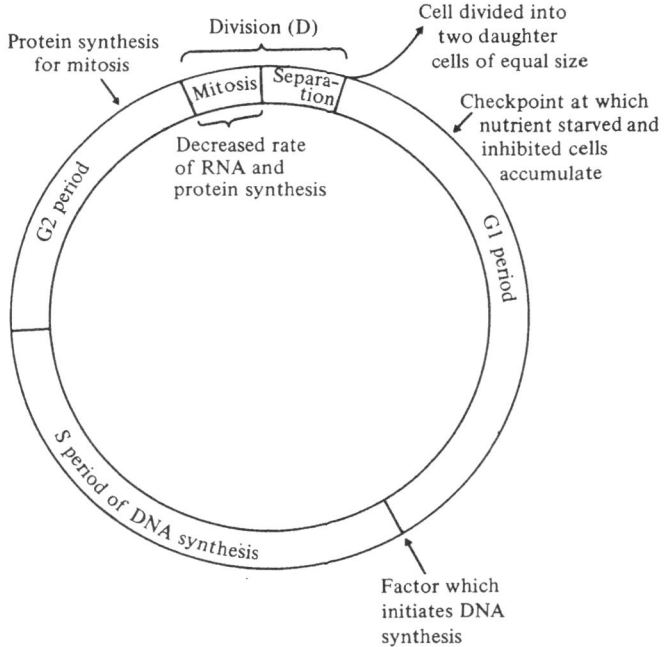

Hartwell, 1977). Thus, in *S. cerevisiae* the G1 checkpoint may be regulated by a gene which is activated by a range of specific (mating type) and non-specific (starvation of various nutrients) factors and prevents advancement to DNA synthesis. Hartwell (personal communication) suggests that this regulatory gene may be involved in the division of the spindle pole body.

Some organisms, e.g. *Amoeba proteus* Leidy and *Physarum polycephalum* Schw. lack a G1 period; it seems that in these organisms DNA synthesis is tightly coupled to the end of mitosis.

The S period

The phase of DNA replication in a cell cycle is called the S period. The S period is composed of a cascade of initiations and replications of the many thousands of replicons present in a eukaryotic nucleus. In *Saccharomyces cerevisiae*, protein synthesis is necessary for initiation but not completion of DNA synthesis (Hereford & Hartwell, 1973; Williamson, 1973).

The G2 period

The G2 period spans the gap between the end of DNA replication and the onset of mitosis. It presumably represents the time required by a cell to synthesize the elements necessary for chromosome condensation and operation of the mitotic apparatus. In mammalian cells, protein synthesis is necessary to complete most of the G2 period but any protein necessary for mitosis is formed before the beginning of prophase.

The division (D) period

The D period represents the time required to complete mitosis (M) and cell separation. At the end of the cell cycle a unicellular, eukaryotic cell is usually divided into two daughter cells of equal size. In cells which divide by the formation of a septum (*Schizosaccharomyces pombe* Lindner) or cell plate (cells of higher plants) the cross-wall is usually formed on a cytoplasmic site previously occupied by the dividing nucleus; indeed in many cells the cross-wall is formed on the cytoplasmic site previously occupied by the metaphase plate. This spatial coupling between the mitotic event and cross-wall formation ensures that each daughter cell contains a nucleus.

Growth during the cell cycle

The G1, S and G2 periods are phases of continuous cell growth during which there is a general increase in all the structural and functional capacities of the cell. In contrast, growth often slows during mitosis; in mammalian cells the rate of RNA synthesis declines during mitosis and the rate of protein synthesis may fall to about 25 % of its original rate. The decline in protein synthesis observed in these cells is caused by a transient incompetence of the ribosomes rather than by any lack of messenger RNA. The rate of volume growth of *Schizosaccharomyces pombe* increases as the cycle progresses, but slows and then ceases near the end of the cycle during septation and cell separation (Mitchison, 1971). However, the mass of the cell increases at a linear rate throughout the cycle.

Variation in cell-cycle length is usually achieved by varying the duration of G1. In contrast, the periods of G2 and particularly of S and D are of more constant duration. For example, Maaløe & Kjeldgaard (1966) observed that the time taken for chromosome replication in *Saccharomyces cerevisiae* is constant and independent of growth rate. A decision of whether or not to proceed to the next cell division is usually made in G1, but in some organisms it may be made in G2 or in both G1 and G2. Thus regulation of the cell cycle operates by stopping the cell at a point in G1 and sometimes in G2 but never in S or D (Prescott, 1976).

Fig. 15.2 shows the cell cycle of *Saccharomyces cerevisiae* and

Fig. 15.2. Cell cycles of (*a*) *Schizosaccharomyces pombe* and (*b*) *Saccharomyces cerevisiae*.

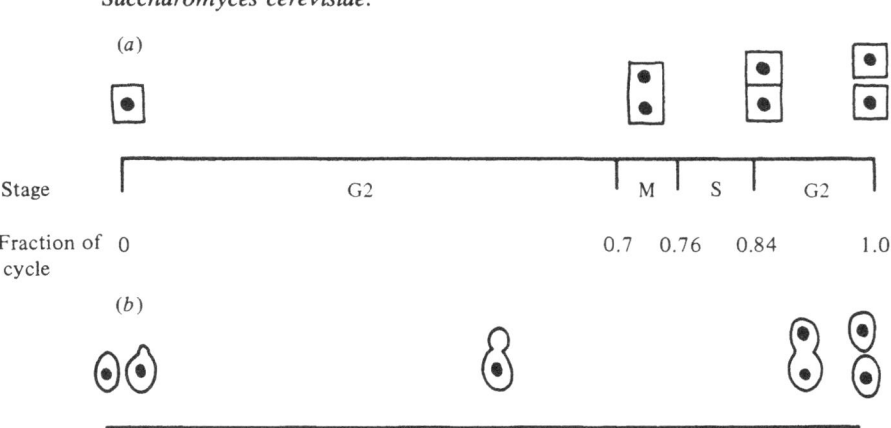

Schizosaccharomyces pombe; the timing of the various events in the cell cycle are shown as a fraction of the total cycle time.

Hyphal morphology

Hyphae are composed of long (up to about 2 mm) apical compartments and relatively short (up to about 250 μm) intercalary compartments. Leading hyphae at the margin of colonies extend at a constant rate and have apical and intercalary compartments of constant *mean* length. However, germ tube and branch hyphae which have an acclerating rate of extension have apical and intercalary compartments which increase in *mean* length. These increases continue until the hyphae eventually attain the linear growth rate characteristic of the strain and conditions (Fiddy & Trinci, 1976*a*). The apical compartment contains one to very many nuclei depending upon the strain and species (Table 15.1). Hyphae are divided into compartments by cross-walls (septa) which are either complete (i.e. lack pores) or have central pores (simple pores or dolipores); septa with unplugged pores allow extensive cytoplasmic continuity between adjacent compartments (see Gull (1978) for a detailed description of the types of septa found in fungi).

Table 15.1. *Number of nuclei per compartment and type of septal pore observed in leading hyphae of various fungi*

Species	Karyotic condition of mycelium	Number of nuclei in		Type of septal pore
		Apical compartment	Intercalary compartments	
Basidiobolus ranarum	Monokaryon	1	1	No pore
Schizophyllum commune	Monokaryon	1	1	Dolipore
Schizophyllum commune	Dikaryon	2	2	Dolipore
Polystictus versicolor	Dikaryon	2	2	Dolipore
Aspergillus nidulans	Coenocyte	*c*. 50	*c*. 4	Simple pore
Alternaria solani	Coenocyte	*c*. 25	*c*. 5	Simple pore
Polyporus arcularius	Coenocyte	*c*. 75	*c*. 2	Dolipore

The duplication cycle in apical compartments of leading hyphae growing linearly

The duration of the duplication cycle in an apical compartment of a mould is given by the interval observed between some discontinuous event (e.g. mitosis) in successive cycles. For example, in *Polyporus arcularius* (Batsch) ex Fr. there was an interval of 2.04 ± 0.22 h* between successive periods of septation in the apical compartment and one of 2.06 ± 0.32 h between successive periods of mitosis (Valla, 1973a). By comparison with the cell cycle of unicellular organisms (Fig. 15.2) it seems reasonable to take septation as the initial event in the duplication cycle of a mould. The interval (2.1 ± 0.2 h) observed between successive septation cycles in an apical compartment of *Aspergillus nidulans* (Eidam) Winter was identical with the mycelial doubling time (2.1 ± 0.2 h) of this organism. Thus the duration of the duplication cycle of this fungus is a direct measure of its doubling time; in this respect the cell (duplication) cycles of unicellular organisms and moulds are identical.

Variation in apical compartment length

Duplication cycles in apical compartments of leading hyphae of *Schizophyllum commune* Fries (monokaryotic and dikaryotic mycelia), *Aspergillus nidulans* and *Polyporus arcularius* are shown in Figs. 15.3 to 15.6. The duplication cycle of *Basidiobolus ranarum* Eidam, which also forms a monokaryotic mycelium, has been described by Trinci (1978a). The cycles illustrated in Figs. 15.3 to 15.6 are basically very similar. In each strain a newly formed apical compartment extends at a linear rate until it has approximately doubled its original length. The length of the apical compartment is then approximately halved by the formation of one to several septa. The regularity of the major morphological events (mitosis and septation) in successive duplication cycles and the precision with which compartments are divided in half by septation are clearly illustrated in Figs. 15.4 to 15.6.

Nuclear migration

Girbardt (1968) has made a careful analysis of the migration of nuclei in dikaryotic, apical compartments of *Polystictus versicolor* Link ex Fries. He divided the migratory behaviour of the nucleus into four phases (Fig. 15.7). During most of the duplication cycle, the nucleus and

---

* Standard deviation of the sample.

Fig. 15.3. Duplication cycle of apical compartment of a leading hypha of a monokaryotic strain of *Schizophyllum commune*.
(*a*) Diagrammatic representation of duplication cycle; (*b*) graph showing mitosis (M), septation (S) and apical compartment length in successive cycles. Adapted from Niederpruem & Jersild (1972).

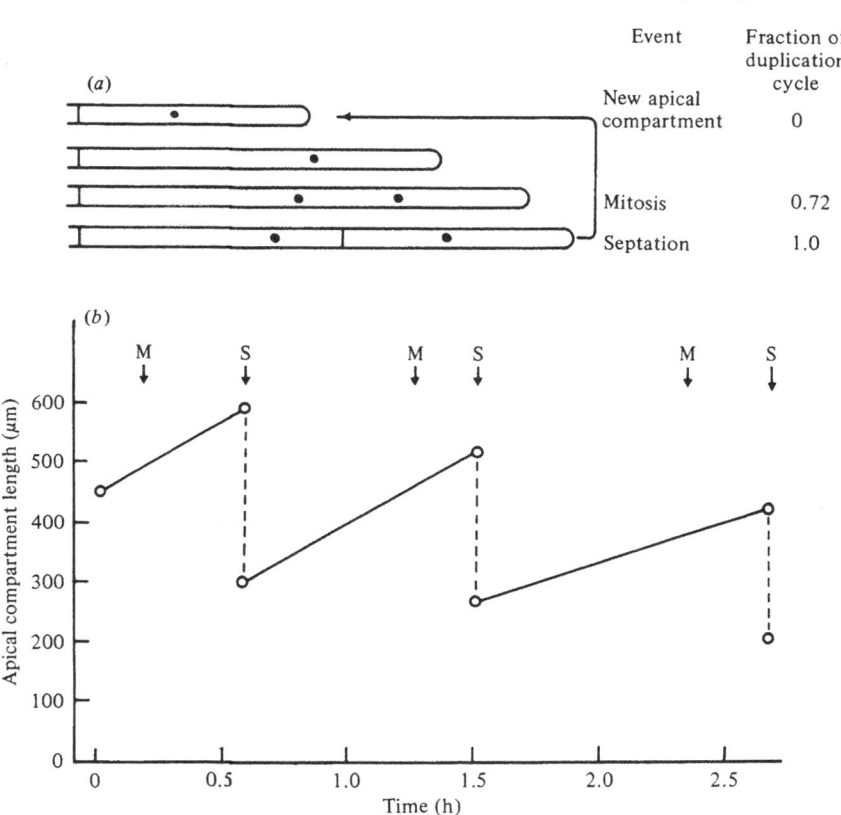

Fig. 15.4. Duplication cycle of an apical compartment of a leading hypha of a dikaryotic strain of *Schizophyllum commune* grown at 25 °C. (*a*) Diagrammatic representation of duplication cycle; (*b*) graph showing mitosis (M, two nuclei dividing synchronously), septation (S, two septa formed per cycle, one in the hypha and one in the clamp connection) and apical compartment length in successive cycles. Adapted from Niederpruem *et al.*, (1971).

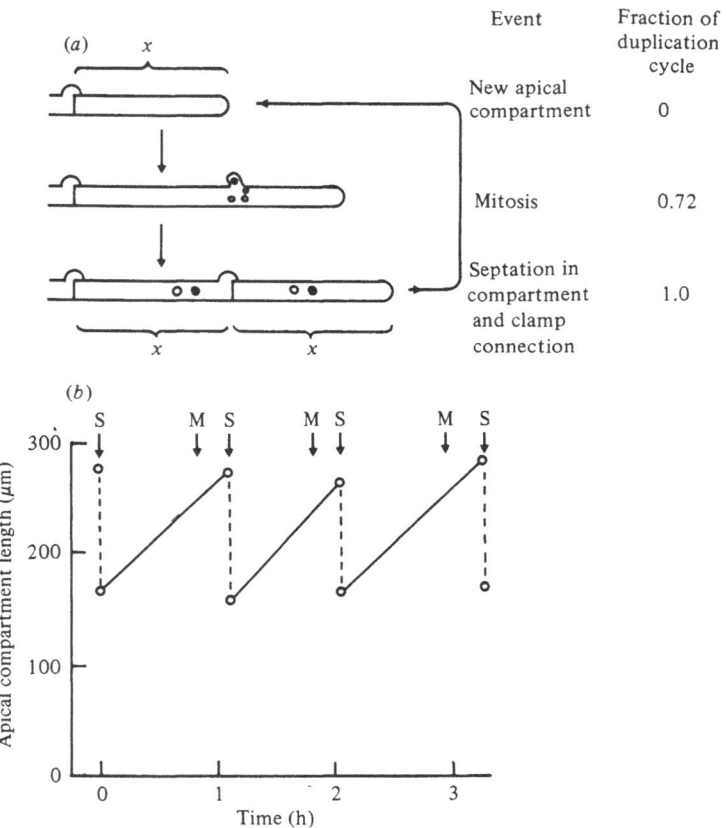

Fig. 15.5. Duplication cycle of an apical compartment of a leading hypha of *Aspergillus nidulans* grown at 25 °C. (*a*) Diagrammatic representation of duplication cycle; (*b*) septation (S, the number of septa produced is indicated) and apical compartment length in successive cycles.

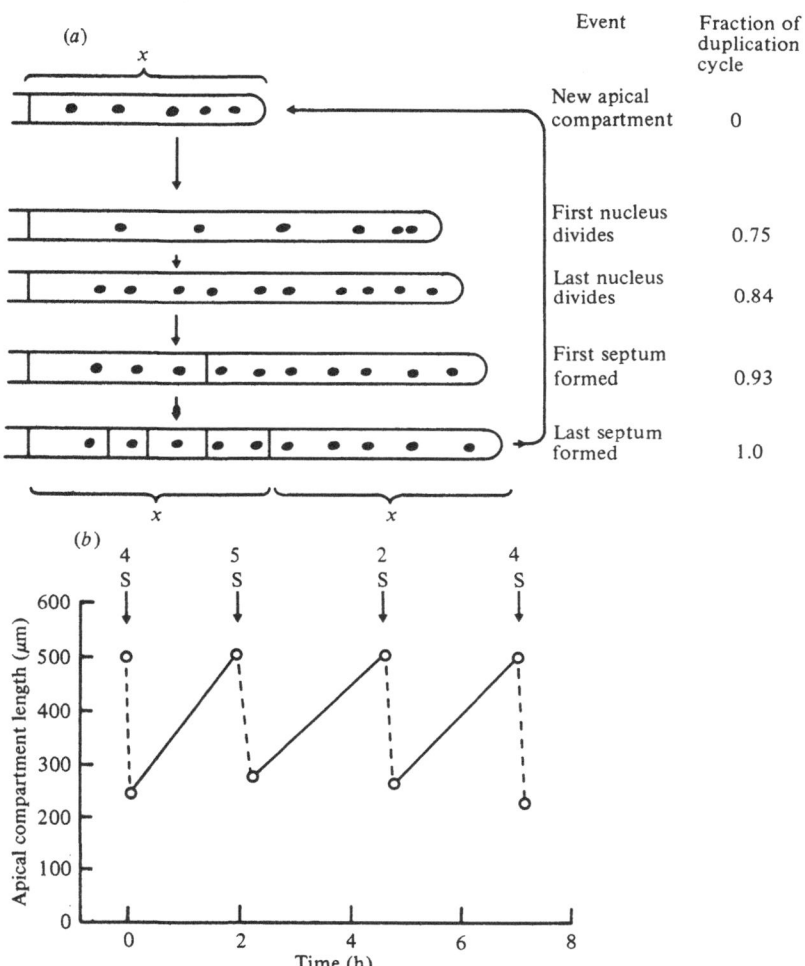

Fig. 15.6. Duplication cycle of an apical compartment of a leading hypha of *Polyporus arcularius* grown at 25 °C. (*a*) Diagrammatic representation; (*b*) initiation of mitosis (M), septation (S, the number of septa produced is indicated) and apical compartment length in successive cycles. Adapted from Valla (1973*b*).

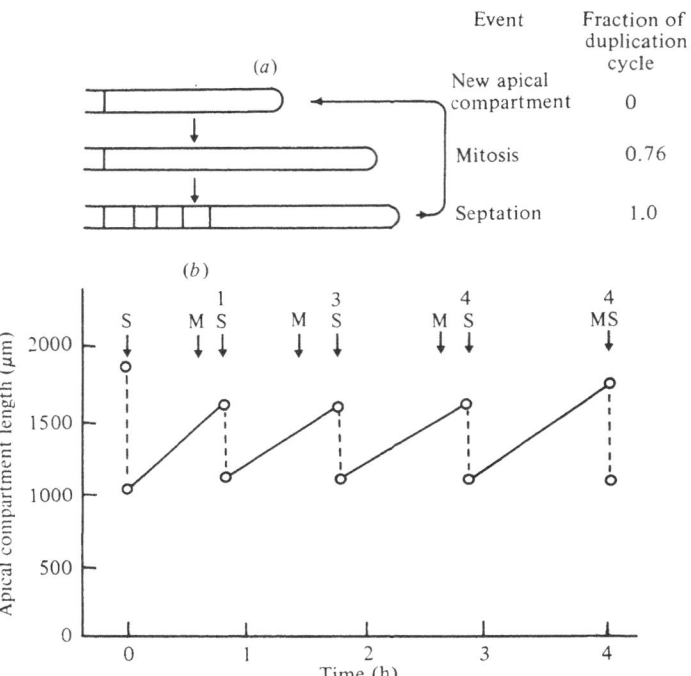

Fig. 15.7. Hyphal extension and nuclear migration in a dikaryotic apical compartment of *Polystictus versicolor* grown at 24 °C. For simplicity the migration of only one of the two nuclei is shown; the curves are mean values of 34 observations and the migratory behaviour of the nucleus is divided into four phases. Adapted from Girbardt (1968).

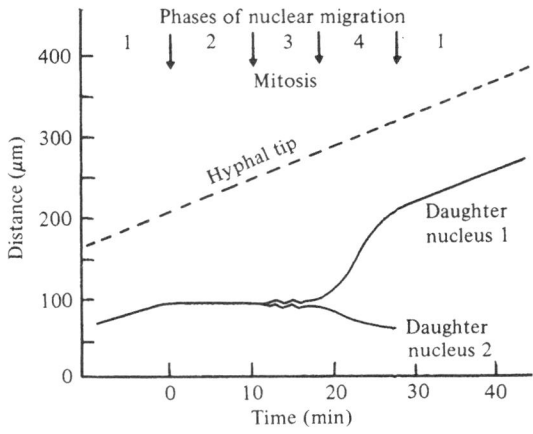

most of the cell contents (granules, mitochondria, vacuoles, etc.) move forward at about the same rate as the growing tip (Phase 1). Three to four minutes before outgrowth of the clamp connection, the nucleus stops migrating, although other cell contents (including at least part of the cytoplasm near the nucleus) continue to move forward; this second phase lasts a total of 10 min. The nucleus divides during the third phase. During the fourth and final phase, one daughter nucleus moves rapidly (at a rate faster than tip extension) forward through the compartment but it eventually assumes the same rate of forward movement as the growing tip (i.e. it returns to the phase 1 condition). The second daughter nucleus moves slowly in the opposite direction, i.e. away from the tip; the slow rate of migration of this nucleus may be related to its having to move against the prevailing current of cytoplasm. Nuclear migration after mitosis restores daughter nuclei to the more or less central position in the apical compartment previously occupied by the parent nuclei.

The nuclei in monokaryotic (*Basidiobolus ranarum* and *Schizophyllum commune*) and dikaryotic (*S. commune*) apical compartments (Niederpruem & Jersild, 1972; Trinci, 1978a) have a migratory behaviour which is very similar to that described in *Polystictus versicolor* by Girbardt (Fig. 15.7). The nuclei in *S. commune* (monokaryotic and dikaryotic mycelia) maintain an approximately central position in the apical compartment during extension growth; the pair of nuclei in the dikaryon migrate together and the distance between them does not change appreciably during most of the duplication cycle (Niederpruem & Jersild, 1972).

It is more difficult to follow nuclear migration in coenocytic fungi than in monokaryons or dikaryons. However, since in coenocytic fungi the distance between the hyphal tip and the first nucleus does not change appreciably during the duplication cycle (Trinci, 1978a), it is clear that at least the leading nuclei migrate through the apical compartments towards the tip at a rate similar to the extension rate of the hypha. During the duplication cycle of *Alternaria solani* the distance between the tip and the first nucleus varies from 55–70 μm just prior to mitosis, to 25–30 μm just after mitosis (King & Alexander, 1969). Robinow & Caten (1969) have observed a differential migratory behaviour of daughter nuclei in *Aspergillus nidulans*; after nuclear division, one daughter nucleus stayed close to the site of mitosis, whilst the other moved rapidly away, proximally or distally to the tip, over distances of about 10–15 μm. In *Aspergillus nidulans* (Fiddy & Trinci, 1976a) and

Alternaria solani (King & Alexander, 1969) the mean distance between adjacent nuclei in the apical compartment increases with distance from the hyphal tip. However, the observation that nuclei are always present throughout the compartment indicates that the pattern of nuclear migration in a coenocyte is probably more complex than that observed in dikaryons. It appears that nuclei do not normally migrate between intercalary and apical compartments during mycelial growth.

The mechanism which regulates the spatial distribution of nuclei in a compartment (Figs. 15.7, 15.8) is not known. Wilson & Aist (1967), Girbardt (1968) and Niederpruem, Jersild & Lane (1971) have suggested that nuclei may be self motile. Indirect evidence for self motility of nuclei in *Coprinus cinereus* (Schaeff.) Fr. is provided by the observation that during dikaryon formation there is no exchange of mitochondria between the mating monokaryons but nuclei migrate from one to the other (Casselton, 1978). The observations of Girbardt (1968) on the kineto-chore equivalent in *Polystictus versicolor* and of Hartwell and his group (personal communication) on the spindle pole body in *Saccharomyces cerevisiae* suggests that these (analogous?) structures are involved in nuclear migration.

Morris (1975) has isolated five temperature-sensitive mutants of *Aspergillus nidulans* which have an abnormal distribution of nuclei after transfer to the restrictive temperature (43 °C). Presumably the mechanism concerned in regulating either nuclear motility or the spatial distribution of nuclei has been affected by these mutations.

Fig. 15.8. Germlings of haploid and diploid strains of *Aspergillus nidulans* which have been stained by the Giemsa method.

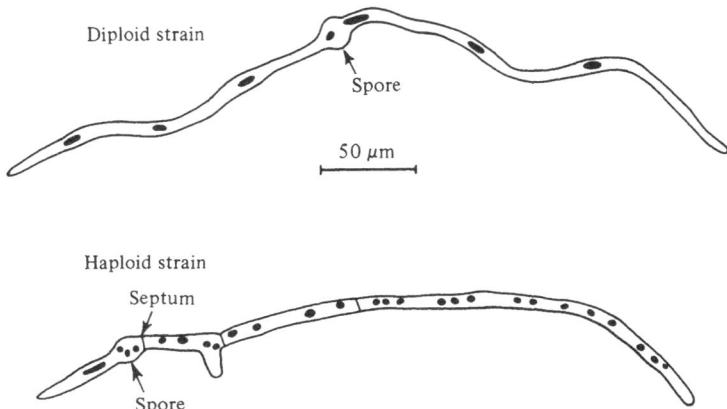

Diploid strain

Spore

50 μm

Haploid strain

Septum

Spore

Mitosis

In all fungi studied to date, mitosis is initiated about three-quarters of the way through the duplication cycle (Table 15.2). The pair of nuclei in dikaryotic compartments of *Schizophyllum commune* divide in perfect synchrony (Fig. 15.4), but the nuclei in coenocytic compartments of *Aspergillus nidulans* (Fig. 15.5), *Alternaria solani* (King & Alexander, 1969) and *Polyporus arcularius* (Fig. 15.6) divide parasynchronously; the degree of mitotic asynchrony in these organisms may be related to apical compartment length since nuclear division is less synchronized (see below) in the long compartments of *P. arcularius* (Valla, 1973*b*) than in the shorter compartments of *A. nidulans* and *A. solani*. The degree of mitotic synchrony is also affected by the growth conditions; Kessel & Rosenberger (1968) found that mitosis was less synchronized in germlings growing at a low specific growth rate than in germlings growing at a high specific growth rate. The periods during which mitosis is initiated in apical compartments of *A. nidulans, A. solani* and *P. arcularius* represent 9 % (average value), 10.5 % (maximum value) and 39 % (average value) of their respective duplication cycles. In *A. solani* and *A. nidulans*, mitosis is usually initiated at the tip

Table 15.2. *Times in successive duplication cycles of a hypha at which mitosis was initiated*

Organisms	Times of successive mitoses	Mean value	Source of data
Schizophyllum commune (monokaryon)	0.73; 0.71	0.72	Niederpruem & Jersild (1972)
Schizophyllum commune (dikaryon)	0.50; 0.73; 0.69; 0.75	0.67 (0.72)[a]	Niederpruem, Jersild & Lane (1971)
Aspergillus nidulans	0.74; 0.83; 0.72	0.76	Clutterbuck (1970)
Polyporus arcularius	0.73; 0.64; 0.79; 0.86	0.76	Valla (1973*a*)
Alternaria solani	0.74 to 0.86[b]		King & Alexander (1969)

The times are indicated as fractions of the total duplication time; septation is taken as the initial and fixed event in the cycles.
[a] Mean value if 0.50 value excluded; there is reason to suspect the validity of this value.
[b] Estimated values.

of the apical compartment and a wave of nuclear division follows progressively basipetally.

Nuclear division in apical compartments (monokaryotic, dikaryotic or coenocytic fungi) may be regulated by a gradual accumulation in the cytoplasm of an initiator which triggers mitosis when it has attained a certain threshold level (see p. 344). If this hypothesis is correct it would be expected that the degree of mitotic synchrony would be influenced both by the efficiency of cytoplasmic mixing in the compartment and the spatial distribution of nuclei; some mitotic asynchrony would be anticipated if cytoplasmic mixing is less than perfect and if the nuclei are dispersed throughout the compartment.

Only in *Basidiobolus ranarum* (Robinow, 1963) has a deceleration been observed in the rate of hyphal extension during mitosis. In other fungi, leading hyphae extend at a constant rate throughout the duplication cycle. The behaviour of *B. ranarum* may differ from the other fungi because unlike them it forms complete septa.

Septation

In *Basidiobolus ranarum* and *Schizophyllum commune* (monokaryotic and dikaryotic strains) there is a relation between the sites where nuclei divide in the cytoplasm of the apical compartment and the subsequent sites of septation; after nuclear division in *S. commune* two septa are formed, one in the clamp connection and one in the apical compartment. In *B. ranarum*, septa are formed on sites in the cytoplasm previously occupied by the metaphase plates of dividing nuclei (Robinow, 1963). This relationship ensures that each newly formed daughter cell or compartment contains a nucleus. Thus in *B. ranarum* and *S. commune* there are temporal (Table 15.3), spatial and quantita-

Table 15.3. *Mean interval between the initiation of mitosis and the initiation of septation in apical compartments of leading hyphae*

Organism	Mean interval[a]	Source of data
Schizophyllum commune (monokaryon)	0.21	Niederpruem & Jersild (1972)
Schizophyllum commune (dikaryon)	0.22	Niederpruem & Jersild (1972)
Aspergillus nidulans	0.18	Clutterbuck (1970)
Polyporus arcularius	0.26	Valla (1973b)
Alternaria solani	0.21	King & Alexander (1969)

[a] Expressed as a fraction of total duplication time.

tive (one septum per mitosis) relations between mitosis and cross-wall formation. However, in *Aspergillus nidulans, Alternaria solani* and *Polyporus arcularius*, although there is a temporal relationship between mitosis and septation (Table 15.3), no obvious spatial or quantitative relationships exist between these two events. For example, although the apical compartment of *A. nidulans* contains about fifty nuclei, their division throughout the compartment is followed by the formation of a maximum of six septa in its distal half (Fig. 15.5). The septa formed after mitosis in a coenocytic compartment are laid down parasynchronously in no fixed order (Fiddy & Trinci, 1976*a*). In the five fungi studied to date, mitosis is initiated at approximately the same point in the duplication cycle (Table 15.2) and the intervals observed between mitosis and septal initiation are also similar (Table 15.3).

Morris (1975) has isolated temperature-sensitive mutants of *Aspergillus nidulans* which do not form septa at the restrictive temperature (42 °C). One of these mutants (*sep-402*) does not produce septa at 37 °C, but has a mycelial doubling time (1.70 h) which is not much greater than the parental strain (1.28 h) grown at the same temperature (A. P. J. Trinci & N. R. Morris, unpublished). When aseptate mycelia of this mutant are transferred from 37 °C to 25 °C (the permissive temperature) there is a lag and then septa are produced in regions of the mycelium which had been formed at 37 °C. These septa are formed parasynchronously and after about 3 h at 25 °C the mycelia are indistinguishable (as far as the number and frequency of cross-walls are concerned) from mycelia grown continuously at 25 °C. Fig. 15.9 shows a mycelium in which 39 septa were formed within about 1.5 h after transfer from 37 °C to 25 °C. These observations suggest that the *sep-402* mutant forms septal 'initials' at 37 °C which can subsequently be recognized at 25 °C. The variation observed in the lag (0.25 to 2.75 h) between transfer of mycelia to 25 °C and initiation of septation may reflect differences in the stage of the duplication cycle of the fungus at which transfer occurred. The method of septation in this mutant could be explained in terms of the hypothesis which Cabib (Chapter 9) has proposed for the regulation of septation in *Saccharomyces cerevisiae*. It is expected that the *sep-402* mutant will be of value in determining how septal sites are determined and how septal growth is regulated in moulds.

Other organelles and organisms

Hawley & Wagner (1967) have reported synchronous division of mitochondria in mycelia of *Neurospora crassa*. It is not known if synchronous mitochondrial development occurs in other organisms, or if other organelles are produced synchronously. Although well defined duplication cycles have been recognized in several fungi, other moulds lack obvious septation cycles, e.g. the septa formed by leading hyphae of *Geotrichum candidum* always divide the apical compartment unequally (Fiddy & Trinci, 1976*b*). It may be significant that *G. candidum*, like *Basidiobolus ranarum*, forms complete septa, i.e. septa which lack a central pore through which cytoplasmic continuity between compartments can occur. It is also not possible to distinguish septation cycles in *Mucor hiemalis* Wehmeyer, another mould which forms complete septa (Fiddy & Trinci, 1976*c*).

Fig. 15.9. Septation in a mycelium of *Aspergillus nidulans sep-402* which had been grown at 37 °C and then transferred to 25 °C. The part of the mycelium formed at 37 °C is shown by the solid lines, subsequent growth at 25 °C is shown by broken lines. The location of septa is shown by lines and the lag between transfer and formation of septa is indicated as follows: A = septa formed 1.0 h after transfer to 25 °C; B = septa formed 1.17 h after transfer to 25 °C; C = septa formed 1.33 h after transfer to 25 °C; D = septa formed 1.62 h after transfer to 25 °C.

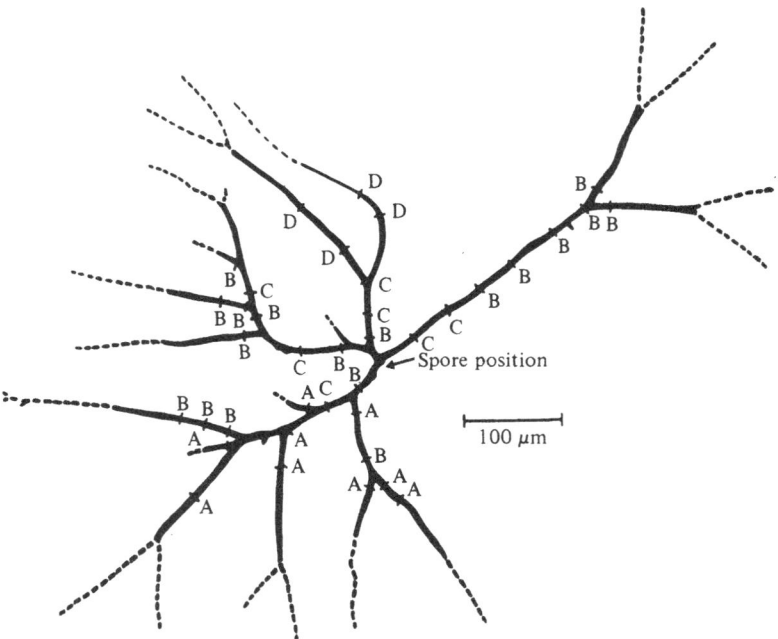

The duplication cycle in (germ tube) hyphae which have an accelerating rate of extension

The rate of extension of a germ tube hypha accelerates until eventually it attains the linear growth rate characteristic of the strain and conditions (Bull & Trinci, 1977). Nuclei are distributed more or less uniformly throughout a germling (Fig. 15.8) and the leading nucleus migrates towards the tip at a rate which is similar to the rate of hyphal extension. The more or less uniform distribution of nuclei in germlings is a further indication of the complex nature of nuclear migration in coenocytic fungi; the nuclei in coenocytic germlings behave in a manner suggesting that they are attached to a cytoplasmic element which stretches or grows uniformly at the same rate as tip extension. There is, however, no evidence for the existence of such a cytoplasic element and it is not thought that nuclei are attached to the protoplasmic membrane.

After spore germination (the conidia are uninucleate), nuclei in a germling of *Aspergillus nidulans* divide more or less synchronously until it contains 8 or 16 nuclei (Rosenberger & Kessel, 1967; Fiddy & Trinci, 1976*a*). There was a two-fold difference in the maximum and minimum lengths of germlings of *A. nidulans* which contained 4, 8, 16 or 32 nuclei (Fiddy & Trinci, 1976*a*); the maximum germling length observed for a particular number of nuclei presumably represents a germling just prior to mitosis, whilst the minimum length observed for the same number of nuclei represents a germling just after mitosis. These data could be interpreted as indicating that mitosis in germlings is initiated when a critical cytoplasmic volume to nuclear number ratio is attained (see discussion below).

Mitosis in apical compartments of germ tube hyphae of *Polyporus arcularius* and *Aspergillus nidulans*, as in leading hyphae, is followed by septation (Fig. 15.10); in *P. arcularius* mitosis occurs 0.78 of the way through the duplication cycle of a germling compared with a value of 0.76 for the initiation of mitosis in leading hyphae (Valla, 1973*b*). Groups of one to nine septa are formed parasynchronously in apical compartments of germ tube hyphae of *A. nidulans* and groups of two to five in germ tube hyphae of *P. arcularius*. The times taken to complete septation in apical compartments of a germ tube and a leading hypha of *A. nidulans* are almost identical (8.6 and 8.5 min respectively at 25 °C). In *P. arcularius* and *A. nidulans*, septation divides the apical compartment of germ tube hyphae unequally and in so doing forms a new apical compartment which is significantly longer than either the apical compartment at the beginning of the previous cycle or the newly formed septated region (Fig. 15.10). This unequal division continues until the germ tube hypha attains a linear rate of extension and its apical compartment attains its maximum *mean* length (Fiddy & Trinci, 1976*a*).

The mean interval (1.8 h) between successive septation cycles in an apical compartment of a germ tube hypha of *Aspergillus nidulans* is very similar to the mycelial doubling time of this organism (2.1 h). Thus in this and other respects, there is a basic similarity in the duplication cycles observed in germ tube and leading hyphae. The unequal division of an apical compartment of a germ tube hypha after septation is probably related to its having an acclerating rate of extension (Fiddy & Trinci, 1976*a*); leading hyphae on the other hand extend at a linear rate and their apical compartments are divided in half by septation. The mechanism which determines the sites of cross-wall formation presum-

Fig. 15.10. Duplication cycles of apical compartments of germlings of *Aspergillus nidulans* (upper) and *Polyporus arcularius* (lower). Initiation of mitosis (M) and septation (S, the numbers indicate the number of septa formed); apical compartment lengths are shown in successive cycles.

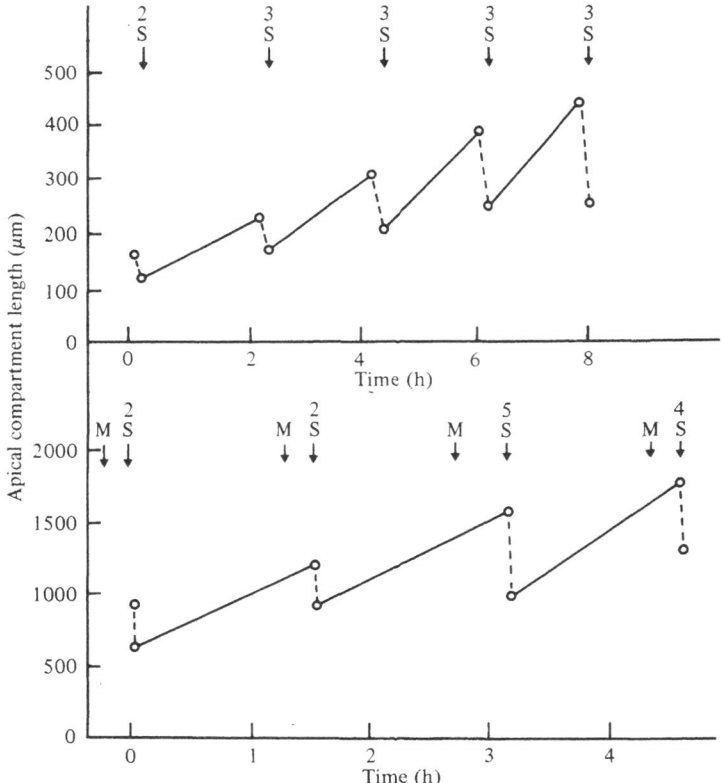

ably operates at a specific point in the cycle. This being so one would expect to observe differences in the location of septa in germ tube and leading hyphae.

Branch hyphae also have an accelerating rate of extension initially and thus they probably exhibit a duplication cycle similar to that observed in germ tube hyphae.

The timing and sequence of events in duplication cycles of germ tube and leading hyphae are probably identical, certainly the timing of the initiation of mitosis in *Polyporus arcularius* is the same in the two types of hyphae (see above).

Decay of synchrony during mycelial growth

For a time after spore germination, nuclei in a germling divide synchronously (Flentje, Stretton & Hawn, 1963; Rosenberger & Kessel, 1967; Crane, 1972; Kumari, Decallone & Meyer, 1975). However, this mitotic synchrony starts to decay when germlings contain more than about eight to sixteen nuclei. The results of Kessel & Rosenberger (1968) indicate that the decay in mitotic synchrony is inversely related to the organism's specific growth rate. In contrast, mitotic synchrony in the coenocytic slime mould *Physarum polycephalum* is maintained more or less indefinitely (Braun *et al.*, 1977). This difference between the behaviour of moulds and slime moulds may be related to differences in protoplasmic mixing since it is thought that in both groups a cytoplasmic substance initiates mitosis when it attains a critical concentration. It is unlikely that the concentration of this substance will vary significantly within a single plasmodium because of the vigorous nature of proto-plasmic mixing in this organism. In contrast, growth in a mould is highly polarized and as a result cytoplasmic mixing within a mycelium is probably less than perfect. In addition the cross-walls formed by many moulds probably impose a further restriction on cytoplasmic mixing, especially if the septa are complete or if their pores have been plugged. Fiddy & Trinci (1976a) have suggested that the decay in mitotic synchrony observed in germlings of *Aspergillus nidulans* is at least, in part, associated with cross-wall formation. Thus in a mycelium of *A. nidulans* it is envisaged that there is intra- but not inter-compartmental mitotic synchrony; a large mycelium may thus be compared with an asynchronous population of a unicellular organism. However, inter-compartmental mitotic synchrony has been reported in *Thanatephorus cucumeris* (Flentje *et al.*, 1963) and *Polyporus arcularius* (Valla, 1973a). In addition, a beautiful example of intercompartmental mitotic synchrony has been observed in the phialides of *Aspergillus flavus* Link ex Fries by Kozakiewicz (1978). Of interest in this particular example is that, although the nuclei in the phialides are dividing synchronously, the nuclei in the metulae (cells which support the phialides) never divide (Fig. 15.11). Thus if a cytoplasmic signal initiates mitosis in this organism, either the signal is confined to the phialides or the nuclei in the metulae may be insensitive to the initiator. The mitotic synchrony observed in the phialides may reflect the natural synchrony with which phialides are differentiated (Trinci, Peat & Banbury, 1968).

The phases of the duplication cycle of moulds

When microconidia of *Fusarium oxysporum* Schlenchtendahl (Kumari *et al.*, 1975) or conidia of *Aspergillus nidulans* (Bainbridge, 1971) germinate, DNA synthesis precedes mitosis, suggesting that the dormant spores are arrested in G1. However, mitosis precedes DNA synthesis during the germination of spores of *Phycomyces blakesleeanus* Burgeff, suggesting that these spores are arrested in G2 (Van Assche & Cartier, 1973). Dormant spores of *Neurospora crassa*, on the other hand, may be arrested in G1 or G2 (Schmit & Brody, 1976). During spore germination, macromolecules are synthesized in *A. nidulans* in the sequence RNA, protein, DNA (Bainbridge, 1971).

Kessel & Rosenberger (1968) and Bainbridge (1971) have investigated the various periods of the duplication cycle of *Aspergillus nidulans* (Table 15.4). Bainbridge found that after the first cycle the duration of the G1 period in a germling was relatively long compared with the duration of G2. A similar observation was made by Kumari *et al.* (1975)

Table 15.4. *Duration of the duplication cycle periods of* Aspergillus nidulans *in relation to carbon source*

Parameter	Carbon source in the medium		
	Glucose	Arabinose	Xylose
Doubling time (min)	87	135	196
G1 period	relatively long		
S period (min)	19	23	20
G2 period	relatively short		
Mitosis (min)	3 to 5	<5	<5

Data from Kessel & Rosenberger (1968) and Bainbridge (1971).

Fig. 15.11. Synchronous nuclear division in phialides of *Aspergillus flavus* viewed by Nomarski interference phase-contrast microscopy. Conidiophore (c), vesicle (v), metula (m), phialide (p), dividing nucleus (dn) and non-dividing nucleus (nn). Crown copyright reserved. Photograph kindly supplied by Miss Z. Kozakiewicz of the Pest Infestation Control Laboratory, Slough.

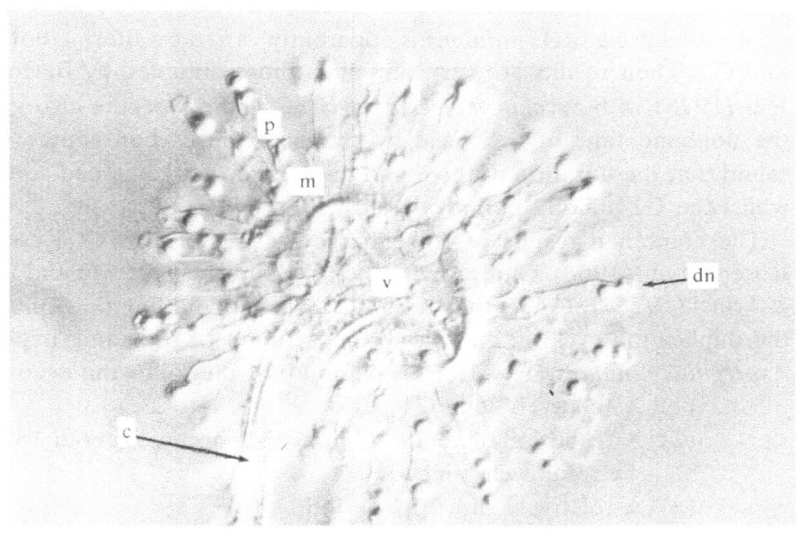

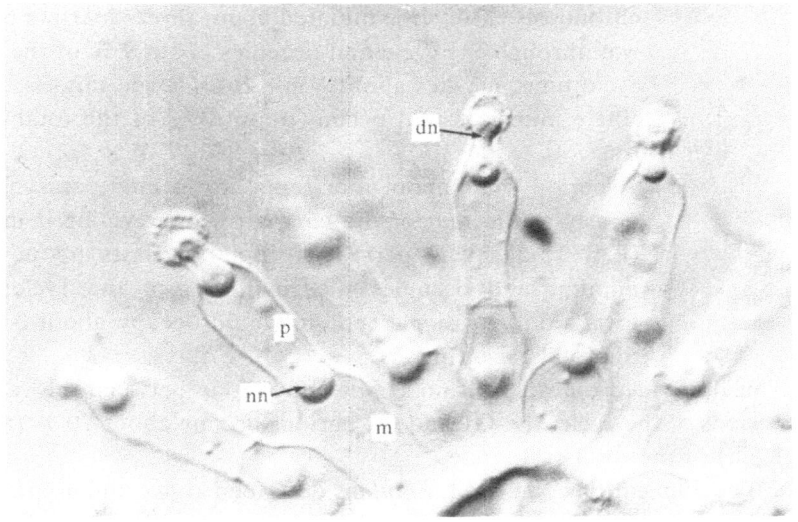

for the respective lengths of the G1 and G2 periods in *Fusarium oxysporum*. Kessel & Rosenberger (1968) varied the doubling time of *A. nidulans* from 1.45 to 3.27 h by changing the carbon source in the medium; they found that the duration of G1 and G2 varied with growth rate whilst the S and mitotic periods remained constant (Table 15.4). Thus, in contrast to some unicellular eukaryotic cells, the duration of the duplication cycle of *A. nidulans* is apparently varied by altering both G1 and G2. Their results are very similar to those obtained by Barford & Hill (1976) with *Saccharomyces cerevisiae*. These workers also varied the doubling time of the yeast by changing the carbon source; they found that the duration of the S and mitotic periods remained constant whilst the G1 and G2 periods varied with growth rate.

The combined results of Rosenberger & Kessel (1967), Kessel & Rosenberger (1968), Clutterbuck (1970), Bainbridge (1971) and Fiddy & Trinci (1976a) suggest that at 25 °C (mycelial doubling time of 2.1 h) the duplication cycle in an apical compartment of a leading hypha of *Aspergillus nidulans* (Fig. 15.5) is made up of the following events:

(1) A relatively long G1 period.

(2) An S period of about 20 min which occupies about 15 % of the total cycle time.

(3) A relatively short G2 period.

(4) A mitotic period during which fifty nuclei divide parasynchronously. Mitosis is initiated about three-quarters of the way through the cycle and occupies about 9 % of the total cycle time; it takes about 4 min for a single mitosis, i.e. a single mitosis would occupy about 3 % of the total cycle time.

(5) A period when about four septa are formed parasynchronously at the rear of the compartment, dividing it in half (Fig. 15.5). This period of septation lasts about nine minutes and occupies the final 9 % of the cycle; the formation of a single septum would occupy about 3 % of the total cycle time.

Thus, provided that there is no significant overlap between the various periods of the cycle, the G1 and G2 periods occupy about 70 % (about 1.5 h) of the total cycle time.

The duplication cycle for *A. nidulans* described above and in Fig. 15.5 may be compared with the cell cycle of *Saccharomyces cerevisiae* and *Schizosaccharomyces pombe* (Fig. 15.2).

Synchronous cultures of mycelia

Information about the cell cycle of an organism may be obtained either from microscopic observations of the growth of individual cells or by studying the behaviour of populations of cells which have been synchronized. It is generally believed that information obtained from studies of populations which have been synchronized by selection methods is more reliable than that obtained from populations synchronized by induction methods (Mitchison, 1971; Prescott, 1976). Unfortunately, since cell separation does not occur in moulds the selection techniques which are commonly used to synchronize populations of unicellular organisms cannot be applied to fungi. However, Padilla, Carter & Mitchison (1975) and Chaffin & Sojin (1976) have shown that the rate of germination (germ tube emergence) of spores of *Schizosaccharomyces pombe* and *Candida albicans* Robin (Berkhout) is related to spore volume. They were thus able to prepare a synchronous population of germinating spores by starting with a population of spores of more or less uniform volume. Clearly this is one selection method for synchrony which could also be applied to moulds. Goodey & Trinci (unpublished) used a sucrose density gradient to separate spores of *Aspergillus nidulans* into different size classes. They found a positive relationship between spore volume and the rate of spore germination, i.e. large spores germinated faster than small spores; a similar observation has previously been made by Bainbridge & Saunders (1973). Goodey & Trinci also showed that when spores isolated from different fractions of a sucrose density gradient were incubated, they germinated more synchronously than spores of the control population, e.g. 50 % of the spores in fractions 8, 10 and 13 (Fig. 15.12) germinated within time periods (15, 23 and 24 min respectively) which were significantly shorter than that observed for the control spores (43 min); Bainbridge (1971) found that 49.5 % of a spore population of *A. nidulans* produced germ tubes within a 60 min period. Wain *et al.* (1976) have used a modification of the spore selection method to obtain synchronous cultures of the hyphal form of the dimorphic yeast *Candida albicans*; the synchronous nature of the hyphal population of this organism was indicated by the stepwise increase observed in the DNA content of the culture. Thus this technique can be successfully applied to moulds.

It has been shown with many organisms that G1 cannot be completed and DNA replication cannot be initiated if a major part of protein or RNA synthesis is inhibited. In *Saccharomyces cerevisiae*, protein synthesis is necessary for the initiation of DNA synthesis but not its continuation (Hereford & Hartwell, 1973; Williamson, 1973). In contrast, in some organisms, e.g. *Physarum polycephalum* (Braun *et al.*, 1977) protein synthesis is essential both for the initiation and continuation of DNA synthesis. Cultures of some prokaryotic and eukaryotic organisms may be synchronized by first inhibiting and then relieving the inhibition of protein synthesis (Ron, Rozenhak & Grossman, 1975; Everhart & Prescott, 1972). Fletcher & Trinci (unpublished) have shown that cultures of a lysine-requiring auxotroph of *Neurospora crassa* may be synchronized (as indicated by the observed stepwise increase in DNA content) by first removing lysine from the medium and then adding it back. In cultures synchronized in this way, the rate of protein synthesis decelerated periodically, and since these periods of

Fig. 15.12. Germination of conidia of *Aspergillus nidulans* of approximately uniform volume. The conidia were separated by centrifugation for 10 min at 2000 rev min^{-1} at room temperature on a linear sucrose gradient (40 to 70 % (w/v)). The graph shows the germination (% germ tube emergence) of conidia from fractions 8, 10, 13 and a control fraction obtained by homogenizing the gradient, (▲, ■, ● and ○ respectively); the fractions were collected from the bottom of the tube. The initial mean volume of the conidial populations (obtained by using a Coulter counter) were as follows: fraction 8, 19.97 μm^3; fraction 10, 16.65 μm^3; fraction 13, 15.22 μm^3; homogenized fraction 16.20 μm^3.

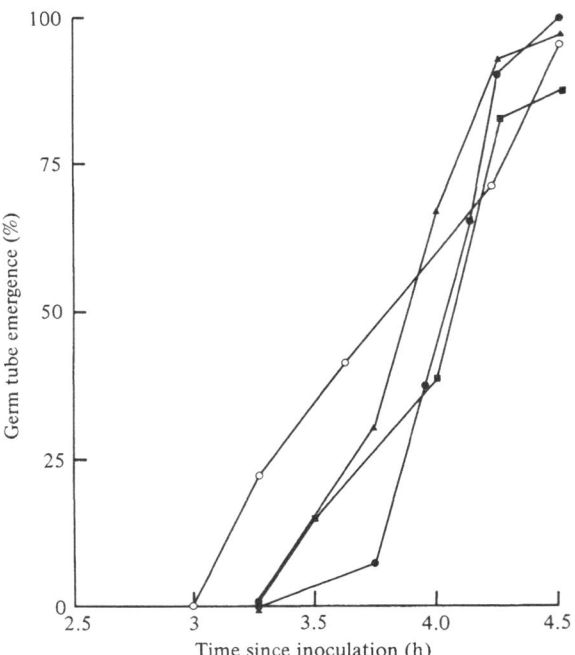

deceleration occurred at intervals which were approximately the same as the mould's doubling time it is possible that in fungi, as in other eukaryotic organisms (see above), there is a deceleration in the rate of protein synthesis during mitosis. Fletcher & Trinci (unpublished) have also shown that DNA synthesis in *N. crassa,* as in *Schizosaccharomyces pombe* (Mitchison & Creanor, 1971) may be synchronized with deoxyadenosine, an inhibitor of DNA synthesis. Until a satisfactory method of synchronizing mycelial cultures has been developed information concerning the duplication cycle of moulds will continue to be largely confined to a description of the morphological events in the cycle.

Regulation of the duplication cycle

Regulatory mechanisms must be present in moulds to control and integrate the ordered synthesis of organelles, macromolecules and enzymes during growth. In particular there must be mechanisms to control the initiation of DNA synthesis, mitosis and cross-wall formation. Mitchison (1971) has suggested that the cell cycle of unicellular eukaryotic organisms may be composed of two cycles that are normally coupled, viz. the 'growth cycle' and the 'DNA-division cycle'. In *Saccharomyces cerevisiae* the events involved in the DNA-division cycle include duplication of the spindle pole body, DNA replication, nuclear division and migration, cytokinesis and cell separation (Johnson, Pringle & Hartwell, 1977). With the exception of cell separation, all these events can be recognized during the DNA-division cycle of moulds.

Three principal types of model have been advanced to explain cell cycle regulation in eukaryotic organisms. There are models in which a particular event is initiated when the organism attains a certain size ('sizer' models); models in which a 'clock' is set at a certain point in the cycle with events occurring at fixed times afterwards; finally there is the 'transition probability' model.

The 'transition probability' theory was first suggested by Smith & Martin (1973). They propose that the cell cycle is composed of an *A*-state (a probabilistic state which occurs somewhere in G1) and a *B*-phase (a deterministic state which includes S, G2 and D periods). Sometime after mitosis the cell enters the *A*-state, in which it is not progressing towards division. The cell may remain in this state for any length of time, throughout which its probability of entering the *B*-phase is constant. The hypothesis was first advanced to explain cell cycle regulation of mammalian cells but it has also been applied to *Saccharomyces cerevisiae* (Shilo, Shilo & Simchen, 1976). However, the conclusions of Shilo *et al.* (1976) have been criticized by Nurse & Fantes (1977). The transition probability hypothesis would account for the observed decay in synchrony of synchronized populations and the decay of mitotic synchrony which Rosenberger & Kessel (1967) observed in germlings of *Aspergillus nidulans*. However, other explanations would also account for these observations (see above). The transition probability model implies a complete absence of the type of 'sizer' control system discussed next.

Hertwig (1908) was probably the first to suggest that a 'sizer' mechanism may be involved in the regulation of cell division; he suggested that cell division was triggered when the cell attained a particular cytoplasmic volume to nuclear ratio. Various 'sizer' mechanisms have subsequently been proposed to explain the timing of DNA synthesis and mitosis. Nurse and his colleagues have studied the behaviour of a *wee* mutant of *Schizosaccharomyces pombe* which divides at half the wild-type size (Nurse, 1975; Nurse & Thuriaux, 1977; Fantes & Nurse, 1977). They suggest that there are two points in the DNA-division cycle at which the cell has to attain a critical size before it can proceed, viz. prior to the initiation of DNA synthesis and prior to the initiation of mitosis. Johnson *et al.* (1977), on the other hand, feel that a single 'sizer' mechanism is present in *Saccharomyces cerevisiae*; they suggest that 'a specific event (called "start") early in G1 cannot be completed until a critical cell size is attained; most, if not all, subsequent events in the DNA-division cycle are independent of continued growth.'

The fact that protein synthesis is necessary for G1 transit is evidence that the initiator protein concept for DNA replication in prokaryotes (Helmstetter *et al.*, 1968) may also operate in eukaryotes. Helmstetter *et al.* (1968) proposed that DNA synthesis is initiated in bacteria when a hypothetical 'initiator' protein reaches a critical concentration in the cytoplasm, and that the protein may be used up in the initiation process so that further initiation only occurs after more protein has been produced. A negative-control mechanism has also been proposed to explain DNA regulation in bacteria (Pritchard, Barth & Collins, 1969); in this model it is suggested that cell growth and division are tied to an inhibitor of DNA synthesis acting as a relaxation oscillator. Recent studies on the replication of plasmids in *Escherichia coli* support the control of DNA replication by an inhibitor (Cabillo, Timmis & Cohen, 1976).

It has also been suggested that initiation of nuclear division, like initiation of DNA synthesis, may be related to the rate of total protein synthesis, or of the rate of synthesis of a particular protein (Sachsen-maier, Remy & Plattner-Shobel, 1972). It is possible that subunits which are assembled into a mitotic structure (or a division protein) are synthesized in the cytoplasm and that completion of this structure triggers mitosis. An alternative model has been considered by Fantes *et al.* (1975). In this model, a cytoplasmic effector molecule binds rever-sibly to a receptor site within the nucleus. The concentration of effector changes continuously throughout the cycle, and, at a critical concentra-tion, the receptor element responds by triggering mitosis. A similar hypothesis has been proposed to explain the regulation of synchronous mitosis in the coenocytic plasmodium of *Physarum polycephalum* (Sachsenmaier *et al.*, 1972).

Hartwell and his colleagues (personal communication) have obtained evidence which suggests that duplication of the spindle pole body in *Saccharomyces cerevisiae* is a crucial step ('start'?) in initiating mitosis. The temperature-sensitive mutants, *cdc-4* and *cdc-28,* accumulate in mitosis at the restrictive temperature; the wild-type product of *cdc-28* gene is a component of the spindle pole body and is apparently essential for its duplication whilst the wild-type product of the *cdc-4* gene is required for the migration of the daughter spindle pole bodies around the nuclear envelope; the mating factors '*a*' and '*a*' in this organism also apparently block the cell cycle of *S. cerevisiae* by preventing division of the spindle pole body. It is thus possible that the normal product of the *cdc-28* gene may be the cytoplasmic initiator which triggers mitosis (see above). Girbardt (1971) observed that the kinetochore equivalent (=spindle pole body?) of *Polystictus versicolor* is composed of two globular elements connected by a flat plate. The globular elements (protein?) in the kinetochore equivalent double in size just prior to mitosis. Girbardt, like Hartwell, suggested that the enlargement of these globular elements might be the trigger which initiates mitosis.

The observation that at mitosis the cytoplasmic volume to nuclear number ratio in *Aspergillus nidulans* is relatively constant for germlings containing up to 32 nuclei (Fiddy & Trinci, 1976*a*) is consistent with the hypothesis that a 'sizer' mechanism(s) may be involved in regulating either DNA synthesis and/or mitosis in this organism. However, the observation that a diploid strain of this organism has a cytoplasmic volume to nuclear number ratio which is approximately double the value observed in a haploid strain (Clutterbuck, 1969; Fiddy & Trinci, 1976*a*; Fig. 15.7) suggests that this 'sizer' mechanism(s) operates at the level of the genome rather than the nucleus. The mechanism may involve the formation of a cytoplasmic initiator protein as described above for eukaryotic cells; certainly the parasynchronous nature of mitosis in apical compartments of many fungi suggests the presence of a cytoplasmic signal.

Studies of the duplication cycle in coenocytic moulds serve to illustrate a particular regulatory mechanism which is not immediately apparent in other systems, viz. that not only must mechanisms exist to integrate the ordered synthesis of cellular components during the cell cycle but that there must also be mechanisms to regulate the spatial distribution of these components. Certainly, proteins, RNA, nuclei and mitochondria are distributed apparently uniformly in hyphae of fungi which are growing exponentially (Nishi, Yanagita & Maruyama, 1968; Fencl, Machek & Novak, 1969). Vigorous protoplasmic streaming serves to randomize the distribution of the various components which make up the plasmodium of *Physarum polycephalum*. However, in fungi at least some organelles (e.g. nuclei) do not appear to be moved within the mycelium in a random manner. Instead, in coenocytic germlings there is apparently a gradual migration of nuclei towards the hyphal apex, whilst at the same time the distance between adjacent nuclei is maintained more or less constant; Fig. 8.6 of Trinci (1978*a*) shows a germling of *Aspergillus nidulans* which contains about 100 nuclei distributed more or less uniformly throughout its length. Further, in diploid strains of *A. nidulans* the nuclei are on average approximately twice as far apart as in haploid strains (Fig. 15.7). Morris (1975) has shown that certain gene products are involved in regulating the spatial distribution of nuclei in mycelia of *A. nidulans*. The mechanisms which control the spatial distribution of protoplasmic components in fungi are not understood.

Branching

In filamentous fungi, cell separation does not follow cross-wall formation (*Geotrichum candidum* is an exception; Caldwell & Trinci, 1973). In this respect moulds differ from unicellular organisms. Indeed some fungi do not regularly produce cross-walls! However, in moulds which do form septa, branching is usually associated with their presence; depending on the species, the events of cross-wall formation and branching may be tightly or loosely coupled.

Since all hyphae eventually attain a linear growth rate, branching is an essential prerequisite for the maintenance of exponential growth in moulds; in an exponential culture the specific rate of branch formation is identical with the specific growth rate of the culture (Caldwell & Trinci, 1973; Trinci, 1974). Thus, branch initiation in moulds is an event which may be regarded as physiologically analogous to cell separation in unicellular organisms.

A general hypothesis for the regulation of branch initiation

I (Trinci, 1974; 1978a) have advanced a general hypothesis to account for the regulation of branch initiation in moulds; this hypothesis is in part based upon the work of Katz, Goldstein & Rosenberger (1972). It is proposed that cytoplasmic components (vesicles, precursors of wall polymers, enzymes which synthesize wall polymers and enzymes which lyse wall polymers during normal wall growth and/or branching) needed for primary wall extension (Trinci, 1978b) are produced at a constant rate throughout the mycelium. They are then transported under the influence of a polarizing mechanism towards the tips of hyphae; I have recently discussed the mechanisms which may be involved in the polarized transport of vesicles (Trinci, 1978b). The specific rate of production of these various cytoplasmic components will be the same as the specific growth rate of the mycelium.

It is proposed that, potentially, a branch may be initiated from any part of a mycelium. However, they actually occur where the cytoplasmic components involved in wall softening and primary wall extension happen to fuse with the rigidified wall of the hypha instead of being transported towards the extending tips. Thus an event which stops or reduces transport of protoplasm within a mycelium is likely to result in branch initiation. The formation of a cross-wall may be such an event. In addition, an apical branch is likely to be initiated when the rate of supply of wall precursors etc. to a tip exceeds the rate at which they can be incorporated into the primary wall of the extension zone.

The growth rate of a newly formed branch accelerates and eventually attains the linear growth rate characteristic of the strain and conditions. The extension rate of a branch during this accelerating phase is proportional to the distal length of hypha which is supplying its tip with the precursors required for primary wall growth (Trinci, 1971; Fiddy & Trinci, 1976a, b); a hypha growing at a constant rate has a constant distal length of hypha supplying it with these components.

Prosser (Chapter 16) has shown that the general hypothesis outlined above can be modelled successfully. However, a successful model does not necessarily validate basic assumptions made in the hypothesis, since models based on alternative hypotheses may also produce realistic simulations. Nevertheless, if a given hypothesis cannot be modelled satisfactorily it is unlikely to be a valid one.

Potential for branch initiation in a mycelium

The observations of Trinci & Collinge (1974), Steele & Trinci (1977a) and Collinge, Fletcher & Trinci (1978) confirm the hypothesis that potentially, a branch can be initiated from any part of the wall of a mycelium, including septa; indeed, the formation of a branch from a septum to produce an intrahyphal hypha is not an uncommon event in moulds (Lowry & Sussman, 1966; Chan & Stephen, 1967; Calonge, 1968). When the temperature-sensitive mutant of *Neurospora crassa, cot-1,* is transferred from 25 °C to 37 °C, it branches profusely from lateral wall regions which would not normally be involved in branching (Fig. 15.13). The sites at which branches are formed in this mutant are clearly determined by a cytoplasmic event.

Fig. 15.13. Septation and branching in mycelia of *Neurospora crassa cot-1* grown at 25 °C and transferred to 37 °C; at 25 °C the mould forms a sparsely branched mycelium. (a) Shows the relationship between septation and branch initiation; (b) shows the progressive production of branches after transfer to 37 °C.

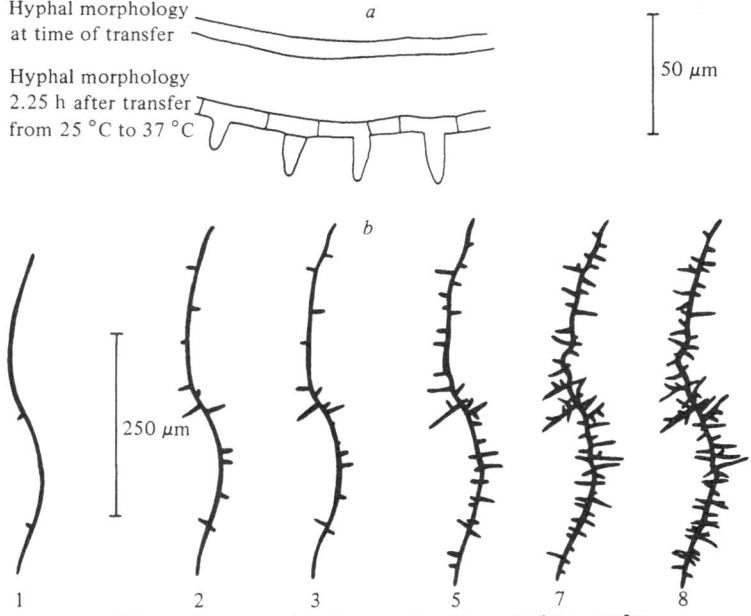

Hyphal morphology at time of transfer

Hyphal morphology 2.25 h after transfer from 25 °C to 37 °C

50 μm

250 μm

1   2   3   5   7   8

Hours after a transfer of a mycelium from 25 °C to 37 °C

A cytoplasmic event involved in branch initiation

I have suggested here and elsewhere (Trinci, 1974) that lateral branches are formed at locations in mycelia where cytoplasmic components (including vesicles) needed for wall softening and primary wall extension fuse with the rigidified wall of the hypha. Vesicles always seem to be implicated in primary wall growth; it is thought that they contain wall precursors and/or the enzymes required for the incorporation of these precursors into the existing wall (Bartnicki-Garcia, 1973). Branching in *Neurospora crassa* (Trinci & Collinge, 1974) and *Achlya ambisexualis* (Nolan & Bal, 1974) is certainly associated with the accumulation of vesicles at the site of branch initiation (Fig. 15.14). Vesicles have been identified which contain enzymes capable of lysing the polymers (cellulose or chitin) which make up the microfibrils of fungal walls (Nolan & Bal, 1974; Rosenberger, this volume Chapter 12). Mahadevan & Mahadkar (1970) and Fèvre (1976) have observed a correlation between the branching frequency in *Neurospora crassa* and *Saprolegnia monoica* and the activity of lysins capable of degrading wall polymers. Rosenberger (Chapter 12) suggests that these lysins are contained in vesicles which become incorporated in an intact form in the wall. He further suggests that the 'trapped' lysins are released when vesicles containing phospholipase fuse with the wall at a site where a

Fig. 15.14. Accumulation of vesicles behind a septum of *Neurospora crassa* prior to the formation of an experimentally induced branch (see Trinci & Collinge (1974) for experimental details). AC = autolysing compartment, V = vesicles, S = septum.

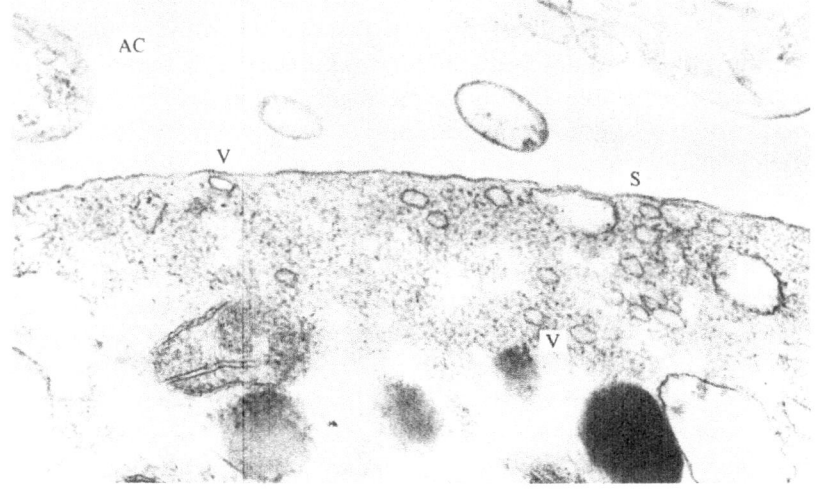

branch is to be initiated. Some of the liberated lysins would soften the wall whilst others (proteases) could activate synthases (e.g. chitin synthase) required for the formation of wall polymers.

Since a branch, potentially, can be formed from any part of the mycelial wall, what determines their observed location? The correlation between septation and branching will be discussed below. However, branching can occur in the absence of septation. Apical branches may be formed when the rate of supply of components required for primary wall growth exceeds the rate at which they can be incorporated into the tip wall. Lateral branching in the absence of septation may occur for a similar reason. I have suggested (Trinci, 1974) that growth of a mycelium may be considered in terms of the duplication of a hypothetical 'growth unit' which consists of a tip associated with a (strain) specific length of hypha. This growth unit has a physiological but not a morphological basis and simply represents the *average* length of hypha associated with each tip. The actual lengths of hypha supporting the growth of the various tips in a mycelium vary over wide limits (Trinci, 1978*a*). The maintenance of a more or less constant hyphal growth unit length (Fig. 15.15) during growth of a mycelium implies that branch initiation is regulated by the changes in cytoplasmic volume which accompany growth. Thus, when the hyphal growth unit of the mycelium exceeds a critical value a branch will be initiated somewhere in the mycelium. In this hypothesis it is assumed that hyphal diameter remains approximately constant. The linear relationship observed between hyphal growth unit and intercalary compartment length in *Neurospora crassa* (Steele & Trinci, 1977*b*) reinforces the hypothesis discussed below that branching may be causally associated with septation. In an aseptate mould, branching will be regulated in a similar way but the sites of branch initiation will be less predictable. Clearly the mechanisms involved in regulating the polarity of hyphal growth and transport of cytoplasm are relevant to any consideration of the regulation of branching in moulds. Unfortunately we know very little about either process.

Branch initiation associated with septation
Lateral branching associated with complete and dolipore septa. Leading hyphae of *Geotrichum candidum* and *Basidiobolus ranarum* form complete septa which may contain micropores (Bracker, 1967) but lack a large central pore; it is unlikely that such septa allow cytoplasmic interchange between compartments. Branches are characteristically initiated in these moulds just behind septa (Fig. 15.16) where presumably the components involved in primary wall growth fuse with the

lateral wall of the hypha; such observations imply that the polarity of the compartment is maintained after it has been isolated from the rest of the hypha. This conclusion is reinforced by the observation of Yanagita (1977) that 'when a single cell (penultimate) of *Coprinus* is cut from the hypha and cultured on a medium it forms a branch just behind the clamp connexion closest to the tip.'

The interval between septation and branch initiation is remarkably constant in some moulds, for example in *Geotrichum candidum* there was a lag of 28 ± 9 min between completion of septation and branch initiation. Thus in these fungi there is both a spatial and temporal relationship between septation and branch initiation. The formation of a second branch from a single compartment of *G. candidum* is associated with the formation of a second septum (Fiddy & Trinci, 1976*b*).

Fig. 15.15. Growth of a young mycelium of *Aspergillus nidulans* at 25 °C on a defined medium. The total mycelial length (O) and the number of branches (□) increase exponentially at the same specific rate. The hyphal growth-unit (●, the ratio between total mycelia length and the number of branches) displays damped oscillations and tends to a constant value. Compare with Fig. 16.5, p. 380.

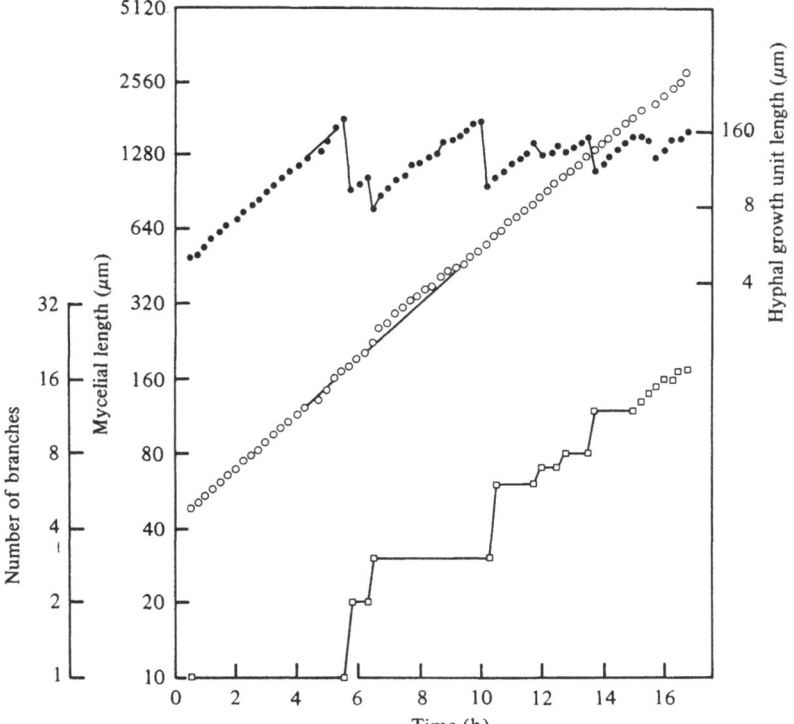

Mucor hiemalis also forms complete septa but in this mould there is no clear spatial or temporal relationship between septation and branch initiation (Fiddy & Trinci, 1976c).

In basidiomycetes, as in *Geotrichum candidum* and *Basidiobolus ranarum*, spatial and temporal relationships are observed between septation and branch initiation (Volz & Niederpruem, 1968; Niederpruem & Jersild, 1972; Buller, 1931; Fig. 15.10b). Thus, as far as branch initiation is concerned, septa with dolipores function in a manner similar to the complete septa of *G. candidum* and *B. ranarum*. The formation of clamp connections in basidiomycetes is also associated with septation. However, these specialized branches are initiated in front of septa (with respect to the tip) and they grow in the opposite direction from the tip.

Lateral branching associated with septa having a single central pore (ascomycete type). The ascomycete septum has a single central pore (Bracker, 1967); in time these pores become plugged (Trinci & Collinge, 1973). A correlation between septation and branch initiation has been observed in *Aspergillus nidulans* (Fiddy & Trinci, 1976a; Fig.

Fig. 15.16. Branch initiation from leading hyphae of (*a*) *Basidiobolus ranarum*; (*b*) a monokaryotic hypha of *Schizophyllum commune*; (*c*) from *Geotrichum candidum*; (*d*) shows branch intiation in a germling of *Aspergillus nidulans*. Note the spatial relationship between branch initiation and septation and the fact that there is usually one branch formed per compartment. (Redrawn from Robinow (1963), Niederpruem & Jersild (1972) and Fiddy & Trinci (1976b); the drawing of *A. nidulans* is original.)

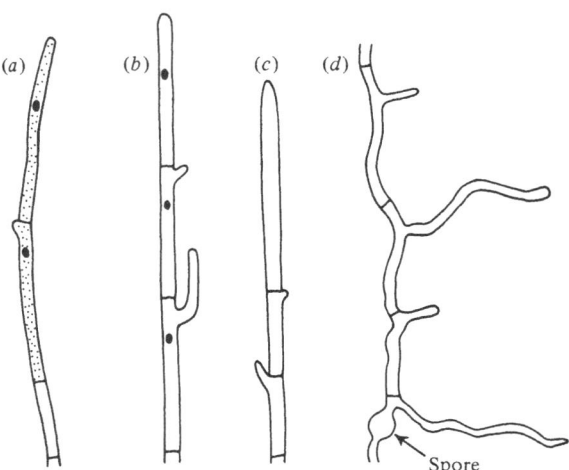

15.16) and *Neurospora crassa* (Steele & Trinci, 1977*b*; Fig. 15.13). Although branches in *A. nidulans* are most commonly formed from the proximal (with respect to the hyphal tip) half of each intercalary compartment (Fig. 15.16) they may also be formed from the distal region of the compartment. Further, the interval between septation and branch initiation is more variable in this fungus than in *Geotrichum candidum*; the lag between septation and branch initiation in *A. nidulans* was 50±40 min (Fiddy & Trinci, 1976*a*). Thus, septation and branch initiation are not as tightly coupled (spatially or temporally) in *A. nidulans* as in *G. candidum*. The ascomycete septum with its central pore may allow greater intercompartmental transport of cytoplasm than other types of septa. Consequently they may not isolate a compartment from the rest of the hypha as rapidly as other septa.

Apical and lateral branching not associated with septation. Branching in moulds is not invariably associated with septation. For example, *Rhizopus stolonifer* (Ehrenberg ex Fries) and *Actinomucor repens* Schostakow do not form septa but branch normally. Further, even in moulds which form septa, e.g. *Aspergillus nidulans* (Fiddy & Trinci, 1976*a*) lateral branching is not always associated with septation and apical branching in *A. nidulans*, *Geotrichum candidum* (Trinci, 1970) and *Neurospora crassa* is never associated with septation. Although the *sep-402* mutant of *A. nidulans* does not produce septa at 37 °C it

Fig. 15.17. Apical branching in *Aspergillus nidulans*. The extension rate of the parent tip and the *combined* rate of extension of the branches (see illustration) are plotted.

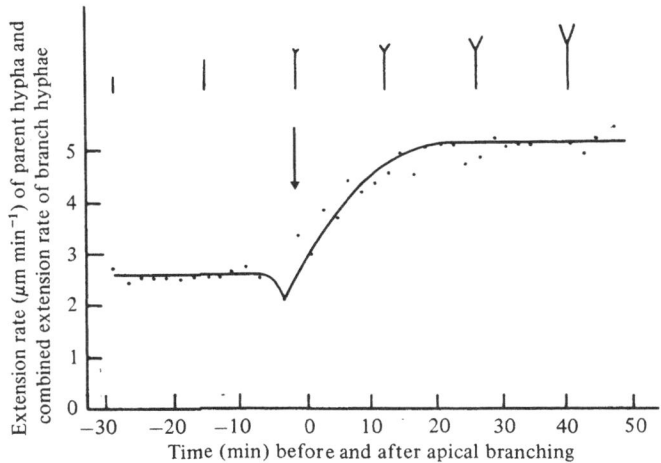

nevertheless branches in a more or less normal manner (Morris, 1975; Trinci & Morris, unpublished). Fig. 15.17 shows the kinetics of apical branching in *A. nidulans*; within 14 to 44 min of branching, each apical branch attains the linear growth rate of the parent hypha. Thus the cytological reorganization of the parent tip to form two extension zones proceeds remarkably rapidly.

Concluding remarks

Burnett (1968) observed in the preface to his book *Fundamentals of Mycology* that 'it is true of fungi as of physics that a science advances as rapidly as its data can be reduced to laws and systems of agreed validity.' I have attempted in this article to show that the growth of mycelia may follow, at least in some respects, the general laws which govern the growth of unicellular organisms. The apparent complexity of mycelial growth may thus conceal basic features which are common to many micro-organisms. However, any significant advance in our understanding of the duplication cycle of moulds will have to await the development of a technique to synchronize mycelial cultures.

Acknowledgement. I wish to thank Dr R. K. Poole for his help during the preparation of this review.

References

Bainbridge, B. W. (1971). Macromolecular composition and nuclear division during spore germination in *Aspergillus nidulans*. *Journal of General Microbiology*, **66**, 319–25.

Bainbridge, B. W. & Saunders, P. T. (1973). A mathematical analysis of conidial germ-tube emergence in *Aspergillus nidulans*. *Journal of General Microbiology*, **75**, xii–xiii.

Barford, J. P. & Hill, R. J. (1976). Estimation of the length of cell cycle phases from asynchronous cultures of *Saccharyomyces cerevisae*. *Experimental Cell Research*, **102**, 276–84.

Bartnicki-Garcia, S. (1973). Fundamental aspects of hyphal morphogenesis. In *Microbial Differentiation* (23rd Symposium of the Society for General Microbiology) eds. J. M. Ashworth & J. E. Smith, pp. 245–67. Cambridge University Press.

Bracker, C. E. (1967). Ultrastructure of fungi. *Annual Review of Phytopathology*, **5**, 347–74.

Braun, R., Hall, L., Schwärzler, M. & Smith, S. S. (1977). The mitotic cycle of *Physarum polycephalum*. In *Cell Differentiation in Microorganisms, Plants and Animals*, eds L. Nover & K. Mothes, pp. 402–3. Amsterdam, New York: North-Holland.

Bull, A. T. & Trinci, A. P. J. (1977). The physiology and metabolic control of fungal growth. *Advances in Microbiol Physiology*, **50**, 1–84.

Buller, A. H. R. (1931). Further observations on the Coprini together with some investigations on social organization and sex in the Hymenomycetes. In *Researches in Fungi*, vol. 4. London, New York: Longmans Green.

Burnett, J. H. (1968). *Fundamentals of Mycology*. First edn. London: Edward Arnold.

Cabillo, F., Timmis, K., & Cohen, S. N. (1976). Replication control in a composite plasmid constructed by *in vitro* linkage of two distinct replicons. *Nature (London)*, **259**, 285–90.

Caldwell, I. Y. & Trinci, A. P. J. (1973). The growth unit of the mould *Geotrichum candidum*. *Archiv für Mikrobiologie*, **88**, 1–10.

Calonge, F. D. (1968). Origin and development of intrahyphal hyphae in *Sclerotinia fructigena*. *Mycologia*, **60**, 932–42.

Campbell, A. (1957). Synchronization of cell division. *Bacteriological Reviews*, **21**, 253–72.

Casselton, L. A. (1978). Dikaryon formation in higher basidiomycetes. In *The Filamentous Fungi*, vol. III, eds J. E. Smith & D. R. Berry, pp. 275–297. London: Edward Arnold.

Chaffin, W. & Sojin, S. J. (1976). Germ tube formation from zonal rotor fractions of *Candida albicans*. *Journal of Bacteriology*, **126**, 771–76.

Chan, S. & Stephen, R. C. (1967). Intrahyphal hyphae in the genus *Linderina*. *Canadian Journal of Botany*, **45**, 1995–8.

Clutterbuck, A. J. (1969). Cell volume per nucleus in haploid and diploid strains of *Aspergillus nidulans*. *Journal of General Microbiology*, **55**, 291–300.

Clutterbuck, A. J. (1970). Synchronous nuclear division and septation in *Aspergillus nidulans*. *Journal of General Microbiology*, **60**, 133–5.

Clutterbuck, A. J. & Roper, J. A. (1966). A direct determination of nuclear distribution in heterokaryons of *Aspergillus nidulans*. *Genetical Research, Cambridge*, **7**, 185–94.

Collinge, A. J. & Trinci, A. P. J. (1974). Hyphal tips of wild-type and spreading colonial mutants of *Neurospora crassa*. *Archiv für Mikrobiologie*, **99**, 353–68.

Collinge, A. J., Fletcher, M. H. & Trinci, A. P. J. (1978). Physiology and cytology of septation and branching in a temperature sensitive colonial mutant (*cot 1*) of *Neurospora crassa*. *Transactions of the British Mycological Society* **77**, 107–20.

Crane, E. (1972). The control of nuclear division and cellular growth in germinating conidia of *Neurospora crassa*. M.Sc. Thesis University of Washington, USA.

Everhart, L. P. & Prescott, D. M. (1972). Reversible arrest of Chinese hamster cells in G1 by partial deprivation of leucine. *Experimental Cell Research*, **75**, 170–4.

Fantes, P. A. & Nurse, P. (1977). Control of cell size at division in fission yeast by growth-modulated size control over nuclear division. *Experimental Cell Research*, **107**, 377–86.

Fantes, P. A., Grant, W. O., Pritchard, R. A., Sudbery, P. E. & Wheals, A. E. (1975). The regulation of cell size and the control of mitosis. *Journal of Theoretical Biology*, **50**, 213–44.

Fencl, Z., Machek, F. & Novak, M. (1969). Kinetics of product formation in multistage continuous culture. In *Fermentation Advances*, ed. D. Perlman, pp. 301–24. New York, London: Academic Press.

Fèvre, M. (1976). Recherches sur le déterminisme de la morphogenèse hyphale. D.Sc. Thesis, Université Claude Bernard, Lyon, France.

Fiddy, C. & Trinci, A. P. J. (1976*a*). Mitosis, septation, branching and the duplication cycle in *Aspergillus nidulans*. *Journal of General Microbiology*, **97**, 169–184.

Fiddy, C. & Trinci, A. P. J. (1976*b*). Nuclei, septation, branching and growth of *Geotrichum candidum*. *Journal of General Microbiology*, **97**, 185–192.

Fiddy, C. & Trinci, A. P. J. (1976*c*). Septation in mycelia of *Mucor hiemalis* and *Mucor rammanianus*. *Transactions of the British Mycological Society*, **68**, 118–20.

Flentje, N. T., Stretton, H. M. & Hawn, E. J. (1963). Nuclear distribution and behaviour throughout the life cycles of *Thanatephorus*, *Waitea* and *Ceratobasidium* species. *Australian Journal of Biological Sciences*, **16**, 450–67.

Girbardt, M. (1968). Ultrastructure and dynamics of the moving nuclei. In *Aspects of Cell Motility* (23rd Symposium of the Society for Experimental Biology), ed. P. L. Miller, pp. 249–59. Cambridge University Press.

Girbardt, M. (1971). Ultrastructure of the fungal nucleus. II. The kinetochore equivalent (KCE). *Journal of Cell Science*, **2**, 453–73.

Gull, K. (1978). Form and function of septa in filamentous fungi. In *The Filamentous Fungi*, vol. III, eds. J. E. Smith & D. R. Berry, pp. 78–93. London: Edward Arnold.

Hartwell, L. H. (1974). *Saccharomyces cerevisiae* cell cycle. *Bacteriological Reviews*, **38**, 164–98.

Hawley, E. S. & Wagner, R. P. (1967). Synchronous mitochondrial division in *Neurospora crassa*. *Journal of Cell Biology*, **35**, 489–99.

Helmstetter, C. E., Cooper, S., Pierucci, O. & Revelas, E. (1968). On the bacterial life sequences. *Cold Spring Harbor Symposium in Quantitative Biology*, **33**, 809–22.

Hereford, L M. & Hartwell, L. H. (1973). Role of protein synthesis in the replication of yeast DNA. *Nature (London), New Biology*, **244**, 129–31.

Hertwig, R. (1908). Über neue probleme der Zellenlebre. *Archiv für Zellforschung*, **1**, 1–32.

Howard, A. & Pelc, S. R. (1953). Synthesis of deoxyribonucleic acid in normal and irradiated cells and its relation to chromosome breakage. *Heredity*, Supplement **6**, 267–73.

Johnson, G. C., Pringle, J. R. & Hartwell, L. H. (1977). Coordination of growth with cell division in the yeast *Saccharomyces cerevisiae*. *Experimental Cell Research*, **105**, 79–98.

Katz, D., Goldstein, D. & Rosenberger, R. F. (1972). Model for branch initiation in *Aspergillus nidulans* based on measurements of growth parameters. *Journal of Bacteriology*, **109**, 1097–1100.

Kessel, M. & Rosenberger, R. F. (1968). Regulation and timing of deoxyribonucleic acid synthesis in hyphae of *Aspergillus nidulans*. *Journal of Bacteriology*, **95**, 2275–81.

King, S. B. & Alexander, L. J. (1969). Nuclear behaviour, septation and hyphal growth of *Alternaria solani*. *American Journal of Botany*, **56**, 249–53.

Kozakiewicz, Z. (1978). Phialide and conidium development in the Aspergilli. *Transactions of the British Mycological Society*, **70**, 175–86.

Kumari, L., Decallone, J. R. & Meyer, J. A. (1975). DNA metabolism and nuclear division during spore germination in *Fusarium oxysporum*. *Journal of General Microbiology*, **88**, 245.

Lowry, R. J. & Sussman, A. S. (1966). Intra-hyphal hyphae in 'clock' mutants of *Neurospora*. *Mycologia*, **58**, 541–48.

Maaløe, O. & Kjeldgaard, N. O. (1966). *Control of Macromolecular Synthesis*. Benjamin: New York.

Mahadevan, P. R. & Mahadkar, U. R. (1970). Role of enzymes in growth and morphology of *Neurospora crassa*: cell wall bound enzymes and their possible role in branching. *Journal of Bacteriology*, **101**, 941–7.

Mitchison, J. M. (1971). *The Biology of the Cell Cycle*. Cambridge University Press.

Mitchison, J. M. & Creanor, J. (1971). Induction synchrony in the fission yeast *Schizosaccharomyces pombe*. *Experimental Cell Research*, **67**, 368–74.

Morris, N. R. (1975). Mitotic mutants of *Aspergillus nidulans*. *Genetical Research*, Cambridge, **26**, 237–54.

Niederpruem, D. J. & Jersild, R. A. (1972). Cellular aspects of morphogenesis in the mushroom *Schizophyllum commune*. *Critical Reviews in Microbiology* **1**, 545–76.

Niederpruem, D. J., Jersild, R. A. & Lane, P. I. (1971). Direct microscopic studies of clamp connection formation in growing hyphae of *Schizophyllum commune*. I. The dikaryon. *Archiv für Mikrobiologie*, **78**, 268–80.

Nishi, A., Yanagita, T. & Maruyama, Y. (1968). Cellular events occurring in growing hyphae of *Aspergillus oryzae* as studied by autoradiography. *Journal of General And Applied Microbiology*, **14**, 171–82.

Nolan, R. A. & Bal, A. K. (1974). Cellulase localization in hyphae of *Achlya ambisexualis*. *Journal of Bacteriology*, **117**, 840–3.

Nurse, P. (1975). Genetic control of cell size at cell division in yeast. *Nature (London)*, **256**, 547–51.

Nurse, P. & Fantes, P. (1977). Transition probability and cell-cycle initiation in yeast. *Nature (London)*, **267**, 647.

Nurse, P. & Thuriaux, P. (1977). Controls over the timing of DNA replication during the cell cycle of fission yeasts. *Experimental Cell Research*, **107**, 365–75.

Padilla, G. M., Carter, B. A. & Mitchison, J. M. (1975). Germinating *Schizosaccharomyces pombe* spores separated by zonal centrifugation. *Experimental Cell Research*, **93**, 325–30.

Prescott, D. M. (1976). *Reproduction of Eukaryotic Cells*. New York, London: Academic Press.

Pritchard, R. H. (1968). Control of DNA synthesis in bacteria. *Heredity (Abstract)*, **23**, 472.

Pritchard, R. H., Barth, P. T. & Collins, J. (1969). Control of DNA synthesis in bacteria. In *Microbial Growth* (19th Symposium of the Society for General Microbiology) eds. P. Meadow & S. J. Pirt, pp. 263–98. Cambridge University Press.

Robinow, C. F. (1963). Observations on cell growth, mitosis and division in the fungus *Basidiobolus ranarum*. *Journal of Cell Biology*, **17**, 123–52.

Robinow, C. F. & Caten, C. E. (1969). Mitosis in *Aspergillus nidulans*. *Journal of Cell Science*, **5**, 403–31.

Ron, E. Z., Rozenhak, S. & Grossman, N. (1975). Synchronization of cell division in *Escherichia coli* by amino acid starvation: strains specificity. *Journal of Bacteriology*, **123**, 374–9.

Rosenberger, R. F. & Kessel, M. (1967). Synchrony of nuclear replication in individual hyphae of *Aspergillus nidulans*. *Journal of Bacteriology*, **94**, 1464–9.

Sachsenmaier, W., Remy, V. & Plattner-Shobel, R. (1972). Initiation of synchronous mitosis in *Physarum polycephalum*. A model of the control of cell division in eukaryotes. *Experimental Cell Research*, **73**, 41–8.

Schmit, J. C. & Brody, S. (1976). Biochemical genetics of *Neurospora crassa* conidial germination. *Bacteriological Reviews*, **40**, 1–41.

Shilo, B., Shilo, V. & Simchen, G. (1976). Cell-cycle initiation in yeast follows first-order kinetics. *Nature (London)*, **264**, 767–70.

Smith, J. A. & Martin, L. (1973). Do cells cycle? *Proceedings of the National Academy of Sciences, USA*, **70**, 1263–7.

Steele, G. W. & Trinci, A. P. J. (1977*a*). Effect of temperature and temperature shift on growth and branching of a wild type and a temperature sensitive colonial mutant (*cot 1*) of *Neurospora crassa*. *Archives of Microbiology*, **113**, 43–8.

Steele, G. W. & Trinci, A. P. J. (1977*b*). Relationship between intercalary compartment length and hyphal growth unit length. *Transactions of the British Mycological Society*, **69**, 156–8.

Trinci, A. P. J. (1970). Kinetics of apical and lateral branching in *Aspergillus nidulans* and

Geotrichum lactis. Transactions of the British Mycological Society, **55**, 17–28.

Trinci, A. P. J. (1971). Influence of the peripheral growth zone on the radial growth rate of fungal colonies. *Journal of General Microbiology*, **67**, 325–44.

Trinci, A. P. J. (1974). A study of the kinetics of hyphal extension and branch initiation of fungal mycelia. *Journal of General Microbiology*, **81**, 225–36.

Trinci, A. P. J. (1978*a*). The duplication cycle and vegetative development in moulds. In *The Filamentous Fungi*, vol. 3, eds. J. E. Smith & D. R. Berry, pp. 132–63. London: Edward Arnold.

Trinci, A. P. J. (1978*b*). Wall and hyphal growth. *Science Progress*, **65**, 75–99.

Trinci, A. P. J. & Collinge, A. J. (1973). Structure and plugging of septa of wild type and spreading colonial mutants of *Neurospora crassa. Archiv für Mikrobiologie*, **91**, 355–64.

Trinci, A. P. J. & Collinge, A. J. (1974). Occlusion of the septal pores of damaged hyphae of *Neurospora crassa* by hexagonal crystals. *Protoplasma*, **80**, 56–67.

Trinci, A. P. J., Peat, A. & Banbury, G. H. (1968). Fine structures of phialide and conidiophore development in *Aspergillus giganteus* Wehmer. *Annals of Botany*, **32**, 241–9.

Valla, G. (1973*a*). Rythmes des division nucléaires et de la septation dans les articles apicaux en croissance du mycelium de *Polyporus arcularius. Comptes rendus Hebdomadaires des Séances de l'Academie des Sciences, Paris*, **279**, 2649–52.

Valla, G. (1973*b*). Division nucléaires, septation et ramification chez l'haplonte de *Polyporus arcularius* (Batsch) Ex Fr. *Naturaliste Canadien*, **100**, 479–92.

Van Assche, J. A. & Cartier, A. R. (1973). The pattern of protein and nucleic acid synthesis in germinating spores of *Phycomyces blakesleeanus. Archiv für Mikrobiologie*, **93**, 129–36.

Volz, P. A. & Niederpruem, D. J. (1968). Growth, cell division and branching patterns of *Schizophyllum commune* Fr. Single basidiospore germlings. *Archiv für Mikrobiologie*, **61**, 232–45.

Wain, W. H., Price, M. F., Brayton, A. R. & Cawson, R. A. (1976). Macromolecular synthesis during the cell cycles of yeast and hyphal phases of *Candida albicans. Journal of General Microbiology*, **97**, 211–17.

Williamson, D. H. (1973). Replication of the nuclear genome in yeast does not require concomitant protein synthesis. *Biochemical and Biophysical Research Communications*, **52**, 731–40.

Wilson, C. L. & Aist, J. R. (1967). Motility of fungal nuclei. *Phytopathology*, **57**, 769–71.

Yanagita, T. (1977). Cellular age in microorganisms. In *Growth and Differentiation in Microorganisms*, eds. T. Ishikawa, Y. Marayama & H. Matsumiya, pp. 1–36. Tokyo: University of Tokyo Press.

16
Mathematical modelling of mycelial growth

J. I. PROSSER

Department of Microbiology, Marischal College, University of Aberdeen, Aberdeen AB9 1AS, UK

Introduction

The application of mathematical modelling techniques to the study of the growth and physiology of bacteria has been developed extensively since the work of Monod (1942). Its use in describing mycelial growth is, however, a relatively recent event and initially involved either direct use of bacteria-based models or modifications of such models, taking into account the mycelial growth form. Models are now appearing which are specific to mycelial growth and it is hoped that these will provide an insight into the fundamental mechanisms of growth.

A mathematical model of a system is a representation of that system in the form of precise mathematical equations. The model immediately provides a quantitative description of the system and the ability to describe complex interactions within the system which might otherwise be conceptually difficult. Mathematical models can be divided into two groups, empirical models and mechanistic models, on the basis of differing philosophical approaches. Empirical models are derived solely by fitting mathematical equations to experimental data. Regression analysis and curve-fitting techniques are used, the final model being that which fits the data most closely. Such models are useful in describing systems within the range of conditions of the experiment, but prediction outside that range is dangerous. Also, such models are only as precise as the experimental data on which they are based and there is always a possibility that a different and even simpler but untried relationship will give a better fit.

Mechanistic models are ideally formed as a subject of experimentation rather than as a result. Such models provide a quantitative hypothesis of a system based on assumptions regarding the fundamental mechanisms controlling the system. Comparison of predicted results with experimental results then provides a means of testing the validity of those assumptions and of discovering the relative importance of different components within the system. Mechanistic models are generally more complex mathematically than empirical models and require more thought by the modeller as to the biology of the system, but predictions outside the range of experimentation may be accepted with more confidence, and the effect of untested factors on a system may be treated by considering their effect on subcomponents of the system if these are known. The quantitative nature of mathematical models and of their predictions provides a much more rigorous test of the assumptions involved than do more usual 'word' hypotheses and previously unsuspected relationships between different factors may be discovered merely by application of mathematical techniques to the equations describing a system. Finally, mechanistic models may focus attention on areas where further study is required by exposing inadequate knowledge of the parameters controlling the system.

Mechanistic models are therefore important in elucidating fundamental mechanisms controlling a system, and can be thought of as explanatory in nature while empirical models provide a description of a system. The distinction between the two groups is, however, not always clear. A mechanistic model may contain an empirically described subcomponent if no suitable mechanism can be proposed for that part of the system, while empirical modelling may indicate areas where mechanistic modelling is required. It is also possible that the relationships described by a successful empirical model may later be found to have a mechanistic basis.

Mathematical models can also be divided into deterministic models, where events occur at predetermined rates, or stochastic models which are statistical in nature and introduce variability and randomness by considering probabilities of events occurring. The former type are more common, because of their simplicity, and the majority of the models discussed below are of this type, consisting of series of differential or algebraic equations. Some, however, also contain stochastic subcomponents and others incorporate 'switches' where a process is initiated by achievement of a threshold value in a certain parameter.

In all models, a balance must be maintained between the desire to model biological systems in all their complexity and the requirement both for predictions that may be tested experimentally and the ability to provide realistic values for all model parameters. This latter is often impossible and estimated values must be used but, as stated above, this can indicate areas where further experimental work is required. It should also be noted that the lack of precision in many biological data makes rigorous testing of models difficult, and as many facets of model behaviour as possible should be tested before acceptance of the assumptions implicit in a model.

Present mathematical models of mycelial growth consider all aspects of growth and adopt both empirical and mechanistic approaches. I have classified these models into those describing mycelial branching patterns, those describing different aspects of growth and models of spore germination. Finally, I will describe a model which attempts to integrate these different aspects of growth in a model which considers the fundamental cellular mechanisms involved in hyphal growth.

Branching patterns

Regular formation of branches by a growing mycelium allows maximum utilization of substrate and particularly of solid substrates. Consequently a point inoculum on a solid substrate typically results in a circular colony containing a mass of branched hyphae. Although analysis of fungal branching has been attempted recently, such analysis is particularly well developed in other fields. Horton (1945) developed a method for quantitative analysis of branching structures in analysing drainage systems of rivers. His method has been modified and applied to a number of systems including branching in trees (Leopold, 1971; Barker, Cumming & Horsfield, 1973; Thornley, 1977).

The analysis involves classification of branches in terms of branch order. First-order branches contain no sub-branches or tributaries. Second-order branches subtend only first-order branches, third-order branches only second-order branches and so on. The length of any second-order branch is that from its origin or branch point to the tip of the longest first-order branch arising from it. Third-order branch length is treated similarly. Leopold (1971) analysed a river drainage network and a number of trees in this way and found a definite linear relationship between the logarithm of the number of branches and branch order and between the logarithm of branch length and branch order. The gradients of these lines are termed branch ratio and length ratio respectively. Branch ratio is the number of branches of a particular order arising from a branch of next highest order while the length ratio represents the ratio of the length of a branch of a particular order to that of a branch of next lowest order.

Values of 3.5 and 2.3 were obtained for branch and length ratios respectively for the river tributary system while branch ratios for a number of trees were in the range 4.7 to 6.5 and length ratios varied from 2.5 to 3.6. Barker *et al.* (1973) carried out a similar analysis on apple and birch trees and tabulated values of branch ratio and length ratio for several systems. They also compared their values with those for arterial and bronchial branching in mechanistic as well as descriptive terms.

Gull (1975) analysed branching patterns of *Thamnidium elegans* Link ex Fries for a third-order system and a fourth-order system. Again, both systems showed a logarithmic relationship between both branch number and branch length when plotted against branch order. Branch ratios of 3.8 and 2.6 were found for the third- and fourth-order systems respectively and length ratios of 4.0 and 2.7. These values are similar to those discussed above for different systems, suggesting an efficiency of branching common to all.

Care must be taken when drawing parallels between such diverse systems. Models of this type are very useful in classifying branching structures and in describing complex systems using just a few parameters. They do not, however, in themselves give information about the mechanisms involved in branching, as branch ratio and length ratio are not related to any particular processes but merely result from experimental data. Indeed it could be argued that a common mechanism is unlikely in systems as diverse as arteries, where resistance to blood flow is an important factor, trees, where attainment of maximum light availability is important, and fungal mycelia, where branching provides a mechanism for utilizing substrate with maximum efficiency. This is illustrated by the development of the above 'state description' to a 'process description' by Thornley (1977) to describe apical bifurcation in trees. This mechanistic model assumes branch formation when apical diameter reaches a certain critical diameter; it is obviously not directly applicable to fungal hyphae, which have constant diameter, although it may be applicable to arterial systems.

However, Uylings (1977) presents an optimization model for lung and vascular tree structures on the basis of power dissipation as a result of frictional resistance of flow through the vessels. This approach may have application to branching of fungal hyphae and this may be one area where empirical models, although not in themselves giving information on mechanisms controlling a system, do suggest possible mechanisms and indicate fruitful areas of study.

A generalized model for generation of branching patterns with direct applicability to mycelial growth is that of Cohen (1967) which provides a link between models concerned merely with overall shape and those concerned with growth. The model defines a number of growth rules. Growth occurs only at the tips of branches, the extent of growth, UG, being given by the equation:

$$UG = \frac{U \cdot DGT - DMIN}{DGT}$$

where U is unit distance and DGT is a limiting density above which no growth can occur. $DMIN$ is the local minimum in the density field which is calculated by sampling at points around the growing tip to determine the proximity of other parts of the pattern, $DMIN$ increasing in proportion to the degree of crowding. Growth angle, GA, is given by the equation:

$$GA = \frac{SA \cdot ST + GRA \cdot GR}{ST + GR}$$

where SA is the angle of the free end of the branch, ST is the inertial factor for growth direction, GRA is the direction of the negative gradient of the local density field and GR is the magnitude of this field. Essentially this introduces autotropism (the tropic response of one hypha to a neighbouring hypha) into the model.

Each branch follows these growth rules, the probability of branch formation being determined by the limiting density for branching (equivalent to DGT above), the distance from the base and the apex of the segment, $DMIN$ and also the result of a random trial, thereby introducing a stochastic element into the model. The branching angle is determined as above (GA) with the addition of a standard branch angle and a persistence factor. Both growth and branching may be biased to introduce directionality.

Continuous iteration of these growth and branching rules produces realistic branching patterns dependent on the model parameters. By varying these parameters, and introducing bias, Cohen (1967) produced a number of different patterns corresponding to tree-like and mycelial structures. Effects of changes in environmental conditions can be investigated by varying parameters during growth. The effects of heterogeneity, introduced for example by a local source of nutrient, can also be investigated. The model, therefore, provides a range of branching patterns on the basis of relatively few parameters while at the same time describing growth. Comparison of quantitative predictions of this model with experimental data on mycelial growth would be valuable.

Models of growth
Hyphal tip growth

Increase in hyphal length proceeds solely as a result of incorporation of material into a portion of the hyphal tip called the extension zone. The rate of wall synthesis within this region and its overall shape have been the subject of a number of mathematical models.

Green & King (1966) developed a model for tip growth in the alga *Nitella,* and Green (1969) and Green (1974) proposed its application to growth of fungal hyphae. Their approach can be explained by reference to Fig. 16.1. The tip is assumed to be hemispherical with radius R, and they consider the change in area and shape of a small, initially circular, area of wall. m represents the length of the meridional arc from the apex to a point on the circle, l the distance around the tip from a reference line and r the length of a perpendicular from the longitudinal axis to the circle. The circle moves from a position near the tip, G, down to a position at H as a result of growth, and will also change shape if rates of expansion in the longitudinal and latitudinal directions are different. This anisotropy is described by the allometric coefficient, K, which is defined as the ratio of the relative rates of growth along the meridian and along the latitudinal direction, i.e.

Fig. 16.1. Diagram showing the parameters of interest during the apparent downward migration, from level G to level H, of a circular region of wall surface on a hemispherical growth zone. (From Green, P. B. & King, A., *Australian Journal of Biological Sciences* 19, 421–437, 1966.)

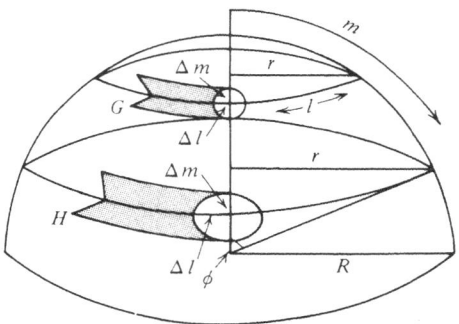

$$\frac{1}{\delta m} \cdot \frac{\mathrm{d}(\delta m)}{\mathrm{d}t} = K \cdot \frac{1}{\delta l} \cdot \frac{\mathrm{d}(\delta l)}{\mathrm{d}t}$$

An expression is then obtained for the rate of increase of distance from the tip of the small circle:

$$\frac{\mathrm{d}m}{\mathrm{d}t} = A\frac{r}{R}K$$

where A is a constant of integration. The rate of increase is therefore proportional to the Kth power of the radial distance from the axis. Integration of this equation gives an expression for m as a function of time:

$$m = 2R \arctan C \exp (At/R)$$

where $\ln C = B/R$.

Green & King (1966) then investigated their model for the case of isotropic expansion (allometric coefficient, K, $= 1$) and for both longitudinal anisotropic expansion ($K = 2$) and latitudinal or transverse anisotropy ($K = 0.5$). The predictions obtained were tested using a simple model consisting of an expanding circular rubber membrane constrained at the rim. Anisotropy was introduced by reinforcing specific regions on the membrane and the processes involved were related to equivalent processes in apical cells of *Nitella*.

Trinci & Saunders (1977) applied this model to tip growth of fungal hyphae and in particular questioned the assumption of a hemispherical extension zone. They measured the extension zones of hyphae of several fungi, with a range of extension rates, and found that the ratio of extension zone length to hyphal radius at the base of the extension zone varied from 2.48 in *Botrytis cinerea* Persoon ex Persoon to 20.45 in sporangiophores of *Phycomyces blakesleeanus* Burgeff. A hemispherical extension zone would give a ratio of 1, so for fungi this assumption is obviously not valid.

Trinci & Saunders (1977) generalized the model of Green & King (1966) by considering a tip of circular cross-section but with otherwise arbitrary shape. For the case of isotropic growth ($K = 1$) they obtained an expression for the specific rate of area expansion:

$$\frac{\mathrm{d}}{\mathrm{d}m}\left(\frac{\mathrm{d}m}{\mathrm{d}t}\right) = \frac{Cb \cos \phi}{(a^2 \sin \phi + b^2 \cos^2 \phi)^{\frac{1}{2}}}$$

where m and t are defined above, C is a constant of integration, a is extension zone length, b is equivalent to R in Fig. 16.1 and ϕ is the angle between the longitudinal axis and the line drawn from the base of the extension zone to the point on the wall being considered (see Fig. 16.1). For a hemispherical extension zone, $a = b$ giving the equation

$$\frac{\mathrm{d}}{\mathrm{d}m}\left(\frac{\mathrm{d}m}{\mathrm{d}t}\right) = \frac{Cb}{a}\cos\phi$$

The specific rate of area expansion is therefore proportional to the cosine of the angle between the longitudinal axis and the point on the tip wall, and this relationship can be obtained from Green & King's (1966) model. For the case of $a^2 \gg b^2$, i.e. an ellipse with large eccentricity, a different relationship is obtained:

$$\frac{\mathrm{d}}{\mathrm{d}m}\left(\frac{\mathrm{d}m}{\mathrm{d}t}\right) = \frac{Cb}{a}\cot\phi$$

The specific rate of area increase is now proportional to the cotangent of ϕ and not the cosine. This predicts a graph of specific rate of area expansion versus distance from the tip apex to be concave-upwards rather than concave-downwards as predicted by a cosine relationship. Substitution of values of a and b similar to those observed for fungal hyphae yields a concave-upwards (cotangent) curve. This fits qualitatively with data on growth of sporangiophores of *Phycomyces blakesleeanus* and on vesicle concentration and incorporation of N-acetylglucosamine in tips of *Neurospora crassa* Shear & Dodge, while the cosine relationship gives no fit. Consideration of the hyphal apex as a half-ellipsoid of revolution rather than a hemisphere improves on the model of Green & King (1966) but still requires modification to give a quantitative fit.

Da Riva Ricci & Kendrick (1972) modelled the physical processes involved in hyphal tip growth. They considered the tip to consist of a cylinder of inelastic set wall capped by a hemisphere of elastic, unset wall capable of being 'blown out' by internal fluid pressure. For normal growth the rate of wall setting must equal that of the blowing out process. Increase in size of a small area of the growing surface occurs as a result of swelling due to internal pressure and is assumed to be proportional to its present area:

$$\frac{\delta\,\mathrm{d}\sigma}{\delta t} = \theta\,\mathrm{d}\sigma$$

where σ is the area under consideration and θ is a coefficient of proportionality. A system of partial differential equations is then derived to describe tip growth based on the expansion function θ. Computer simulation of these equations represents growth as a series of lines describing tip shape at successive time intervals, deriving each line from information on the line at the previous time interval. Two cases are investigated, that in which θ remains constant and that in which it decreases with increasing distance from the tip apex. The former case results in formation of a hypha whose diameter decreases with time. The latter generates a hypha with constant diameter, suggesting a steady decrease in unsetness and intussusception as a particular point moves from the apex to its final position on the hyphal wall.

The mechanism for tip growth proposed here, i.e. expansion of unset material as a result of internal fluid pressure, may be correct even though hemispherical tip shape is assumed, but the situation may be simpler than that described in all the models above. The models of Green & King (1966) and Trinci & Saunders (1977) depend on a parameter K, the allometric coefficient, which has no obvious biological meaning and the model of Da Riva Ricci & Kendrick (1972) introduces an expansion factor, θ, which must vary with distance from the tip in order to give realistic predictions. However, Saunders (1978) points out that uniform hydrostatic pressure applied to a membrane with uniform isotropic elasticity will result in a paraboloid of revolution. If the wall of the extension zone is assumed to consist of unset material with uniform isotropic elasticity then hydrostatic pressure generated by protoplasmic flow towards the tip may in itself give the observed shapes of fungal hyphal tips which can be approximated by paraboloids of revolution as closely as by ellipsoids.

Phototropic growth

Medina & Cerdá-Olmedo (1977) proposed a quantitative model, based on previous qualitative models, for growth of sporangiophores of *Phycomyces blakesleeanus*, which are negatively geotropic and positively phototropic. The growing zone of the sporangiophore tip has constant length and each point on the surface of the growing zone rotates around the longitudinal axis, with angular velocity at a maximum at the tip apex and decreasing to zero where the wall sets at the base of the zone. A symmetrical variation in light intensity results in a transient change in growth velocity, termed photomecism. This can be explained by the existence of an adaptation level, A, within the system which is continuously compared with the external light intensity I. If I remains constant, A approaches I exponentially, stimulation ceasing when $I - A$ is less than the system noise.

Medina & Cerdá-Olmedo (1977) quantify this model by assuming that the growing zone rotates with constant angular velocity and contains photoreceptors (flavin molecules) aligned parallel to the sporangiophore wall and connected to two sets of linear transducers separated by an adaptation mechanism. The first set of transducers produces a signal EI (where E is a constant) which is proportional to the incoming light intensity, I. This is then compared with the adaptor level, A, to give an instantaneous subjective stimulus $S_i = (EI - A)/A$. The adaptation level meanwhile approaches E at a rate given by the equation

$$\frac{\mathrm{d}A}{\mathrm{d}t} = (EI - A)/b$$

where b is a constant. The second set of transducers produces a growth-promoting signal, M, proportional to S, i.e. $M = KS$, where K is a constant and S represents total subjective stimulus.

To describe changes in light intensity within the sporangiophore, the growing zone is assumed to be cylindrical with height m, diameter n and at an angle θ to the horizontal light beam. Total light flux, F, falling on the sensitive zone can be calculated by the equation $F = Imn \sin \theta$. Some of this is scattered but most is refracted to give two bright bands on the opposite surface of the sporangiophore. The intensity at the bands, I_n, is given by the equation

$$I_n = \frac{0.63 \, (\ln \sin \theta)}{2a}$$

where a is the band width.

On the basis of this model the instantaneous subjective stimulus, the stimulus when the photoreceptor hits the first bright band, is calculated and from this the subjective stimulus is given by either of two equations, depending on the relative values of EI_s and A_d:

$$S = 0.63 \, \pi \sin \theta / jC \qquad \text{if } EI_s > A_d$$
$$S = 0.63 \, EI \sin \theta / jA_d \qquad \text{if } EI_s < A_d$$

where I_s is the scattered light intensity and A_d the minimum adaptation level. S therefore depends on light intensity only if I_s is lower than a critical value $I_c = \pi A_d / EC$ and the response is almost entirely due to the first band. Production of growth-promoting substances results in differential growth, the direction of which is determined by the latency of the response and the time taken to travel to the point furthest from the light source. The difference in growth velocity between two sides of the sporangiophore is proportional to the size of the response and also the geotropic response, G, and is given by:

$$\omega = K(P \sin \theta - G \cos \theta) \qquad \text{when } I > I_c$$
$$\omega = K(QI \sin \theta - G \cos \theta) \qquad \text{when } I < I_c$$

where $P = 0.63 \, K_1 \, \pi/jC$, $Q = 0.63EK_1/jA_d$, j is angular velocity and K_1 a constant.

Model predictions are then tested by comparison with class 1 mutants, where geotropic behaviour is normal but phototropism reduced, apparently due to reduced efficiency of an early phototransducer, and class 2 mutants, where photomecisms are smaller than those of the wild type and tropisms much slower. The first type are explained either by assuming a reduction (h) in efficiency of the first linear transducer and replacing E by E/h or by increasing the minimum value of the adaptation level, A_d, by a factor h. Class 2 mutants are explained by replacing K by K_h or by increasing the minimum value of the normal photogeotropic equilibria but taking longer to reach them. Variation of parameters in this way is stated to give predictions in good agreement with experimental results and the model provides a quantitative description of previous qualitative models.

Colony growth

Models of mycelial growth in terms of increase in biomass or total hyphal length tend to be divided into those describing growth in liquid submerged culture and those describing surface growth. Both types of model have been reviewed by Righelato (1975). To a large extent, models of growth in submerged culture are derived from studies on bacterial growth kinetics. Thus, growth of fungi in batch culture follows the same sort of growth curve exhibited by bacterial cultures and continuous culture growth is described by Monod kinetics with added terms for maintenance. The filamentous form provides practical problems, e.g. by increasing viscosity of the culture medium, rather than theoretical problems and these are discussed by Bull & Bushell (1976).

However, the formation of pellets, uncommon in cultures of non-filamentous unicellular organisms, is a common occurrence during growth of some fungi and requires special treatment. This was provided by Pirt (1966) and stated that initial growth was exponential but that eventually growth at the centre became limited by diffusion of substrate into the pellet. The pellet then consisted of an exponentially growing shell surrounding a central non-growing sphere into which nutrients could not diffuse. Pellet growth is now described by the equations:

$$\frac{dx}{dt} = \mu x_p$$

$$\frac{dr}{dt} = \mu \omega$$

where x is pellet mass, t is time, μ is specific growth rate, r is pellet radius and ω and x_p the depth and mass respectively of the outer growing shell. These equations can be combined and modified to give:

$$x^{\frac{1}{3}} = \left(\frac{3}{4\pi d} \right)^{-\frac{1}{3}} \mu \omega t + x_0^{\frac{1}{3}}$$

where d is pellet density and x_0 the initial biomass. The model, therefore, predicts cube root kinetics once the radius becomes greater than ω and this explained such observations by Emerson (1950) and other workers. Pirt (1966) also calculated the depth of the outer peripheral growth zone on the assumption that oxygen was the most likely limiting substrate.

Pirt (1967) also developed a model for surface growth which he applied to bacterial colonies and Trinci (1971) applied this to fungal colonies. Again initial growth is exponential following the equation:

$$\ln r_t = \ln r_0 + \frac{\mu t}{2}$$

where r_t and r_0 are colony radii at times t and zero respectively and μ is specific growth rate. Eventually, diffusion of nutrient into the centre of the colony will limit growth and the colony can then be considered as a non-growing central disc surrounded by an exponentially growing annulus. Growth is now described by the equation:

$$r_t = a\mu t + r_0$$

where a is the width of the peripheral growth zone. This model was tested experimentally by Trinci (1971) and was found to describe the growth of a number of fungi. It also explains the need for care in using colony radial expansion as a parameter of mould growth as certain factors may alter peripheral growth zone width, and hence colony expansion, without affecting specific growth rate (Bull & Trinci, 1977).

Koch (1975) devised a model to describe the kinetics of mycelial growth which was based on the logistic equation. This equation forms the basis for many models of population growth in microbial and higher organism ecology. It has the form:

$$\frac{\mathrm{d}N}{\mathrm{d}t} = \lambda N - \frac{\lambda N^2}{K}$$

which can be solved analytically to give

$$N = N_0\, \mathrm{e}^{\lambda t}\left(\frac{K}{K - N_0 + N_0 \mathrm{e}^{\lambda t}}\right)$$

where N is the number of organisms, t is the time since there were N_0 organisms, λ the maximum growth rate and K the maximum population or carrying capacity of the system. The model is based on the assumption that specific growth rate decreases as a linear function of population size and ceases when the population size equals the carrying capacity. The equation describes best those situations where lack of space or crowding is the factor limiting specific growth rate.

Koch (1975) modified this equation to describe mycelial growth by replacing N, the number of organisms, by $\mathrm{d}W/\mathrm{d}V$, the weight of protoplasm per unit volume element, and N_0 by S the initial mycelial density that inoculated the volume of space $\mathrm{d}V$. t is replaced by $T - t$ where T is the time elapsed since growth of the colony was initiated and t is now the time at which the particular volume element being considered became infected. By further substitution for $\mathrm{d}V$ he obtained the following expression for the case of growth in two dimensions:

$$W = \int_0^T \frac{SK\mathrm{e}^{\lambda(T-t)}}{K - S + S\mathrm{e}^{\lambda(T-t)}} \cdot 2\pi h a^2 t\, \mathrm{d}t$$

where a is the constant rate of increase in colony size.

The most important factor controlling model behaviour is found to be the value of K/S, the ratio of the final hyphal density of a particular volume element to the initial hyphal density inoculating that volume. Simulation of the model for the case of two dimensional growth gives a linear increase in log W with time followed by a curvilinear relationship. The cube root of W also increased linearly for a significant region and the square root of W shows a linear relationship when growth is limited by circular expansion of the colony. The region of cube root linearity appears to represent a transition between exponential and square root kinetics.

Koch (1975) fitted this model to the data of Deppe (1973) on two-dimensional surface growth of several fungi and of Trinci (1970) on three-dimensional growth of mycelial pellets of *Aspergillus nidulans* (Eidam) Winter in submerged culture. The former gave the best fit with values of K/S greater than 100 while the latter was described best with a value of K/S approximately equal to 1. This can be explained by the fact that growth on a surface initially requires contact with the surface and is then followed by growth above the surface which is limited by transport of nutrients upwards. The final hyphal density will consequently be much greater than that initially inoculating a particular area and K/S will be large. The volume surrounding a pellet is invaded simultaneously by hyphae from the whole pellet surface and then has no source of nutrients, so that initial and final hyphal densities are similar and K/S will be approximately 1. The model corresponds to that of Pirt (1966) described above if his growing cortex is considered analogous to the area behind the growing front that has not yet reached carrying capacity. In this respect the model of Koch (1975) is more realistic in that growth within this zone is not constant but shades from growing to non-growing areas. The model is also valuable in providing a unified description of different forms of mycelial growth and in explaining the different kinetics of growth observed under different conditions.

Growth has also been modelled from a biochemical basis by Alberghina & Martegani (1977) and in this volume (Chapter 14). The model is concerned particularly with protein and DNA synthesis and relates growth of *Neurospora crassa* to a model of the dynamics of the cell cycle in mammalian cells (Alberghina, 1977).

Spore germination

Waggoner & Parlange (1974*a*) presented an empirical model for spore germination in *Alternaria solani* Sorauer, the causal agent of potato and tomato early blight, and further developed it in Waggoner & Parlange (1974*b*, 1975). The model is based on three parameters: H, the proportion of the spore population which germinates, $t_{\frac{1}{2}}$, the time taken for half the population to germinate and s^2, the variance of the individual arrival times around $t_{\frac{1}{2}}$. The model was derived from a series of experiments in which *A. solani* spores at 25 °C were subjected to a temperature shock at 45 °C. This introduced three further parameters: t_{45}, the length of exposure to 45 °C, t_d, the delay before heating to 45 °C and t_b, the time between successive exposures when the maltreatment was repeated.

Results for germination of control spores, not subjected to maltreatment, were fitted to a normal cumulative distribution and the effect of maltreatment on H, $t_\frac{1}{2}$ and s_2 was found to fit equations:

$$H = H_0 - t_{45}(0.15 + 0.6t_d)$$

$$t_\frac{1}{2} = t_0 + t_b = t_{45}(15 + 8t_d)$$

$$s^2 = \frac{t_\frac{1}{2}}{12} + t_{45}(2.5 + 10t_d)$$

where H_0 and t_0 are the control values for H and t. Waggoner & Parlange (1974*a*) then construct a 'box' model where spores pass through a number of hypothetical boxes or stages, representing different environmental conditions or different stages in germination, until germination is achieved. Progress through stages is described by equations of the type:

$$\frac{\mathrm{d}C_n}{\mathrm{d}t} = P(C_{n-1} - C_n) - BC_n$$

where C_n is the number of spores in the nth stage, C_{n-1} the number in the $(n-1)$th stage, P the relative rate of passage between stages and B the relative rate of death in a particular stage. The rates of passage and death of spores at 45 °C, P_{45} and B_{45}, are assumed to be constant during the period of exposure but affect spores at different stages of development to different extents. Permanent damage as a result of exposure is modelled by assuming a new value for P_{25} on return to 25 °C (P'_{25}). This model takes the form:

$$P_{25}t_d + P_{45}t_{45} + P'_{25}(t_\frac{1}{2} - t_d - t_{45}) = f$$

$$s^2 = f/P'^{\,2}_{25}$$

$$H = H_0 \exp(-t_{45}B_{45})$$

where f is the number of stages and subscripts 25 and 45 represent values at 25 °C and 45 °C. If spore 'memory' is short and exposure to 45 °C produces no lasting effect, then the first of these equations may be simplified to give:

$$P_{45}\,t_{45} + P_{25}(t_\frac{1}{2} - t_{45}) = f.$$

This second, conceptual, model was then compared with the first, purely empirical, model and the fit between the two suggested that in fact spore memory was relatively short and the simplified equation could be used.

In Waggoner & Parlange (1974*b*) and Waggoner & Parlange (1975) the model is used to investigate the effects of variable moderate temperatures, in the range 4 °C to 29 °C, and of changes between moderately warm (35 °C) and moderately cool (15 °C) temperature. It was found that below 30 °C the rate of spore germination increased linearly with temperature and then decreased from 30 °C to 40 °C, higher temperatures having a lethal effect. The length of the germination path, i.e. the value of *f*, was found to increase when changing temperature between the two temperature ranges with the rate of germination, *P*, being greater at the higher temperature. Despite these discontinuities the model was able to describe spore germination under constant and transient conditions and the second model provides a basis for a mechanistic model which could possibly be applied to other types of maltreatment and to other species.

Lapp & Skoropad (1976) describe an empirical model for conidial germination and appressorial formation in another plant pathogen, *Colletotrichum graminicola* (Cesati) Wilson, the causal agent of anthracnose of grasses. Their model is based on the Gompertz equation:

$$y = AB^{r^x}$$

where *y* is the mortality at time *x* and *A*, *B* and *r* are constants. This equation has been used frequently to describe a variety of biological systems, values for the parameters being obtained by fitting the equation to experimental data. Lapp & Skoropad (1976) used a modification of this equation:

$$y = K \exp(-e^{a-bx})$$

where *y* is the number of spores that germinated or formed appressoria at time *x*; *K*, *a* and *b* are constants. *K* represents the maximum number of spores in the population which have germinated or appressoria which have formed while *a* and *b* control the rate of germination or formation but have no obvious biological significance. By substituting for *y*, on the basis that *a* and *b* define the time taken to reach 99 % of the maximum, Lapp & Skoropad (1976) obtain the expression:

$$x_K = (a - \ln \ln 1.0101)/b$$

where x_K is the time taken to reach the maximum, *K*.

This equation was then fitted using non-linear regression analysis to data on the effect of different day-lengths of low-level light intensities on appressorial formation and conidial germination in *Colletotrichum graminicola* giving values for *a*, *b*, x_K and *K* for each treatment. The equation fitted the data well and it was found that the longest light regime (24 h) reduced the proportion of conidia that can germinate but did not alter the time taken for the maximum number to germinate while there was no significant effect on appressorial formation.

This model demonstrates the usefulness of empirical models in analysing experimental data such that it can be described by a few parameters, making comparisons easy. The parameters do not necessarily have any biological significance, however, and here, for example, do not indicate the mechanism by which conidial germination may be affected by light. This may not be important in areas such as plant pathology, and the models of Lapp & Skoropad (1976) and Waggoner & Parlange (1974a) may, by their simplicity, be more useful in predicting, for example, the effect of application of a fungicide on a disease than a more complex mechanistic model. Empirical models, however, are limited to the range of experimental conditions on which they are based and a successful mechanistic model may be capable of simplification for routine use and can be applied with some confidence outside its tested range.

Integrated model of mycelial growth

None of the models described above considers the fundamental mechanisms involved in hyphal growth at a cellular level. The following is an attempt to construct such a model and to relate cytological events within hyphae to colony growth kinetics. Essentially the model quantifies qualitative theories on hyphal growth proposed by Bartnicki-Garcia (1973), Collinge & Trinci (1974) and Fiddy & Trinci (1976a, 1976b). These can be summarized as the production throughout the mycelium of vesicles containing wall precursors and/or enzymes required for new wall synthesis. These vesicles travel to the hyphal tip where they fuse, thereby increasing the surface area of the hypha and giving elongation. There also exists within the hypha a duplication cycle equivalent to the cell cycle observed in unicellular micro-organisms, with branch initiation analogous to cell division (see Chapter 15). The model is described in more detail by Prosser & Trinci (1979).

Description of the model

In order to model a system as complex as the fungal hypha, a finite-difference model was constructed in which a hypha is considered to consist of two types of hypothetical segments (Fig. 16.2). Tip segments are initially of negligible length and do not produce vesicles but absorb them. The number of vesicles absorbed in time interval δt is given by the equation:

$$R_v = \frac{R_a \delta t V_0}{V_0 + K_v}$$

where R_v is the number absorbed, R_a is the maximum rate of absorption, V_0 is the number of vesicles in the tip compartment and K_v is equivalent to a 'saturation constant' for absorption, being the vesicle concentration at which the rate of absorption is half the maximum. This is similar to the Monod (1942) model for substrate utilization, and merely states that the rate of vesicle absorption is proportional to vesicle concentration at low concentrations and increases to a maximum at high concentrations. Absorption of vesicles results in elongation, E, of the tip segment given by:

$$E = (V_{0(t-\delta t)} - V_{0(t)}) R_1$$

where $V_{0(t)}$ and $V_{0(t-\delta t)}$ are the vesicle concentrations at times t and $t - \delta t$ respectively and R_1 is the increase in hyphal length resulting from absorption of a single vesicle. The tip segment elongates in this fashion until its length is equal to that of a hyphal segment, when a new hyphal segment is formed and the tip segment returns to having negligible length. This process is illustrated in Fig. 16.3.

Hyphal segments have constant length, l_h, and do not absorb vesicles but produce them at a constant rate R_p. The terms apical and intercalary compartments have their usual meanings, the former representing the region from the hyphal tip to the first septum and the latter a length of hypha separated by two septa. The model considers each of these to consist of a number of hyphal segments.

Vesicles produced in hyphal segments flow to the tip segment by direct transfer from one hyphal segment to the adjacent proximal

Fig. 16.2. Division of hypha into compartments and hypothetical segments. The hypha has constant diameter.

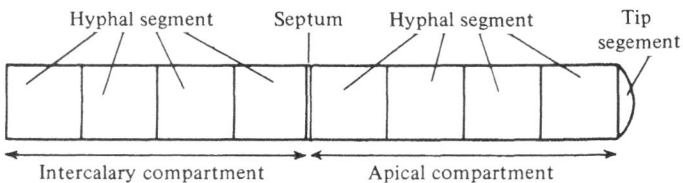

segment. Flow rate is controlled by altering relative numbers of two steps represented by equations of the type:

Step 1

$$\begin{cases} V_{2(t)} = V_{2(t-\delta t)} + R_p \cdot \delta t \\ V_{0(t)} = V_{0(t-\delta t)} - \dfrac{R_a \cdot \delta t \cdot V_{0(t-\delta t)}}{V_{0(t-\delta t)} + K_v} \end{cases}$$

Step 2

$$\begin{cases} V_{2(t)} = V_{3(t-\delta t)} + R_p \cdot \delta t \\ V_{0(t)} = V_{0(t-\delta t)} - \dfrac{R_a \cdot \delta t \cdot V_{0(t-\delta t)}}{V_{0(t-\delta t)} + K_v} + V_{1(t-\delta t)} \end{cases}$$

Subscripts 1, 2 and 3 represent values in hyphal segments 1, 2 and 3. The first equation for each step illustrates the changes in vesicle concentration in the second hyphal segment and the second equation describes changes in the tip segment. In the first step there is no flow

Fig. 16.3. Elongation of a tip segment to give a new hyphal segment.

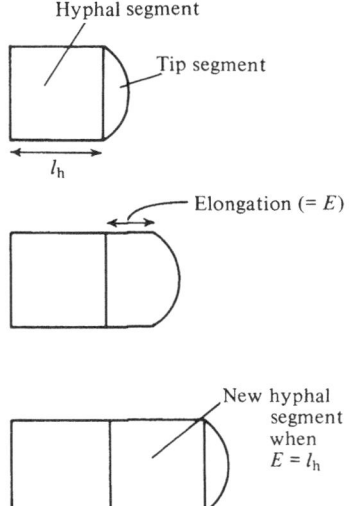

Hyphal segment

Tip segment

l_h

Elongation (= E)

New hyphal segment when $E = l_h$

l_h

and accumulation and depletion of vesicles occur in the hyphal and tip segments respectively. In the second step flow occurs from hyphal segments to adjacent proximal segments, the tip segment being supplied with vesicles from the first hyphal segment. Flow rate can be increased or decreased by decreasing or increasing the ratio of steps 1 to steps 2.

The duplication cycle hypothesis (see Chapter 15) is introduced into the model by consideration of the concentration of nuclear material in the apical compartment. In the absence of nuclear division, this concentration will decrease as elongation proceeds. The model states that nuclear division will occur when the concentration of nuclear material in the apical compartment falls to a critical value. A septum then forms midway in the apical compartment such that the nuclear material is distributed equally to the new intercalary and apical compartments.

The significance of septum formation for this model is its effect on vesicle flow. Thus, septation is represented as a linear decrease in the proportion of vesicles flowing between hyphal segments separated by a developing septum. For complete septa, which lack central pores, or plugged septa, whose central pores have been occluded, this proportion decreases to zero. For septa with unplugged pores a proportion of the vesicles from one intercalary compartment is allowed to flow to the next intercalary compartment.

Branch initiation occurs in the model when vesicles accumulate in a tip or hyphal segment above a certain critical level. Accumulation in a tip segment will result from a rate of supply of vesicles faster than the rate of absorption, and will give rise to apical branching. Accumulation in a hyphal segment may result from prevention or reduction of flow by the presence of a septum. Either case results in formation of a new tip segment and hyphal segment from the segment of accumulation. These then develop as described above, although provision is made for flow of a proportion of vesicles from the parent segment into the branch.

Spore germination is modelled as a means of initiating the whole process. A spore is considered to consist of a hyphal segment which produces vesicles until the concentration reaches a critical value. This results in germination, which is represented by formation of a tip segment. The germ tube hypha may therefore be considered to be a branch from the spore.

Model predictions

The model is capable of predicting changes in hyphal length and numbers and positions of branches and septa on the basis of changes in vesicle and nuclear concentrations. In order to generate such predictions, values must be supplied for all the model parameters; some of these can be calculated or measured, others of them must be estimated. This has been attempted for *Geotrichum candidum* Link ex Persoon from data of Trinci (1974) and Fiddy & Trinci (1976b). This requires slight modification of the model, as septation here usually results in formation of an intercalary compartment half the length of the new apical compartment.

A value of R_a, the maximum rate of vesicle absorption, can be calculated by dividing the rate of increase in hyphal protoplasmic surface area during linear growth by vesicle surface area and this gives a value of 490 vesicles min^{-1}. R_l, the increase in hyphal length resulting from absorption of a single vesicle, was calculated by assuming the hypha to be cylindrical and then taking R_l to be the ratio of vesicle surface area to hyphal circumference:

$$R_1 = \frac{4\pi r_v^2}{2\pi r_h}$$

where r_v is vesicle radius and r_h hyphal radius. This gave a value for *G. candidum* of 0.0048 μm. The rate of vesicle production per segment (R_p) was calculated by first dividing R_a, the rate of absorption during linear growth, by the length of the peripheral growth zone, i.e. the region contributing to growth at the tip. This gave a value of 1.1 vesicles μm^{-1} min^{-1}, which was then multiplied by l_h (hyphal segment length) to give the rate of production per segment.

Experimental data are not available on rates of flow of vesicles nor on K_v, the saturation constant for absorption, nor on V_{max}, the vesicle concentration at which a branch is initiated. However, model predictions were relatively insensitive to changes in these three parameters so that their precise values may not be important. A flow rate of 40μm min^{-1} was used, and values of $K_v = 200$ vesicles and $V_{max} = 500$ vesicles.

Model predictions are illustrated in Fig. 16.4 as increase in total mycelial length and individual branch length and in Fig. 16.5 where total hyphal length, number of branches and hyphal growth unit (defined as total length divided by number of branches) are plotted on log scales. Total hyphal length increases exponentially, while the germ tube hypha and individual branches initially grow exponentially but eventually attain a common linear growth rate. These features were observed experimentally by Trinci (1974) and the linear growth rates and exponential growth rates equal those from experimental data. Initiation of the first branch occurs before the germ tube hypha attains its linear growth rate and branches attain their linear growth rate more rapidly than the germ tube hypha as observed by Fiddy & Trinci (1976*b*).

Number of branches increases exponentially at a rate approximately equal to the specific growth rate and the hyphal growth unit oscillates before reaching a steady value. The nature of the oscillations and the

Fig. 16.4. Increase in total hyphal length (+) and individual branch lengths (■) predicted by the model for *Geotrichum candidum*.

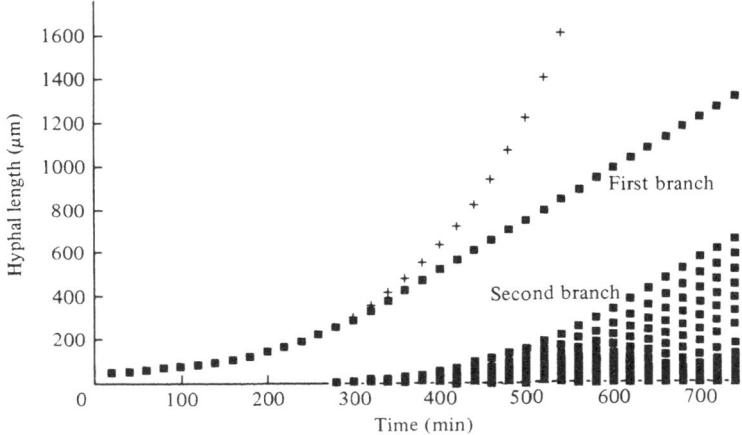

Fig. 16.5. Predicted increases in total hyphal length (O), number of branches (□) and hyphal growth unit (●) for *Geotrichum candidum*.

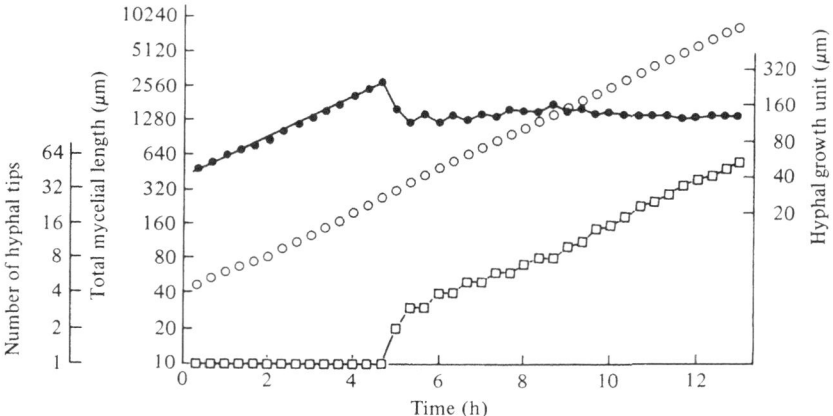

final value reached are in good agreement with observations of Trinci (1974).

The branching pattern predicted by the model is illustrated in Fig. 16.6 assuming a branch angle of 90° and that branching occurs on alternate sides of the parent hypha. The pattern resembles that observed experimentally, in that branches are formed behind septa and branch length and the number of secondary branches decrease towards the tip of the growing germ tube hypha.

This model demonstrates that despite the use of simplifying assumptions and the lack of complete and precise values for parameters of the model, predictions can be obtained in good agreement with experimental data. While not proving the assumptions implicit in the model to be correct, this does indicate that they are not incompatible with observed results on a quantitative basis and that the vesicular hypothesis can account for both the spatial distribution of hyphae within a mycelium and the kinetics of branching. The formulation of such a model highlights areas where precise quantitative data is absent and

Fig. 16.6. Predicted branching pattern for *Geotrichum candidum* assuming a branch angle of 90° and branching on alternate sides of parent hyphae.

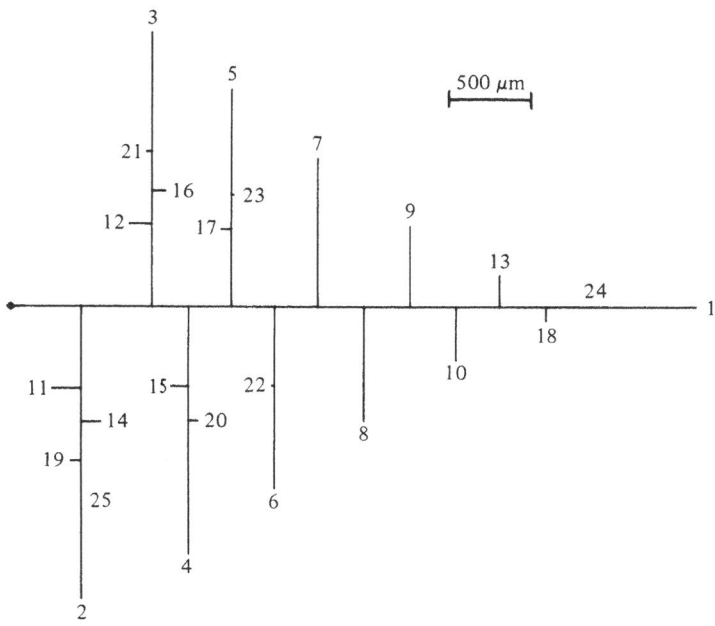

where information is required and assesses the relative importance of these different areas. Thus, while there are not experimental values for K_v, the model is insensitive to changes in K_v, within a reasonable range, suggesting that further work initially be concentrated in other areas.

Conclusion

Mathematical models are seen to operate at several levels of mycelial growth. Their main use has been to quantify previously qualitative hypotheses and, in so doing, to state the assumptions implicit in these hypotheses in a more precise fashion. This results in a more critical appraisal of such assumptions and a more rigorous test of hypotheses. Such testing can be limited by two factors. The first is a fault of the theoretician in introducing concepts or parameters or generating predictions which are impossible to measure experimentally. The second limitation is the lack of precision in many biological data which often results in an inability to distinguish between two different models. This is unimportant in the case of empirical models, where any number of equations may theoretically be valid. However, where two different mechanistic models are proposed for the same process, a distinction must be made on the basis of predictions to determine which process is correct. In this situation modelling can suggest critical experiments to distinguish alternatives and areas in which further work should be concentrated.

The integration of models representing the different levels and aspects of mycelial growth has not yet been achieved, although an attempt has been made to describe colony growth as a function of cellular mechanisms. It is conceivable, however, that low levels of model be used to determine the important factors which must be considered by models at a higher level. For example, models describing hyphal tip growth may indicate important parameters for colony growth.

The models described generally serve one of two functions. The first, which is represented by the models of spore germination, may be considered a 'technological' function, where the model exists only to predict the effect of certain environmental conditions. The second function may be termed scientific, in that the model serves as a hypothesis to be tested. Ideally such models should be constructed initially and then used to provide 'rule of thumb' relationship for practical use if required. Mechanistic modelling is more difficult both experimentally and theoretically but provides a powerful technique for the study of quantitative aspects of mycelial growth and physiology.

Acknowledgements. I would like to acknowledge receipt of a NERC Postdoctoral Research Fellowship during the period of this work.

References

Alberghina, L. (1977). Dynamics of the cell cycle in mammalian cells. *Journal of Theoretical Biology,* **69,** 633–43.

Alberghina, L. & Martegani, E. (1977). Modeling *Neurospora* growth. *Neurospora Newsletter,* **24,** 10–11.

Barker, S. B., Cumming, G. & Horsfield, K. (1973). Quantitative morphometry of the branching structure of trees. *Journal of Theoretical Biology,* **40,** 33–43.

Bartnicki-Garcia, S. (1973). Fundamental aspects of hyphal morphogenesis. In *Microbial Differentiation* (23rd Symposium of the Society for General Microbiology) eds. J. M. Ashworth & J. E. Smith, pp. 245–67. Cambridge University Press.

Bull, A. T. & Bushell, M. E. (1976). Environmental control of fungal growth. In *Filamentous Fungi,* vol. 2, eds. J. E. Smith & D. R. Berry, pp. 1–13. London: Edward Arnold.

Bull, A. T. & Trinci, A. P. J. (1977). The physiology and metabolic control of fungal growth. *Advances in Microbial Physiology,* **15,** 1–84.

Cohen, D. (1967). Computer simulation of biological pattern generation processes. *Nature (London),* **216,** 246–8.

Collinge, A. J. & Trinci, A. P. J. (1974). Hyphal tips of wild-type and spreading colonial mutants of *Neurospora crassa. Archives of Microbiology,* **99,** 353–68.

Da Riva Ricci, D. & Kendrick, B. (1972). Computer modelling of hyphal tip growth in fungi. *Canadian Journal of Botany,* **50,** 2455–62.

Deppe, C. S. (1973). Protein degradation in *Schizophyllum commune.* Ph.D. thesis, University of Harvard, USA.

Emerson, S. (1950). The growth phase in *Neurospora* corresponding to the logarithmic phase in unicellular organisms. *Journal of Bacteriology,* **60,** 221–3.

Fiddy, C. & Trinci, A. P. J. (1976*a*). Mitosis, septation, branching and the duplication cycle in *Aspergillus nidulans. Journal of General Microbiology,* **97,** 169–84.

Fiddy, C. & Trinci, A. P. J. (1976*b*). Nuclei, septation, branching and growth of *Geotrichum candidum. Journal of General Microbiology,* **97,** 185–92.

Green, P. B. (1969). Cell morphogenesis. *Annual Review of Plant Physiology,* **20,** 365–94.

Green, P. B. (1974). Morphogenesis of the cell and organ axis – biophysical models. *Brookhaven Symposia in Biology,* **25,** 166–90.

Green, P. B. & King, A. (1966). A mechanism for the origin of specifically oriented texture in development with special reference to *Nitella* wall texture. *Australian Journal of Biological Sciences,* **19,** 421–37.

Gull, K. (1975). Mycelium branch patterns of *Thamnidium elegans. Transactions of the British Mycological Society,* **64,** 321–4.

Horton, R. E. (1945). Erosional development of streams and their drainage basins: hydrophysical approach to quantitative morphometry. *Bulletin of the Geographical Society of America,* **56,** 275–370.

Koch, A. L. (1975). The kinetics of mycelial growth. *Journal of General Microbiology,* **89,** 209–16.

Lapp, M. S. & Skoropad, W. P. (1976). A mathematical model of conidial germination and appressorial formation for *Colletotrichum graminicola. Canadian Journal of Botany,* **54,** 2239–42.

Leopold, L. B. (1971). Trees and streams: the efficiency of branching patterns. *Journal of Theoretical Biology,* **31,** 339–54.

Medina, J. R. & Cerdá-Olmedo, E. (1977). A quantitative model of *Phycomyces* phototropism. *Journal of Theoretical Biology*, **69**, 709–19.

Monod, J. (1942). *Recherches sur la Croissance des cultures Bacteriennes*. Paris: Hermann et Cie.

Pirt, S. J. (1966). A theory of the mode of growth of fungi in the form of pellets in submerged culture. *Proceedings of the Royal Society, London*, **B166**, 369–73.

Pirt, S. J. (1967). A kinetic study of the mode of growth of surface colonies of bacteria and fungi. *Journal of General Microbiology*, **47**, 181–97.

Prosser, J. I. & Trinci, A. P. J. (1979). A model for hyphal growth and branching. *Journal of General Microbiology*. **111**, 153–64.

Righelato, R. C. (1975). Growth kinetics of mycelial fungi. *Filamentous Fungi*, vol. 1, eds. J. E. Smith & D. R. Berry, pp. 79–103. London: Edward Arnold.

Saunders, P. T. (1978). Models of tip growth. *Bulletin of the British Mycological Society*. **12**, 117.

Thornley, J. H. M. (1977). A model of apical bifurcation applicable to trees and other organisms. *Journal of Theoretical Biology*, **65**, 165–76.

Trinci, A. P. J. (1970). Kinetics of growth of mycelial pellets of *Aspergillus nidulans*. *Archiv für Mikrobiologie*, **73**, 353–67.

Trinci, A. P. J. (1971). Influence of the width of the peripheral growth zone on the radial growth rate of fungal colonies on solid media. *Journal of General Microbiology*, **67**, 325–44.

Trinci, A. P. J. (1974). A study of the kinetics of hyphal extension and branch initiation of fungal mycelia. *Journal of General Microbiology*, **81**, 225–36.

Trinci, A. P. J. & Saunders, P. T. (1977). Tip growth of fungal hyphae. *Journal of General Microbiology*, **103**, 243–8.

Uylings, H. B. M. (1977). Optimisation of diameters and bifurcation angles in lung and vascular tree structures. *Bulletin of Mathematical Biology*, **39**, 509–20.

Waggoner, P. E. & Parlange, J.-Y. (1974a). Mathematical model for spore germination at changing temperature. *Phytopathology*, **64**, 605–10.

Waggoner, P. E. & Parlange, J.-Y. (1974b). Verification of a model of spore germination at variable, moderate temperatures. *Phytopathology*, **64**, 1192–6.

Waggoner, P. E. & Parlange, J.-Y. (1975). Slowing of spore germination with changes between moderately warm and cool temperatures. *Phytopathology*, **65**, 551–3.

17
The kinetics of mycelial growth

R.C.RIGHELATO
Tate & Lyle Group Research and Development, PO Box 68, Reading, RG6 2BX, UK

Introduction

A fungal mycelium is composed of a system of branching hyphae, each of which consists of a linear series of compartments bounded by septa. Each compartment usually contains a number of nuclei (i.e. it is a coenocyte) and cytoplasmic continuity between adjacent compartments may be maintained via pores in the septa; these pores may become plugged in time.

Few people would now dispute that fungi, like other microbes, grow exponentially when the medium contains an excess of the substrates necessary for growth and when inhibitors are not present. Exponential growth means that the rate of increase in biomass is a constant function of the biomass present, the function being the specific growth rate i.e. $dx/dt = \mu x$. It implies that all, or a constant proportion, of the cells contribute to growth, and do so homogeneously with respect to time. Whilst it is easy to accept this for unicellular microbes, whose mass increases approximately twofold and which then divide into two similar smaller cells, one can less readily accept the dogma of homogeneous cultures and homogeneous exponential growth for filamentous fungi. Microscopic examination, even of so-called homogeneous cultures, reveals an array of hyphal fragments varying in size and cytoplasmic contents. In this contribution I shall review briefly the mechanisms by which it is supposed that exponential growth occurs and then move on to the constraints on it, imposed largely by the hyphal growth habit. These constraints are significant in the design and interpretation of experiments on fungal physiology and are of over-riding importance in the fermentation industry, which relies heavily on products derived from filamentous fungi.

Various aspects of the kinetics of fungal growth have recently been reviewed by Righelato (1975), Bull & Bushell (1976) and Bull & Trinci (1977).

Exponential growth

Exponential increase in biomass has been observed in most groups of fungi when the organisms are grown at relatively low concentrations in excess of nutrients. In some cases studied in detail, the increase in biomass is a paralleled by an increase in the total length of the hyphal system and in the number of growing tips (Caldwell & Trinci, 1973). In batch cultures it is difficult to observe exponential increases in biomass for more than about five doublings but the fact that it is possible to obtain steady-state continuous cultures of fungi demonstrates that exponential growth can be maintained indefinitely. The rates of growth achieved by fungi in batch culture under conditions of excess nutrients can be quite high, only a little lower than those achieved by some of the fastest-growing bacteria under similar conditions (Table 17.1).

Table 17.1. *Growth rates of some fungi in batch culture (Forage & Righelato 1979)*

Species	Maximum specific growth rate, μ_{max} (h^{-1})	Doubling time (h)	Incubation temperature (°C)
Neurospora crassa Shear & Dodge	0.35	1.98	30
Fusarium graminearum Schwabe	0.28	2.48	30
Aspergillus nidulans (Eidam) Winter	0.36	1.93	37
Aspergillus niger van Tiegham	0.30	2.31	30
Geotrichum candidum Link ex Persoon	0.35	1.98	25
Penicillium chrysogenum Thom.	0.28	2.48	25
Achlya bisexualis Coker	0.80	0.87	24
Saccharomyces cerevisiae Hansen	0.45	1.54	30
Candida utilis (Henneberg) Loder et Kreger-van Rij	0.40	1.73	30

In defining the specific growth rate (μ) as $1/x \cdot dx/dt$ the implicit simplification is made that all of the mycelial mass (x) contributes equally to growth. However, microscopic examination shows that hyphae of *Penicillium chrysogenum* grown at low specific growth rates exhibit more vacuolation and other signs of differentiation than do hyphae grown at high specific growth rates (Righelato *et al.*, 1968). The protein content of the mycelia fell and the carbohydrate content increased at lower specific growth rates (Righelato, 1967), consistent with a decrease in the ratio of cytoplasm:hyphal wall. In addition, hyphal compartments showing signs of lysis were always present in the cultures of *P. chrysogenum* that we studied (Trinci & Righelato, 1970). The assumption of equal contribution of all hyphal compartments to growth cannot therefore be supported, and other models must be considered. Two examples are shown below using, for simplicity, continuous culture kinetics.

(1) Constant death rate (k) when all hyphal compartments have an equal chance of death and all viable compartments grow at the same rate:

$$Dx = \mu x^* \text{ and } Dx^* = \mu x^* - kx^*$$

Proportion of viable compartments

$$\frac{x^*}{x} = \frac{D}{D+k}$$

Specific growth rate of viable compartments

$$\mu = D + k$$

where D = dilution rate, x = total number of compartments, x^* = number of viable compartments.

(2) Cells die at age t, all viable compartments grow at same rate:

Overall productivity

$$Dx = \mu x(1 - e^{-Dt})$$

Proportion of viable compartments

$$\frac{x^*}{x} = (1 - e^{-Dt})$$

Specific growth rate of viable compartments

$$\mu = D\frac{x}{x^*}$$

These models imply that the proportion of non-viable hyphal compartments increases with decreasing dilution rate or 'overall specific growth rate'. More complex models could be generated in which the specific growth rate varies with cell age. However, it is difficult to test such models, as the proportion of viable material in a hypha cannot readily be measured. Moreover, at low specific growth rates, differentiation may occur. Conidiation has been observed in *Penicillium chrysogenum* and *Aspergillus nidulans* (Righelato *et al.*, 1968; Ng, Smith & McIntosh, 1973). In *P. chrysogenum*, conidiation was induced at growth rates between 0 and 0.011 h^{-1} and the rate of accumulation of conidia in the medium was related to the specific growth in that range and to the previous growth rate in 'shift-down' experiments (Righelato *et al.*, 1968).

It is thought that exponential growth of a mycelium is maintained by assimilation of substrates and synthesis of macromolecules and wall precursors over all or a large part of the mycelium, whilst extension only occurs at hyphal tips; the number of tips increases to maintain a constant ratio of tips to hyphal length (G, the hyphal growth unit) in the mycelium. The hypothesis of Katz, Goldstein & Rosenberger (1972) based on observations of *Aspergillus nidulans* provides a useful model. They concluded that:

   (i) the rate of extension at the tips of short hyphae is proportional to both the hyphal length and the specific growth rate;

   (ii) each tip can extend at a rate between a low value and a maximum value characteristic of the organism and independent of the mass specific growth rate;

   (iii) a new tip is formed when the capacity of the mycelium to extend exceeds that of the existing tips.

The specific growth rate can be described as the ratio of the mean tip extension rate, E ($\mu\text{m h}^{-1}$) and the hyphal growth unit, G (μm), the mean length of hypha per growing tip

$$\mu = \frac{E}{G}$$

The conclusions of Katz *et al.* (1972) were derived from studies of branching of mycelia grown on media supporting three different specific growth rates. The use of different media in these experiments permitted a complicating possibility of the maximum rate of extension of the tips being influenced by the medium, as well as the maximum specific growth rate of the mycelia. A more rigid test would be the use of chemostat continuous culture to vary the growth rate without altering medium composition. Limited data with *Penicillium chrysogenum* appear to support the hypothesis that branching frequency, rather than tip extension rate is the prime variant with growth rate (Morrison & Righelato 1974). However, Metz (1976), with another strain of *P. chrysogenum* in chemostat culture, found that the total length of hyphal fragments increased with growth rate but the hyphal growth unit did not vary.

When the specific growth rate is varied by temperature, the hyphal growth unit remains constant (Caldwell & Trinci, 1973; Trinci, 1973*a*). Trinci (1973*a*) also found with *Neurospora crassa* Shear & Dodge that several inhibitors of specific growth rate had little effect on the hyphal growth unit, but considerably reduced the colony radial growth rate. In these cases presumably the *rate of tip extension is the main variant* with growth rate. One might also expect there to be conditions which restrict the tip extension rate but not specific growth rate, in which case branching frequency would be the main variant with growth rate. The presence of sorbose (Trinci & Collinge, 1973; Galpin, Jennings & Thornton, 1977) and the mutations affecting the spreading colonial morphology of *N. crassa* (Trinci, 1973*b*) and the hyphal growth unit of *P. chrysogenum* (Morrison & Righelato, 1974) perhaps fall into this category. The observation by Pirt (1973), that colony radial growth rate, hence presumably tip extension rate, decreased with decreasing substrate concentration on agar, is difficult to reconcile with the observations of Katz *et al.* (1972) and Morrison & Righelato (1974) in submerged culture. It is however, in accord with the observation of Metz (1976) that the hyphal growth unit was independent of specific growth rate. It appears that either hyphal tip extension rate or branching frequency may vary with growth rate, depending on the strain and conditions.

The effect of morphology on growth kinetics in submerged culture

The morphology of fungi in submerged culture varies from unicells to long, infrequently branched filaments to dense pellets formed of tightly branched hyphae. Often the same strain can grow in both the pellet and the filamentous form; the environmental conditions which induce growth in one form or the other are poorly understood (Whittaker & Long, 1973; Metz, 1976). The morphological factors which influence the macroscopic morphology of hyphal flocs are probably branching frequency, hyphal rigidity and fragmentation; high degrees of branching and rigidity and low rates of fragmentation favouring the pellet form. Aggregation of hyphal flocs into larger units may also play a part in determining the development of pellets (Galbraith & Smith, 1969). The morphology of the hyphal flocs influences growth kinetics in submerged culture in two main ways: by determining the rheology of the culture and through diffusion limitation of substrate assimilation within the flocs.

Rheology

Rheology is normally expressed in terms of viscosity (resistance to flow), flow behaviour (the relationship between viscosity and shear rate) and yield stress (the force required to make the static fluid flow). Cultures of microbes usually consist of a continuous aqueous phase, which alone behaves like water (that is, it has low viscosity, Newtonian flow behaviour and no significant yield-stress); and a discontinuous organism phase, which determines the gross rheology of the culture. Cultures containing approximately spherical microbial bodies, yeast-like and pelletted mycelia, exhibit low viscosities and more or less Newtonian flow-behaviour (Fig. 17.1). By contrast, mycelia growing in long filaments cause high viscosities and exhibit shear-thinning (pseudoplastic) flow behaviour (Fig. 17.1). Filamentously growing cultures also exhibit a yield stress. Both apparent viscosity (i.e. the viscosity at any

Fig. 17.1. Approximate relationship between shear rate and apparent viscosity for filamentous and pelleted broths.

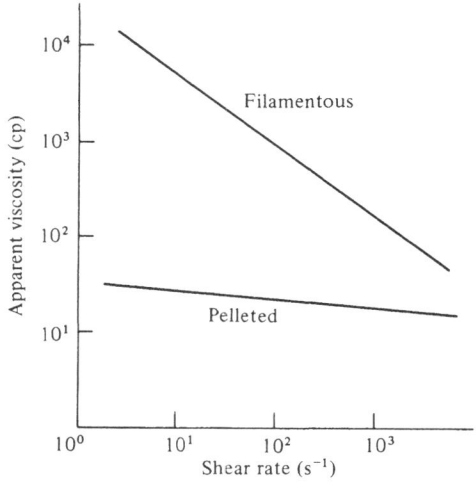

specified shear rate) and yield stress are a function of mycelial concentration. Roels, van den Berg & Voncken (1974) found that yield stress of filamentously growing *Penicillium chrysogenum* was proportional to (mycelium concentration)$^{2\cdot5}$ and a similar relationship was found by Metz (1976). Several authors have found that the apparent viscosity of cultures of filamentously-growing moulds is approximately related to the square of the mycelium concentration (Carilli *et al.*, 1961; Solomons & Weston, 1961; Roels *et al.*, 1974; Metz, 1976). Metz (1976), in an excellent study of the rheology of cultures of *P. chrysogenum*, was able to relate the rheological parameters to morphological characteristics of the hyphal units. The yield stress was related to the longest length of hypha in the units and the viscosity was related to the branching frequency:

$$\tau_0 = K_1 \cdot x^{2\cdot5} \cdot l^{0\cdot8}$$

$$\eta_\infty = (K_2 \cdot x \cdot G^{0\cdot6})^2$$

where τ_0 = yield stress; x = mycelium concentration; l = length of main hypha divided by diameter; η_∞ = viscosity at infinitely high shear rate, extrapolated from Casson equation; G = hyphal growth unit; K_1 and K_2 are constants. Metz (1976) also showed that the rigidity of the hyphae influenced viscosity and yield stress.

The transfer of oxygen into cultures is markedly influenced by the broth rheology; the oxygen transfer coefficient falls with increasing viscosity (Fig. 17.2) and filamentously growing cultures can quickly become oxygen-limited (Banks, 1977; Taguchi, 1971). The poor oxygen transfer is probably a result of low rates of formation of bubbles or increased rates of bubble coalescence, and decreased flow rates of the broth around the fermenter. When a culture becomes oxygen limited, growth is stoichiometrically related to the quantity of oxygen transferred. In a batch culture, the increasing mycelium concentration causes a rapid decline in oxygen transfer and the growth rate quickly slows to zero (Fig. 17.3). In an oxygen-limited continuous culture, because the total quantity of oxygen transferred per unit of medium decreases with decreasing residence time, the mycelium concentration falls at higher dilution rates (Fig. 17.4). To overcome the effects of oxygen limitation in industrial fermentations, water may be added, diluting the broth and hence decreasing the viscosity and increasing oxygen transfer rate (Taguchi, 1971).

In addition to the generally low rates of oxygen transfer, the

Fig. 17.2. Effect of mycelium concentration in cultures of *Penicillium chrysogenum* (a) on oxygen transfer coefficient (K_La), and (b) apparent viscosity. (a) after Deindoerfer & Gaden (1955), (b) after Bongenaar *et al.* (1973).

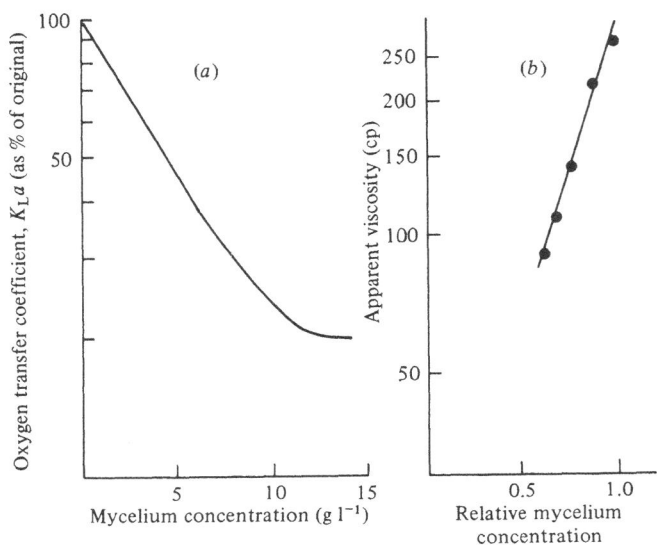

Fig. 17.3. Oxygen-limitation in batch cultures.

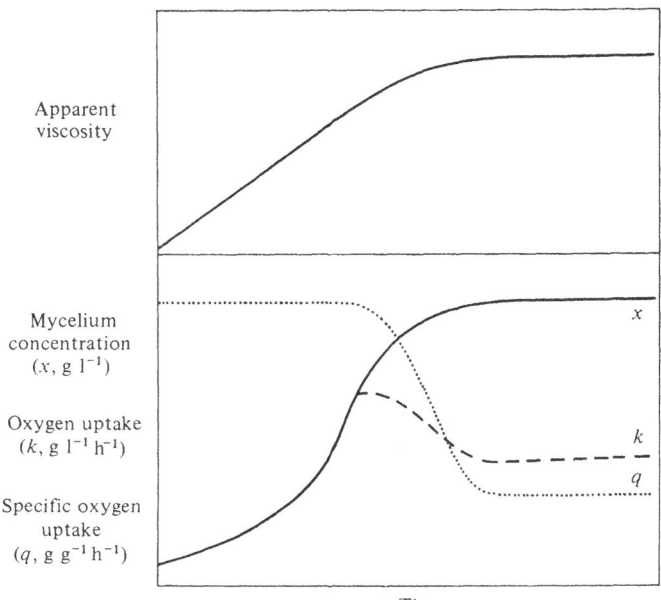

pseudoplastic flow behaviour and the yield stress can cause considerable heterogeneity in the mixing conditions in the fermenter. In the regions of high shear, close to the impellers, the broth exhibits relatively low viscosities and turbulent flow; moving out of the high shear region the viscosity increases, the flow becomes laminar, and close to surfaces such as baffles and vessel walls, stationary zones occur. Individual units of mycelium reside for variable periods in the well aerated zone and the relatively stagnant zone which can often be deficient in oxygen. In addition to the oxygen deficiency, stagnant layers around heat-transfer surfaces impede removal of heat. In continuous cultures, mycelium in the poorly mixed zones may be less likely to be washed out, hence anomalous kinetics may be observed (Fig. 17.5). These problems may be more severe on a laboratory scale than in larger reactors, as the ratio of surfaces to volume is larger. To overcome these mixing problems, high-power inputs and mixers designed for high pumping rates are used.

Diffusion limitation

Mixing in a fermenter is applied, *inter alia*, to bring substrates as close as possible to the cells through turbulent flow. Aggregates of

Fig. 17.4. Oxygen-limitation in continuous culture. Broken line indicates cell concentration and oxygen uptake rate in oxygen-sufficient cultures.

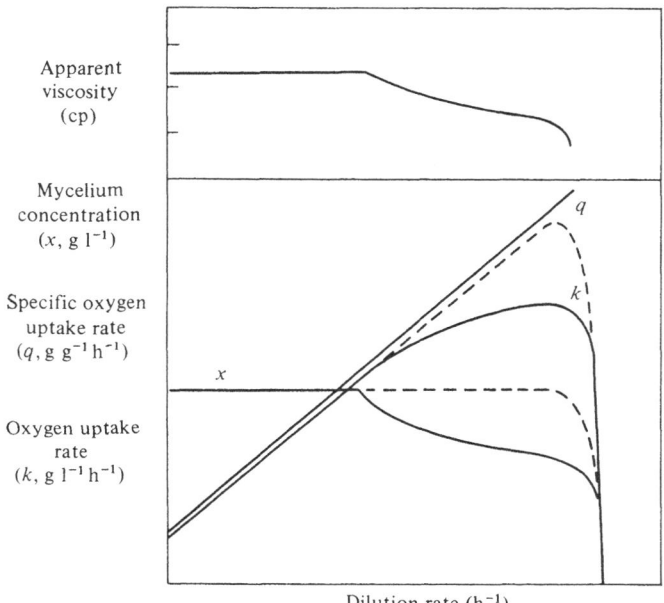

mycelia prevent turbulent flow of the medium within them and hence the availability of substrates to hyphae within flocs is controlled by diffusion. Two types of mycelial flocs are observed: loose, unstable aggregates of filamentously growing hyphae and the denser, more stable, pellet growth form. Metz (1976) observed that aggregation of filamentously growing hyphae into larger flocs occurred within about 0.5 s of removal from a high shear region. The flocs appeared 10^2 to 10^3 times larger than the component hyphae. Within such a floc, turbulence is unlikely to contribute significantly to substrate transfer, which would therefore be determined by diffusion until the floc is again disaggregated in a high-shear region such as the impeller zone. The pellet growth form produces flocs which do not disaggregate in fermenters and hence substrate transfer to cells within a pellet is always diffusion-limited. Pirt (1966) showed that for pellets of such a size that the rate of diffusion of substrate inwards was lower than the demand of the mycelium for the substrate, only a peripheral zone of the pellet would contribute to growth and a cube-root growth law would apply:

$$x^{\frac{1}{3}} = x_0^{\frac{1}{3}} + K\mu wt$$

where μ is specific growth rate, w is the peripheral growth zone of the pellet, x is the biomass after t hours and x_0 the biomass at the start;

Fig. 17.5 Cell concentration as a function of dilution rate for perfectly mixed cultures (continuous line), for imperfectly mixed cultures (broken line) and for cultures with wall growth (dotted line).

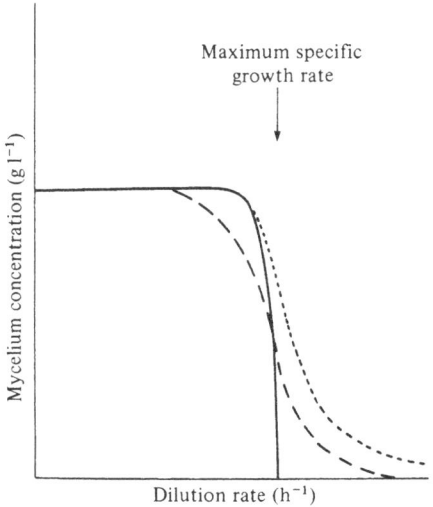

$K = (4\pi\rho n/3)^{\frac{1}{3}}$, where ρ is biomass density and n the number of pellets, assumed to be of equal diameters.

Where the substrate affinity constant is small, as is generally the case for substrates such as oxygen and glucose, the hyphae deeper into the pellet than the peripheral growth zone would be virtually starved of substrate. Hence mycelium in the centre of pellets is frequently seen to show signs of autolysis.

Oxygen is the substrate most commonly found to be limiting, as it is sparingly soluble in water so that the maximum concentrations that can be achieved outside the flocs are low, *c*. 0.25 mM with air at STP. The relation between oxygen concentration and respiration rate for pellets of different sizes and for homogenized pellets is shown in Fig. 17.6. The larger the pellet size the higher is the concentration necessary to achieve a given respiration rate and the lower the maximum rate achieved. Thus, although the filamentous fungi have high affinity oxygen uptake mechanisms – for *Aspergillus niger* $K_s = 3$ μM (Kobayashi, van Dedem & Moo-Young, 1973) and for *Penicillium chrysogenum c.* 6 μM (Mason & Righelato, 1976) – much higher dissolved oxygen concentrations are required to avoid oxygen limitation under fermentation conditions.

Fig. 17.6. Oxygen uptake rates of pellets of *Aspergillus niger*, redrawn from Kobayashi, van Dedem & Moo-Young (1973).

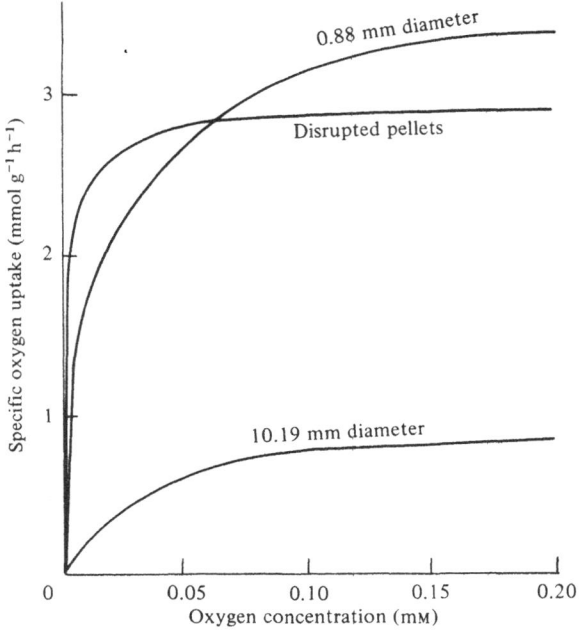

Pellet or floc diameter, biomass density and oxygen demand determine the onset of oxygen limitation. Pirt (1966) calculated that for dense pellets, growing exponentially at a rate of $0.1\ h^{-1}$, oxygen diffusion would become limiting at a diameter of 0.2 mm at an external oxygen concentration of 0.2 bar; cube root growth would ensue.

The assumption has been made here that oxygen transfer into pellets relies entirely on molecular diffusion. However, Miura & Miyamoto (1977) suggested that pellets should not necessarily be regarded as rigid spheres and that deformation of pellets could account for the high effective diffusivity of oxygen which they had observed.

Fragmentation of mycelium

Hyphal extension, branching and pellet formation have been studied in many laboratories, but very little is known of the mechanisms which cause hyphae to fragment. To obtain stable prolonged batch cultures or continuous cultures, new hyphal fragments or new pellets must be formed from existing flocs.

Fragmentation of pellets is thought to involve two shear-dependent processes: chipping of pieces of hyphae from the outside of the pellet and total rupture of the pellet (Taguchi, 1971). In a non-growing system chipping resulted in a progressive decrease in diameter which was related to the impeller tip velocity and the pellet diameter. Rupture, which also occurred, caused a first-order decrease in pellet numbers. Taguchi (1971) hypothesized that rupture was a function of passage through the high-shear region of the impeller. The Reynolds stress at the impeller was estimated to be of the same order as the tensile stress of the pellets, measured as $0.1\ kg\ cm^{-2}$. The tensile strength decreased with repeated loading; the number of exposures to stress before rupture being (local maximum shear stress/tensile strength of pellet)$^{-5.77}$. Loss of pellets may also result from a change in the morphology from tightly branched to less branched filamentous growth. Clark (1962) reported such a change resulting from oxygen deficiency. Whatever the mechanisms of breakdown of pellets, fragmentation and reformation in such a way as to permit steady-state continuous cultures has not been reported.

Fragmentation of filamentously growing mycelia occurs in submerged culture and produces mycelia with a wide spread, both of length and hyphal growth unit (Metz, 1976). It has been suggested that fragmentation in submerged culture is due to shear forces imparted by the impeller (e.g. Caldwell & Trinci, 1973). Whilst shear forces may be involved, it is unlikely that they are sufficient to account for the observed breakup of hyphae. Similar mycelial lengths are observed in shake-flask and stirred fermenters, although the shear rates are presumably much smaller in the former. Although somewhat smaller growth units were observed in shake flasks than chemostat cultures of *Penicillium chrysogenum* (Morrison & Righelato, 1974), the small differences could be accounted for by a difference in hyphal diameter. Metz (1976) observed only a small effect of shear on the length of the longest hyphae of units of another strain of *P. chrysogenum*. In some species, such as *Geotrichum candidum* (Caldwell & Trinci, 1973) and *Cephalosporium acremonium* Corda (Drew, Winstanley & Demain, 1976) dimorphism occurs, apparently under the control of nutrients or inhibitors. In such cases the organisms break up into similarly sized yeast-like units, without the imposition of high rates of shear. The mechanisms of dimorphism have been discussed extensively (see Smith & Galbraith (1971) for a review). In species which do not possess such mechanisms, fragmentation may be the result of autolysis of some hyphal compartments and degradation of the wall of the lysed compartment. Occasional compartments showing advanced signs of autolysis were observed in the middle of intact, apparently growing, hyphae of *P. chrysogenum* (Righelato *et al.*, 1968; Trinci & Righelato, 1970). Control of the fragmentation process would be of considerable practical value, since hyphae fragmented into shorter units produce less viscous cultures which, therefore, have higher oxygen transfer rates (Table 17.2). Moreover, cultures of short hyphal fragments are less likely to be subject to diffusion limitation of growth than cultures of pellets or other dense flocs.

Growth kinetics in surface culture

The basic kinetics of hyphal extension and branching which have been discussed above are probably similar in surface and sub-

Table 17.2. *Effect of morphology of* Penicillium chrysogenum *on viscosity and* $K_L a$ *(Carilli* et al., *1961)*

		Dry weight $(g\ l^{-1})$	Viscosity (arbitrary units)	$K_L a$ (h^{-1})
Strain 1	Short filaments 50 to 100 μm	2.1	4.7	210
Strain 2	Long filaments 200 to 400 μm	2.3	8.6	120

merged culture. However, the relative ease with which fungi can be grown on agar surfaces makes an understanding of colony growth kinetics valuable for designing simple physiological experiments.

Linear growth of colony diameter is a well known feature of growth on agar; the problem has been relating this character to the exponential growth observed in submerged culture. Using similar reasoning to that applied to the growth of pellets in submerged culture, Pirt (1967) developed a general model for colony growth and applied it to bacterial colonies. This was refined by Trinci (1971) for growth of filamentous fungi. It is assumed that initially the growth of the mass of a colony is exponential, i.e.

$$x_t = x_0 \, e^{\mu t} = H\pi dr_t^2$$

$$= (\pi H dr_0^2)e^{\mu t} \tag{1}$$

This assumes that the colony grows as a disc of radius r, height H and density d. Taking logarithms, equation (1) becomes:

$$\ln x_t = \ln x_0 + \mu t = \ln \, (r_t^2 \pi H d) = \ln(r_0^2 \pi H d) + \mu t$$

or:

$$\ln r_t = \ln r_0 + \frac{\mu t}{2} \tag{2}$$

Once a colony reaches such a size that the diffusion of nutrient into its centre becomes smaller than the demand for the nutrient, exponential growth would cease at the centre. The mycelium in the annulus outside the substrate-limited centre would continue to grow at the maximum specific growth rate but inside the annulus the transition to zero growth rate is likely to be rapid because small changes in the substrate concentration at low growth-limiting levels have a large effect on growth rate. Thus growth of the colony can be regarded as being solely due to exponential growth of the mycelium in the outer annulus. Let the annulus have a depth a which is constant, and a mass x_a:

$$\frac{dx}{dt} = \mu x_a$$

x_a approximates to $2 \, rHad$ where a is small compared with r, and:

$$\frac{dr}{dt} = \mu a \tag{3}$$

$$r_t = a\mu t + r_0$$

As *a* and μ are constants, equation (3) predicts linear growth of the colony. These kinetics are in agreement with the observations that the diameters of colonies of *Chaetomium* sp. (Plomley, 1959) and *Aspergillus nidulans* (Trinci, 1971) initially grew exponentially (equation 2) and then became linear at about 0.2 mm (equation 3).

In order to calculate the specific growth rate (μ), *a* must be measured as well as the increase in colony diameter. This can be done by cutting the colony along a chord and observing the growth rate of the colony edge on the outside of the cut (Trinci, 1971). Where the distance between the cut and the colony edge is less than *a* the radial growth rate would decrease; where the cut is deeper into the colony than *a* there would be no effect on the radial growth rate. On the basis of this measurement of *a*, the peripheral growth zone, Trinci (1971) calculated the specific growth rate of nine species of fungi and found that the values obtained were very close to the specific growth rate in submerged culture on the same medium. He also observed a substantial variation of *a* between strains even when the specific growth rates were similar. This may be explained by variations in hyphal diameter or degree of branching, both of which would affect the distance covered by growing hyphae. Clearly the less the branching and the narrower the diameter, the faster would be apical extension rate for a fixed mass growth rate. For instance, Bainbridge & Trinci (1970) noted that a more branched mutant of *Aspergillus nidulans* had a substantially lower colony radial growth rate and a smaller peripheral growth zone than the wild-type but an almost identical specific growth rate in submerged culture. Similar observations have been made with a series of strains of *Penicillium chrysogenum* with different branching frequences (Morrison & Righelato, 1974). These results show that colony growth rate measurements can be related to morphology and specific growth rate in submerged culture. Morrison & Righelato (1974) showed that *a*, the peripheral growth zone would be described by *G*, the branching frequency or hyphal growth unit, times a constant, *k*:

$$\frac{dr}{dt} = \mu G k$$

Although still a relatively crude technique, colony radial growth measurements could be very useful in scanning the effects of large numbers of conditions or some characters of large numbers of strains (Trinci & Gull, 1970; Okanishi & Gregory, 1970; Trinci, 1973*b*).

References

Bainbridge, B. W. & Trinci, A. P. J. (1970). Colony and specific growth rates of normal and mutant strains of *Aspergillus nidulans*. *Transactions of the British Mycological Society*, **53**, 473–5.

Banks, G. T. (1977). Aeration of mould and streptomycete culture fluids. In *Topics in Enzyme and Fermentation Biotechnology*, vol. 1, ed. A. Wiseman, pp. 73–110. Chichester: Ellis Horwood.

Bongenaar, J. J. T. M., Kossen, N. W. F., Metz, B. & Meijboom, F. W. (1973). A method for characterising the rheological properties of viscous fermentation broths. *Biotechnology and Bioengineering*, **15**, 201–6.

Bull, A. T. & Bushell, M. E. (1976). Environmental control of fungal growth. In *The Filamentous Fungi* vol. 2, eds. J. E. Smith & D. R. Berry, pp. 1–31. London: Edward Arnold.

Bull, A. T. & Trinci, A. P. J. (1977). The physiology and metabolic control of fungal growth. *Advances in Microbial Physiology*, **15**, 1–84.

Caldwell, I. Y. & Trinci, A. P. J. (1973). The growth unit of the mould *Geotrichum candidum*. *Archiv für Mikrobiologie*, **88**, 1–10.

Carilli, A., Chain, E. B., Gualandi, G. & Morisi, G. (1961). Aeration studies III. Continuous measurement of dissolved oxygen during fermentation in large fermenters. *Scientific Reports of the Istituto Superiore di Sanità*, **1**, 177–89.

Clark, D. S. (1962). Submerged citric acid fermentation of ferrocyanide-treated beet molasses: morphology of pellets of *Aspergillus niger*. *Canadian Journal of Microbiology*, **8**, 133–6.

Deindoerfer, F. H. & Gaden, E. L. (1955). Effects of liquid physical properties on oxygen transfer in penicillin fermentation. *Applied Microbiology*, **3**, 253–7.

Drew, S. W., Winstanley, D. J. & Demain, A. L. (1976). Effect of norleucine on mycelial fragmentation in *Cephalosporium acremonium*. *Applied and Environmental Microbiology*, **31**, 143–5.

Forage, A. J. & Righelato, R. C. (1979). Microbial biomass from carbohydrates. In *Economic Microbiology*, ed. A. H. Rose. London, New York: Academic Press. (In press).

Galbraith, J. C. & Smith, J. E. (1969). Filamentous growth of *Aspergillus niger* in submerged shake culture. *Transactions of the British Mycological Society*, **52**, 237–46.

Galpin, M. F. J., Jennings, D. H. & Thornton, J. D. (1977). Hyphal branching in *Dendryphiella salina*: effect of various compounds and further elucidation of the effect of sorbose and the role of cyclic AMP. *Transactions of the British Mycological Society*, **69**, 175–82.

Katz, D., Goldstein, D. & Rosenberger, R. F. (1972). Model for branch initiation in *Aspergillus nidulans* based on measurements of growth parameters. *Journal of Bacteriology*, **109**, 1097–1100.

Kobayashi, T., van Dedem, G. & Moo-Young, M. (1973). Oxygen transfer into mycelial pellets. *Biotechnology and Bioengineering*, **15**, 27–45.

Mason, H. R. S. & Righelato, R. C. (1976). Energetics of fungal growth. *Journal of Applied Chemistry and Biotechnology*, **26**, 145–52.

Metz, B. (1976). From Pulp to Pellet. Ph.D. thesis, Technical University of Delft, The Netherlands.

Miura, Y. & Miyamoto, K. (1977). Oxygen transfer within fungal pellets. *Biotechnology and Bioengineering*, **19**, 1407–9.

Morrison, K. B. & Righelato, R. C. (1974). The relationship between hyphal branching, specific growth rate and colony radial growth rate in *Penicillium chrysogenum*. *Journal of General Microbiology*, **81**, 517–20.

Ng, A. M. L., Smith, J. E. & McIntosh, A. F. (1973). Conidiation of *Aspergillus niger* in continuous culture. *Archiv für Mikrobiologie,* **88,** 119–26.

Okanishi, M. & Gregory, K. F. (1970). Isolation of mutants of *Candida tropicalis* with increased methionine content. *Canadian Journal of Microbiology,* **16,** 1139–43.

Pirt, S. J. (1966). A theory of the mode of growth of fungi in the form of pellets in submerged culture. *Proceedings of the Royal Society,* **B166,** 369–73.

Pirt, S. J. (1967). A kinetic study of the mode of growth of surface colonies of bacteria and fungi. *Journal of General Microbiology,* **47,** 181–97.

Pirt, S. J. (1973). Estimation of substrate affinities (K_s values) of filamentous fungi from colony growth rates. *Journal of General Microbiology,* **75,** 245–7.

Plomley, N. J. B. (1959). Formation of the colony in the fungus *Chaetomium. Australian Journal of Biological Sciences,* **12,** 53–64.

Righelato, R. C. (1967). Studies on the penicillin fermentation. Ph.D. thesis, University of London.

Righelato, R. C. (1975). Growth kinetics of mycelial fungi, in *The Filamentous Fungi,* vol. 1, p. 79, eds. J. E. Smith & D. R. Berry. London: Edward Arnold.

Righelato, R. C., Trinci, A. P. J., Pirt, S. J. & Peat, A. (1968). The influence of maintenance energy and growth rate on the metabolic activity, morphology and conidiation of *Penicillium chrysogenum. Journal of General Microbiology,* **50,** 399–412.

Roels, J. A., van den Berg, J. & Voncken, R. M. (1974). The rheology of mycelial broths. *Biotechnology and Bioengineering,* **16,** 181–208.

Smith, J. E. & Galbraith, J. C. (1971). Biochemical and physiological aspects of differentiation in the fungi. *Advances in Microbial Physiology,* **5,** 45–134.

Solomons, G. L. & Weston, G. O. (1961). The prediction of O_2 transfer rates in the presence of mould mycelium. *Biotechnology and Bioengineering,* **3,** 1–6.

Taguchi, H. (1971). The nature of fermentation fluids. *Advances in Biochemical Engineering,* **1,** 1–30.

Trinci, A. P. J. (1971). Influence of the width of the peripheral growth zone on the radial growth rate of fungal colonies on solid media. *Journal of General Microbiology,* **67,** 325–44.

Trinci, A. P. J. (1973a). The hyphal growth unit of wild type and spreading colonial mutants of *Neurospora crassa. Archiv für Mikrobiologie,* **91,** 127–36.

Trinci, A. P. J. (1973b). Growth of wild type and spreading colonial mutants of *Neurospora crassa* in batch culture and on agar medium. *Archiv für Mikrobiologie,* **91,** 113–26.

Trinci, A. P. J. & Collinge, A. J. (1973). Influence of L-sorbose on the growth and morphology of *Neurospora crassa. Journal of General Microbiology,* **78,** 179–92.

Trinci, A. P. J. & Gull, K. (1970). Effect of actidione, griseofulvin and triphenyltin acetate on the kinetics of fungal growth. *Journal of General Microbiology,* **60,** 286–92.

Trinci, A. P. J. & Righelato, R. C. (1970). Changes in the constituents and ultrastructure of hyphal compartments during autolysis of glucose-starved *Penicillium chrysogenum. Journal of General Microbiology,* **60,** 239–49.

Whittaker, A. & Long, P. A. (1973). Fungal Pelleting. *Process Biochemistry,* **8,** 27–31.

Index of specific names

'p' means *passim*, scattered references

Subject index

The numbers in **bold type** refer to figures, legends to figures or tables on the page indicated.

The abbreviation t-s, means temperature sensitive and the term *passim*, means scattered references.

N-acetylglucosamine, 27–31 *passim*; and protoplast reversion, 54; aryl-*β*- , 267; chitin and, 171, 203, 207; chitin synthase activation and, 217, 222; chitinase activity and, 176, 177, 181, 182; cycle, 216; incorporation, 256, 366; incorporation into regenerated walls, 62, 65, 66, 83; oligosaccharide as primer, 197

N-acetylglucosamine kinase, 207

N-acetylglucosamine 6-phosphate, 207

N-acetyl-*β*-glucosaminidase, 216; and chitin degradation, 174, 182, 183

acid phosphatases, location of, 231, **232**

actidione, and t-s mutants, 87

activators (activating factor): of chitin synthase, 191, 192, 196, 197, 198; of glucan synthase, 246, 247

adenosine monophosphate, *see* cAMP

adenosine triphosphate, *see* ATP

adenosine triphosphatase, *see* ATPase

adenylosuccinase, in t-s mutants, 87

aecidiospores, germ pores in, 135

aglycone, 182

amino acids: biosynthetic pathway repression by, 312; coremia initiation and, 94, 105, 106; in walls, 3, **4**, 27–9 *passim*, 32, 55; rhizomorph initiation and, 94, 100, 101; transport of, 284; *see also individual amino acids*

cAMP: and branching, 287; and membrane transport, 286, 289; involved in apical control of flow, in hypha, 291

anisotropy, 364, 365

antibiotics, and wall polymer formation, 58, 61; *see also* chloramphenicol, cycloheximide, griseofulvin *and* polyoxin D

apex, hyphal: cytology of, 253–5, **257**;

diameter at, and branching, 362; distribution of potassium and sodium in, 281, 287; flow of ions in, 288, **291** (*see also* hyphal tip); growth in coremia, 102, 103; growth in rhizomorphs, 99; growth of, and polyoxin D, 257; hydrolases in, 235; lytic enzymes at, 15, 272, 273; wall polymers at, 6, 7; *see also* compartments, apical; appressoria, model for formation of, 374, 375

arthrospores, 117, **143**; plasmalemmal grooves on, **139**; wall formation in, 129

ascomycetes, 102, 117, 135, 137; chitin synthase activation in, 198; septa and lateral branching in, 352–3

ascospores: formation and wall development in, 117–19, 121; germ pores in, 135; secondary walls of, 131

ascus, and ascospore formation, 117, 118, 121

ATP: and chitin synthesis, 220, 221; and electrogenic pump for potassium ions, 284, 287, 288; effect of cyanide on, 285, 286; effect on potential difference across membrane, 289

ATPase: and wall growth, 225; distribution of, 287, 289; magnesium dependence of, 243; plasma membrane marker, **244**, 245

autolysis: and lysin–wall association, 271–3 *passim*; cell wall, 236; 246; in pellets, 395; mycelial fragmentation and, 397; site of, 237, 238; sugars released by, 238

balloon formation (gross hyphal swellings), 73, **78**, **79**; composition of walls in, 80, 82; in t-s mutants, 76–88; model for, 85–7

basidiomycetes: chitin synthase activation in, 198; mycelial strands in, 95; septation